D0022066

Linear algebra

linear algebra

AN INTRODUCTION WITH CONCURRENT EXAMPLES

A. G. HAMILTON

Department of Computing Science, University of Stirling

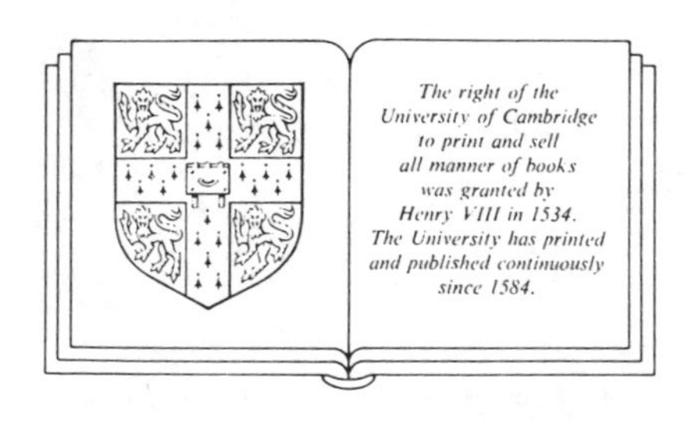

CAMBRIDGE UNIVERSITY PRESS

Cambridge

New York Port Chester Melbourne Sydney

Published by the Press Syndicate of the University of Cambridge
The Pitt Building, Trumpington Street, Cambridge CB2 1RP
40 West 20th Street, New York, NY 10011, USA
10 Stamford Road, Oakleigh, Melbourne 3166, Australia

© Cambridge University Press 1989

First published 1989

Printed in Great Britain at the University Press, Cambridge

British Library cataloguing in publication data
Hamilton, A. G. (Alan G.)
Linear algebra.
1. Linear algebra.
I. Title II. Hamilton, A. G. (Alan G.)
Linear algebra: an introduction with
concurrent examples
512′.5

Library of Congress cataloguing in publication data
Hamilton, A. G., 1943–
Linear algebra: an introduction with concurrent examples / A. G. Hamilton
p. cm.
Includes index.
1. Algebras, Linear. I. Title.
QA184.H362 1989
512.5—dc19 88-31177CIP

ISBN 0 521 32517 X hard covers
ISBN 0 521 31042 3 paperback

MC

CONTENTS

Part 2

PREFACE

My earlier book, *A First Course in Linear Algebra with Concurrent Examples* (referred to below as the First Course), was an introduction to the use of vectors and matrices in the solution of sets of simultaneous linear equations and in the geometry of two and three dimensions. As its name suggests, that much is only a start. For many readers, such elementary material may satisfy the need for appropriate mathematical tools. But, for others, more advanced techniques may be required, or, indeed, further study of algebra for its own sake may be the objective.

This book is therefore in the literal sense an extension of the First Course. The first eleven chapters are identical to the earlier book. The remainder forms a sequel: a continuation into the next stage of the subject. This aims to provide a practical introduction to perhaps the most important applicable idea of linear algebra, namely eigenvalues and eigenvectors of matrices. This requires an introduction to some general ideas about vector spaces. But this is not a book about vector spaces in the abstract. The notions of subspace, basis and dimension are all dealt with in the concrete context of n-dimensional real Euclidean space. Much attention is paid to the diagonalisation of real symmetric matrices, and the final two chapters illustrate applications to geometry and to differential equations.

The organisation and presentation of the content of the First Course were unusual. This book has the same features, and for the same reasons. These reasons were described in the preface to the First Course in the following four paragraphs, which apply equally to this extended volume.

'Learning is not easy (not for most people, anyway). It is, of course, aided by being taught, but it is by no means only a passive exercise. One who hopes to learn must work at it actively. My intention in writing this book is not to teach, but rather to provide a stimulus and a medium through which a reader can learn. There are various sorts of textbook with widely differing approaches. There is the encyclopaedic sort, which tends to be unreadable but contains all of the information relevant to its subject. And at the other extreme there is the work-book, which leads the reader in a progressive series of exercises. In the field of linear algebra there are already enough books of the former kind, so this book is aimed away from that end of the spectrum. But it is not a work-book, neither is it comprehensive. It is a book to be worked through, however. It is intended to be read, not referred to.

'Of course, in a subject such as this, reading is not enough. Doing is also necessary. And doing is one of the main emphases of the book. It is about methods and their application. There are three aspects of this provided by this book: description, worked examples and exercises. All three are important, but I would stress that the most important of these is the exercises. You do not know it until you can do it.

'The format of the book perhaps requires some explanation. The worked examples are integrated with the text, and the careful reader will follow the examples through at the same time as reading the descriptive material. To facilitate this, the text appears on the right-hand pages only, and the examples on the left-hand pages. Thus the text and corresponding examples are visible simultaneously, with neither interrupting the other. Each chapter concludes with a set of exercises covering specifically the material of that chapter. At the end of the book there is a set of sample examination questions covering the material of the whole book.

'The prerequisites required for reading this book are few. It is an introduction to the subject, and so requires only experience with methods of arithmetic, simple algebra and basic geometry. It deliberately avoids mathematical sophistication, but it presents the basis of the subject in a way which can be built on subsequently, either with a view to applications or with the development of the abstract ideas as the principal consideration.'

Last, this book would not have been produced had it not been for the advice and encouragement of David Tranah of Cambridge University Press. My thanks go to him, and to his anonymous referees, for many helpful comments and suggestions.

Part 1

Examples

1.1 Simple elimination (two equations).

$$2x+3y=1$$
$$x-2y=4.$$

Eliminate x as follows. Multiply the second equation by 2:

$$2x+3y=1$$
$$2x-4y=8.$$

Now replace the second equation by the equation obtained by subtracting the first equation from the second:

$$2x+3y=1$$
$$-7y=7.$$

Solve the second equation for y, giving $y=-1$. Substitute this into the first equation:

$$2x-3=1,$$

which yields $x=2$. Solution: $x=2$, $y=-1$.

1.2 Simple elimination (three equations).

$$x-2y+\ z=5$$
$$3x+\ y-\ z=0$$
$$x+3y+2z=2.$$

Eliminate z from the first two equations by adding them:

$$4x-y=5.$$

Next eliminate z from the second and third equations by adding twice the second to the third:

$$7x+5y=2.$$

Now solve the two simultaneous equations:

$$4x-\ y=5$$
$$7x+5y=2$$

as in Example 1.1. One way is to add five times the first to the second, obtaining

$$27x=27,$$

so that $x=1$. Substitute this into one of the set of two equations above which involve only x and y, to obtain (say)

$$4-y=5,$$

so that $y=-1$. Last, substitute $x=1$ and $y=-1$ into one of the original equations, obtaining

$$1+2+z=5,$$

so that $z=2$. Solution: $x=1$, $y=-1$, $z=2$.

1

Gaussian elimination

We shall describe a standard procedure which can be used to solve sets of simultaneous linear equations, no matter how many equations. Let us make sure of what the words mean before we start, however. A *linear* equation is an equation involving unknowns called x or y or z, or x_1 or x_2 or x_3, or some similar labels, in which the unknowns all occur to the first degree, which means that no squares or cubes or higher powers, and no products of two or more unknowns, occur. To *solve* a set of simultaneous equations is to find all values or sets of values for the unknowns which satisfy the equations.

Given two linear equations in unknowns x and y, as in Example 1.1, the way to proceed is to *eliminate* one of the unknowns by combining the two equations in the manner shown.

Given three linear equations in three unknowns, as in Example 1.2, we must proceed in stages. First eliminate one of the unknowns by combining two of the equations, then similarly eliminate the same unknown from a different pair of the equations by combining the third equation with one of the others. This yields two equations with two unknowns. The second stage is to solve these two equations. The third stage is to find the value of the originally eliminated unknown by substituting into one of the original equations.

This general procedure will extend to deal with n equations in n unknowns, no matter how large n is. First eliminate one of the unknowns, obtaining $n-1$ equations in $n-1$ unknowns, then eliminate another unknown from these, giving $n-2$ equations in $n-2$ unknowns, and so on until there is one equation with one unknown. Finally, substitute back to find the values of the other unknowns.

There is nothing intrinsically difficult about this procedure. It consists of the application of a small number of simple operations, used repeatedly.

1.3 The Gaussian elimination process.

$$2x_1 - x_2 + 3x_3 = 1 \qquad (1)$$
$$4x_1 + 2x_2 - x_3 = -8 \qquad (2)$$
$$3x_1 + x_2 + 2x_3 = -1 \qquad (3)$$

Stage 1:
$$x_1 - \tfrac{1}{2}x_2 + \tfrac{3}{2}x_3 = \tfrac{1}{2} \qquad (1)\div 2$$
$$4x_1 + 2x_2 - x_3 = -8 \qquad (2)$$
$$3x_1 + x_2 + 2x_3 = -1 \qquad (3)$$

Stage 2:
$$x_1 - \tfrac{1}{2}x_2 + \tfrac{3}{2}x_3 = \tfrac{1}{2} \qquad (1)$$
$$4x_2 - 7x_3 = -10 \qquad (2) - 4\times(1)$$
$$\tfrac{5}{2}x_2 - \tfrac{5}{2}x_3 = -\tfrac{5}{2} \qquad (3) - 3\times(1)$$

Stage 3:
$$x_1 - \tfrac{1}{2}x_2 + \tfrac{3}{2}x_3 = \tfrac{1}{2} \qquad (1)$$
$$x_2 - \tfrac{7}{4}x_3 = -\tfrac{5}{2} \qquad (2)\div 4$$
$$\tfrac{5}{2}x_2 - \tfrac{5}{2}x_3 = -\tfrac{5}{2} \qquad (3)$$

Stage 4:
$$x_1 - \tfrac{1}{2}x_2 + \tfrac{3}{2}x_3 = \tfrac{1}{2} \qquad (1)$$
$$x_2 - \tfrac{7}{4}x_3 = -\tfrac{5}{2} \qquad (2)$$
$$\tfrac{15}{8}x_3 = \tfrac{15}{4} \qquad (3) - \tfrac{5}{2}\times(2)$$

Stage 5:
$$x_1 - \tfrac{1}{2}x_2 + \tfrac{3}{2}x_3 = \tfrac{1}{2} \qquad (1)$$
$$x_2 - \tfrac{7}{4}x_3 = -\tfrac{5}{2} \qquad (2)$$
$$x_3 = 2. \qquad (3)\div\tfrac{15}{8}$$

Now we may obtain the solutions. Substitute $x_3 = 2$ into the second equation.

$$x_2 - \tfrac{7}{2} = -\tfrac{5}{2}, \quad \text{so } x_2 = 1.$$

Finally substitute both into the first equation, obtaining

$$x_1 - \tfrac{1}{2} + 3 = \tfrac{1}{2}, \quad \text{so } x_1 = -2.$$

Hence the solution is $x_1 = -2$, $x_2 = 1$, $x_3 = 2$.

These include multiplying an equation through by a number and adding or subtracting two equations. But, as the number of unknowns increases, the length of the procedure and the variety of different possible ways of proceeding increase dramatically. Not only this, but it may happen that our set of equations has some special nature which would cause the procedure as given above to fail: for example, a set of simultaneous equations may be *inconsistent*, i.e. have no solution at all, or, at the other end of the spectrum, it may have many different solutions. It is useful, therefore, to have a standard routine way of organising the elimination process which will apply for large sets of equations just as for small, and which will cope in a more or less automatic way with special situations. This is necessary, in any case, for the solution of simultaneous equations using a computer. Computers can handle very large sets of simultaneous equations, but they need a routine process which can be applied automatically. One such process, which will be used throughout this book, is called *Gaussian elimination*. The best way to learn how it works is to follow through examples, so Example 1.3 illustrates the stages described below, and the descriptions should be read in conjunction with it.

Stage 1 Divide the first equation through by the coefficient of x_1. (If this coefficient happens to be zero then choose another of the equations and place it first.)

Stage 2 Eliminate x_1 from the second equation by subtracting a multiple of the first equation from the second equation. Eliminate x_1 from the third equation by subtracting a multiple of the *first* equation from the third equation.

Stage 3 Divide the second equation through by the coefficient of x_2. (If this coefficient is zero then interchange the second and third equations. We shall see later how to proceed if neither of the second and third equations contains a term in x_2.)

Stage 4 Eliminate x_2 from the third equation by subtracting a multiple of the second equation.

Stage 5 Divide the third equation through by the coefficient of x_3. (We shall see later how to cope if this coefficient happens to be zero.)

At this point we have completed the elimination process. What we have done is to find another set of simultaneous equations which have the same solutions as the given set, and whose solutions can be read off very easily. What remains to be done is the following.

Read off the value of x_3. Substitute this value in the second equation, giving the value of x_2. Substitute both values in the first equation, to obtain the value of x_1.

1.4 Using arrays, solve the simultaneous equations:

$$x_1+x_2-x_3=4$$
$$2x_1-x_2+3x_3=7$$
$$4x_1+x_2+x_3=15.$$

First start with the array of coefficients:

$$\begin{array}{rrrr} 1 & 1 & -1 & 4 \\ 2 & -1 & 3 & 7 \\ 4 & 1 & 1 & 15 \\ \hline 1 & 1 & -1 & 4 \\ 0 & -3 & 5 & -1 \\ 0 & -3 & 5 & -1 \\ \hline 1 & 1 & -1 & 4 \\ 0 & 1 & -\frac{5}{3} & \frac{1}{3} \\ 0 & -3 & 5 & -1 \\ \hline 1 & 1 & -1 & 4 \\ 0 & 1 & -\frac{5}{3} & \frac{1}{3} \\ 0 & 0 & 0 & 0 \\ \hline \end{array} \qquad \begin{array}{l} \\ \\ \\ \\ \\ (2)-2\times(1) \\ (3)-4\times(1) \\ \\ \\ (2)\div -3 \\ \\ \\ \\ \\ (3)+3\times(2) \end{array}$$

See Chapter 2 for discussion of how solutions are obtained from here.

1.5 Using arrays, solve the simultaneous equations:

$$3x_1-3x_2+x_3=1$$
$$-x_1+x_2+2x_3=2$$
$$2x_1+x_2-3x_3=0.$$

What follows is a full solution.

$$\begin{array}{rrrr} 3 & -3 & 1 & 1 \\ -1 & 1 & 2 & 2 \\ 2 & 1 & -3 & 0 \\ \hline 1 & -1 & \frac{1}{3} & \frac{1}{3} \\ -1 & 1 & 2 & 2 \\ 2 & 1 & -3 & 0 \\ \hline 1 & -1 & \frac{1}{3} & \frac{1}{3} \\ 0 & 0 & \frac{7}{3} & \frac{7}{3} \\ 0 & 3 & -\frac{11}{3} & -\frac{2}{3} \\ \hline 1 & -1 & \frac{1}{3} & \frac{1}{3} \\ 0 & 3 & -\frac{11}{3} & -\frac{2}{3} \\ 0 & 0 & \frac{7}{3} & \frac{7}{3} \\ \hline \end{array} \qquad \begin{array}{l} \\ \\ \\ (1)\div 3 \\ \\ \\ \\ (2)+(1) \\ (3)-2\times(1) \\ \\ \left.\begin{array}{l} \\ \\ \end{array}\right\} \text{interchange rows} \\ \\ \end{array}$$

Notice that after stage 1 the first equation is not changed, and that after stage 3 the second equation is not changed. This is a feature of the process, however many equations there are. We proceed downwards and eventually each equation is fixed in a new form.

Besides the benefit of standardisation, there is another benefit which can be derived from this process, and that is brevity. Our working of Example 1.3 includes much that is not essential to the process. In particular the repeated writing of equations is unnecessary. Our standard process can be developed so as to avoid this, and all of the examples after Example 1.3 show the different form. The sets of equations are represented by arrays of coefficients, suppressing the unknowns and the equality signs. The first step in Example 1.4 shows how this is done. Our operations on equations now become operations on the rows of the array. These are of the following kinds:

- interchange rows,
- divide (or multiply) one row through by a number,
- subtract (or add) a multiple of one row from (to) another.

These are called *elementary row operations*, and they play a large part in our later work. It is important to notice the form of the array at the end of the process. It has a triangle of 0s in the lower left corner and 1s down the diagonal from the top left.

Now let us take up two complications mentioned above. In stage 5 of the Gaussian elimination process (henceforward called the GE process) the situation not covered was when the coefficient of x_3 in the third equation (row) was zero. In this case we divide the third equation (row) by the number occurring on the right-hand side (in the last column), if this is not already zero. Example 1.4 illustrates this. The solution of sets of equations for which this happens will be discussed in the next chapter. What happens is that either the equations have no solutions or they have infinitely many solutions.

The other complication can arise in stage 3 of the GE process. Here the coefficient of x_2 may be zero. The instruction was to interchange equations (rows) in the hope of placing a non-zero coefficient in this position. When working by hand we may choose which row to interchange with so as to make the calculation easiest (presuming that there is a choice). An obvious way to do this is to choose a row in which this coefficient is 1. Example 1.5 shows this being done. When the GE process is formalised (say for computer application), however, we need a more definite rule, and the one normally adopted is called *partial pivoting*. Under this rule, when we interchange rows because of a zero coefficient, we choose to interchange with the row which has the coefficient which is numerically the *largest* (that

1	−1	$\frac{1}{3}$	$\frac{1}{3}$	
0	1	$-\frac{11}{9}$	$-\frac{2}{9}$	(2)÷3
0	0	1	1	(3)÷$\frac{7}{3}$

From here, $x_3=1$, and substituting back we obtain

$$x_2-\tfrac{11}{9}=-\tfrac{2}{9},\quad \text{so } x_2=1.$$

Substituting again:

$$x_1-1+\tfrac{1}{3}=\tfrac{1}{3},\quad \text{so } x_1=1.$$

Hence the solution sought is: $x_1=1$, $x_2=1$, $x_3=1$.

1.6 Using arrays, solve the simultaneous equations:

$$\begin{aligned} x_1+x_2-x_3&=-3\\ 2x_1+2x_2+x_3&=0\\ 5x_1+5x_2-3x_3&=-8. \end{aligned}$$

Solution:

1	1	−1	−3	
2	2	1	0	
5	5	−3	−8	
1	1	−1	−3	
0	0	3	6	(2)−2×(1)
0	0	2	7	(3)−5×(1)
1	1	−1	−3	
0	0	1	2	(2)÷3
0	0	2	7	
1	1	−1	−3	
0	0	1	2	
0	0	0	3	(3)−2×(2)

Next, and finally, divide the last row by 3. How to obtain solutions from this point is discussed in Chapter 2. (In fact there are no solutions in this case.)

1.7 Solve the simultaneous equations:

$$\begin{aligned} 2x_1-2x_2+x_3-3x_4&=2\\ x_1-x_2+3x_3-x_4&=-2\\ -x_1-2x_2+x_3+2x_4&=-6\\ 3x_1+x_2-x_3-2x_4&=7. \end{aligned}$$

Convert to an array and proceed:

2	−2	1	−3	2
1	−1	3	−1	−2
−1	−2	1	2	−6
3	1	−1	−2	7

is, the largest when any negative signs are disregarded). This has two benefits. First, we (and more particularly, the computer) know precisely what to do at each stage and, second, following this process actually produces a more accurate answer when calculations are subject to rounding errors, as will always be the case with computers. Generally, we shall not use partial pivoting, since our calculations will all be done by hand with small-scale examples.

There may be a different problem at stage 3. We may find that there is no equation (row) which we can choose which has a non-zero coefficient in the appropriate place. In this case we do nothing, and just move on to consideration of x_3, as shown in Example 1.6. How to solve the equations in such a case is discussed in the next chapter.

The GE process has been described above in terms which can be extended to cover larger sets of equations (and correspondingly larger arrays of coefficients). We should bear in mind always that the form of the array which we are seeking has rows in which the first non-zero coefficient (if there is one) is 1, and this 1 is to the *right* of the first non-zero coefficient in the preceding row. Such a form for an array is called *row-echelon form.* Example 1.7 shows the process applied to a set of four equations in four unknowns.

Further examples of the GE process applied to arrays are given in the following exercises. Of course the way to learn this process is to carry it out, and the reader is recommended not to proceed to the rest of the book before gaining confidence in applying it.

Summary

The purpose of this chapter is to describe the Gaussian elimination process which is used in the solution of simultaneous equations, and the abbreviated way of carrying it out, using elementary row operations on rectangular arrays.

$$
\begin{array}{rrrrrll}
1 & -1 & \frac{1}{2} & -\frac{3}{2} & 1 & & (1)\div 2 \\
1 & -1 & 3 & -1 & -2 & & \\
-1 & -2 & 1 & 2 & -6 & & \\
3 & 1 & -1 & -2 & 7 & & \\
\hline
1 & -1 & \frac{1}{2} & -\frac{3}{2} & 1 & & \\
0 & 0 & \frac{5}{2} & \frac{1}{2} & -3 & & (2)-(1) \\
0 & -3 & \frac{3}{2} & \frac{1}{2} & -5 & & (3)+(1) \\
0 & 4 & -\frac{5}{2} & \frac{5}{2} & 4 & & (4)-3\times(1) \\
\hline
1 & -1 & \frac{1}{2} & -\frac{3}{2} & 1 & & \\
0 & -3 & \frac{3}{2} & \frac{1}{2} & -5 & \rbrace & \text{interchange rows} \\
0 & 0 & \frac{5}{2} & \frac{1}{2} & -3 & & \\
0 & 4 & -\frac{5}{2} & \frac{5}{2} & 4 & & \\
\hline
1 & -1 & \frac{1}{2} & -\frac{3}{2} & 1 & & \\
0 & 1 & -\frac{1}{2} & -\frac{1}{6} & \frac{5}{3} & & (2)\div -3 \\
0 & 0 & \frac{5}{2} & \frac{1}{2} & -3 & & \\
0 & 4 & -\frac{5}{2} & \frac{5}{2} & 4 & & \\
\hline
1 & -1 & \frac{1}{2} & -\frac{3}{2} & 1 & & \\
0 & 1 & -\frac{1}{2} & -\frac{1}{6} & \frac{5}{3} & & \\
0 & 0 & \frac{5}{2} & \frac{1}{2} & -3 & & \\
0 & 0 & -\frac{1}{2} & \frac{19}{6} & -\frac{8}{3} & & (4)-4\times(2) \\
\hline
1 & -1 & \frac{1}{2} & -\frac{3}{2} & 1 & & \\
0 & 1 & -\frac{1}{2} & -\frac{1}{6} & \frac{5}{3} & & \\
0 & 0 & 1 & \frac{1}{5} & -\frac{6}{5} & & (3)\div \frac{5}{2} \\
0 & 0 & -\frac{1}{2} & \frac{19}{6} & -\frac{8}{3} & & \\
\hline
1 & -1 & \frac{1}{2} & -\frac{3}{2} & 1 & & \\
0 & 1 & -\frac{1}{2} & -\frac{1}{6} & \frac{5}{3} & & \\
0 & 0 & 1 & \frac{1}{5} & -\frac{6}{5} & & \\
0 & 0 & 0 & \frac{49}{15} & -\frac{49}{15} & & (4)+\frac{1}{2}\times(3) \\
\hline
1 & -1 & \frac{1}{2} & -\frac{3}{2} & 1 & & \\
0 & 1 & -\frac{1}{2} & -\frac{1}{6} & \frac{5}{3} & & \\
0 & 0 & 1 & \frac{1}{5} & -\frac{6}{5} & & \\
0 & 0 & 0 & 1 & -1 & & (4)\div \frac{49}{15} \\
\hline
\end{array}
$$

From this last row we deduce that $x_4=-1$. Substituting back gives:

$$
\begin{aligned}
x_3-\tfrac{1}{5}&=-\tfrac{6}{5}, && \text{so } x_3=-1, \\
x_2+\tfrac{1}{2}+\tfrac{1}{6}&=\tfrac{5}{3}, && \text{so } x_2=1, \quad \text{and} \\
x_1-1-\tfrac{1}{2}+\tfrac{3}{2}&=1, && \text{so } x_1=1.
\end{aligned}
$$

Hence the solution is: $x_1=1$, $x_2=1$, $x_3=-1$, $x_4=-1$.

Exercises

Using the Gaussian elimination process, solve the following sets of simultaneous equations.

(i) $\begin{aligned} x-y&=2\\ 2x+y&=1. \end{aligned}$

(ii) $\begin{aligned} 3x+2y&=0\\ x-y&=5. \end{aligned}$

(iii) $\begin{aligned} x_1+x_2+x_3&=2\\ 2x_1+x_2-2x_3&=4\\ -x_1-2x_2+3x_3&=4. \end{aligned}$

(iv) $\begin{aligned} 3x_1-x_2-x_3&=6\\ x_1-x_2+x_3&=0\\ -x_1+2x_2+2x_3&=-2. \end{aligned}$

(v) $\begin{aligned} 2x_1-4x_2+x_3&=2\\ x_1-2x_2-2x_3&=-4\\ -x_1+x_2&=-1. \end{aligned}$

(vi) $\begin{aligned} -x_1+2x_2-x_3&=-2\\ 4x_1-x_2-2x_3&=2\\ 3x_1-4x_3&=-1. \end{aligned}$

(vii) $\begin{aligned} 2x_2-x_3&=-5\\ x_1-x_2+2x_3&=8\\ x_1+2x_2+2x_3&=5. \end{aligned}$

(viii) $\begin{aligned} x_1+5x_2-2x_3&=0\\ 3x_1-x_2+10x_3&=0\\ -x_1-2x_2+7x_3&=0. \end{aligned}$

(ix) $\begin{aligned} x_1-3x_2-x_3&=0\\ 2x_1-4x_2-7x_3&=2\\ 7x_1-13x_2&=8 \end{aligned}$

(x) $\begin{aligned} 2x_1-x_2-x_3&=-2\\ 3x_2-7x_3&=0\\ -3x_1+x_2-4x_3&=3 \end{aligned}$

(xi) $\begin{aligned} x_1-2x_2+x_3-x_4&=1\\ -x_1-x_2+2x_3+2x_4&=-5\\ 2x_1+x_2+x_3+3x_4&=2\\ x_1+3x_2-3x_3+3x_4&=2. \end{aligned}$

(xii) $\begin{aligned} x_1+x_2-3x_3+x_4&=-2\\ 2x_1+2x_2+x_3+3x_4&=0\\ x_1+2x_2-2x_3+2x_4&=-2\\ -3x_2+4x_3-x_4&=1. \end{aligned}$

Examples

2.1 Find all values of x and y which satisfy the equations:

$$x+2y=1$$
$$3x+6y=3.$$

GE process:

1	2	1	
3	6	3	
1	2	1	
0	0	0	$(2)-3\times(1)$

Here the second row gives no information. All we have to work with is the single equation $x+2y=1$. Set $y=t$ (say). Then on substitution we obtain $x=1-2t$. Hence all values of x and y which satisfy the equations are given by:

$$x=1-2t, \quad y=t \quad (t\in\mathbb{R}).$$

2.2 Find all solutions to:

$$x_1+x_2-\ x_3=\ \ 5$$
$$3x_1-x_2+2x_3=-2.$$

The GE process yields:

1	1	-1	5
0	1	$-\frac{5}{4}$	$\frac{17}{4}$

Set $x_3=t$. Substituting then gives

$$x_2-\tfrac{5}{4}t=\tfrac{17}{4}, \quad \text{so } x_2=\tfrac{17}{4}+\tfrac{5}{4}t, \quad \text{and}$$
$$x_1+(\tfrac{17}{4}+\tfrac{5}{4}t)-t=5, \quad \text{so } x_1=\tfrac{3}{4}+\tfrac{1}{4}t.$$

Hence the full solution is:

$$x_1=\tfrac{3}{4}+\tfrac{1}{4}t, \quad x_2=\tfrac{17}{4}+\tfrac{5}{4}t, \quad x_3=t \quad (t\in\mathbb{R}).$$

2.3 Solve the equations:

$$x_1-\ x_2-4x_3=0$$
$$3x_1+\ x_2-\ x_3=3$$
$$5x_1+3x_2+2x_3=6.$$

The GE process yields:

1	-1	-4	0
0	1	$\frac{11}{4}$	$\frac{3}{4}$
0	0	0	0

In effect, then, we have only two equations to solve for three unknowns. Set $x_3=t$. Substituting then gives

$$x_2+\tfrac{11}{4}t=\tfrac{3}{4}, \quad \text{so } x_2=\tfrac{3}{4}-\tfrac{11}{4}t, \quad \text{and}$$
$$x_1-(\tfrac{3}{4}-\tfrac{11}{4}t)-4t=0, \quad \text{so } x_1=\tfrac{3}{4}-\tfrac{5}{4}t.$$

Hence the full solution is: $x_1=\frac{3}{4}-\frac{5}{4}t$, $x_2=\frac{3}{4}-\frac{11}{4}t$, $x_3=t$ $(t\in\mathbb{R})$.

2

Solutions to simultaneous equations 1

Now that we have a routine procedure for the elimination of variables (Gaussian elimination), we must look more closely at where it can lead, and at the different possibilities which can arise when we seek solutions to given simultaneous equations.

Example 2.1 illustrates in a simple way one possible outcome. After the GE process the second row consists entirely of zeros and is thus of no help in finding solutions. This has happened because the original second equation is a multiple of the first equation, so in essence we are given only a single equation connecting the two variables. In such a situation there are infinitely many possible solutions. This is because we may specify *any* value for one of the unknowns (say y) and then the equation will give the value of the other unknown. Thus the customary form of the solution to Example 2.1 is:

$$y=t, \quad x=1-2t \quad (t\in\mathbb{R}).$$

These ideas extend to the situation generally when there are fewer equations than unknowns. Example 2.2 illustrates the case of two equations with three unknowns. We may specify any value for one of the unknowns (here put $z=t$) and then solve the two equations for the other two unknowns. This situation may also arise when we are originally given three equations in three unknowns, as in Example 2.3. See also Example 1.4.

2.4 Find all solutions to the set of equations:

$$x_1 + x_2 + x_3 = 1$$
$$x_1 + x_2 + x_3 = 4.$$

This is a nonsensical problem. There are no values of x_1, x_2 and x_3 which satisfy both of these equations simultaneously. What does the GE process give us here?

1	1	1	1	
1	1	1	4	
1	1	1	1	
0	0	0	3	(2) − (1)
1	1	1	1	
0	0	0	1	(2) ÷ 3

The last row, when transformed back into an equation, is

$$0x_1 + 0x_2 + 0x_3 = 1.$$

This is satisfied by *no* values of x_1, x_2 and x_3.

2.5 Find all solutions to the set of equations:

$$x_1 + 2x_2 + x_3 = 1$$
$$2x_1 + 5x_2 - x_3 = 4$$
$$x_1 + x_2 + 4x_3 = 2.$$

GE process:

1	2	1	1	
2	5	−1	4	
1	1	4	2	
1	2	1	1	
0	1	−3	2	(2) − 2 × (1)
0	−1	3	1	(3) − (1)
1	2	1	1	
0	1	−3	2	
0	0	0	3	(3) + (2)
1	2	1	1	
0	1	−3	2	
0	0	0	1	(3) ÷ (3)

Because of the form of this last row, we can say straight away that there are no solutions in this case (indeed, the last step was unnecessary: a last row of 0 0 0 3 indicates inconsistency immediately).

Here, then, is a simple-minded rule: if there are fewer equations than unknowns then there will be infinitely many solutions (if there are solutions at all). This rule is more usefully applied *after* the GE process has been completed, because the original equations may disguise the real situation, as in Examples 2.1, 2.3 and 1.4.

The qualification must be placed on this rule because such sets of equations may have no solutions at all. Example 2.4 is a case in point. Two equations, three unknowns, and no solutions. These equations are clearly *inconsistent* equations. There are no values of the unknowns which satisfy both. In such a case it is obvious that they are inconsistent. The equations in Example 2.5 are also inconsistent, but it is not obvious there. The GE process automatically tells us when equations are inconsistent. In Example 2.5 the last row turns out to be

$$0 \quad 0 \quad 0 \quad 1,$$

which, if translated back into an equation, gives

$$0x_1 + 0x_2 + 0x_3 = 1,$$

i.e.

$$0 = 1.$$

When this happens, the conclusion that we can draw is that the given equations are inconsistent and have no solutions. See also Example 1.6. This may happen whether there are as many equations as unknowns, more equations, or fewer equations.

2.6 Find all solutions to the set of equations:

$$\begin{aligned} x_1 + x_2 &= 2 \\ 3x_1 - x_2 &= 2 \\ -x_1 + 2x_2 &= 3. \end{aligned}$$

GE process:

1	1	2	
3	−1	2	
−1	2	3	
1	1	2	
0	−4	−4	(2) − 3 × (1)
0	3	5	(3) + (1)
1	1	2	
0	1	1	(2) ÷ −4
0	3	5	
1	1	2	
0	1	1	
0	0	2	(3) − 3 × (2)

Without performing the last step of the standard process we can see here that the given equations are inconsistent.

2.7 Find all solutions to the set of equations:

$$\begin{aligned} x_1 - 4x_2 &= -1 \\ 2x_1 + 2x_2 &= 8 \\ 5x_1 - x_2 &= 14. \end{aligned}$$

GE process:

1	−4	−1	
2	2	8	
5	−1	14	
1	−4	−1	
0	10	10	(2) − 2 × (1)
0	19	19	(3) − 5 × (1)
1	−4	−1	
0	1	1	(2) ÷ 10
0	0	0	(after two steps)

Here there is a solution. The third equation is in effect redundant. The second row yields $x_2 = 1$. Substituting in the first gives:

$$x_1 - 4 = -1, \quad \text{so } x_1 = 3.$$

Hence the solution is: $x_1 = 3$, $x_2 = 1$.

Example 2.6 has three equations with two unknowns. Here there are more equations than we need to determine the values of the unknowns. We can think of using the first two equations to find these values and then trying them in the third. If we are lucky they will work! But the more likely outcome is that such sets of equations are inconsistent. Too many equations may well lead to inconsistency. But not always. See Example 2.7.

We can now see that there are three possible outcomes when solving simultaneous equations:

(i) there is no solution,

(ii) there is a unique solution,

(iii) there are infinitely many solutions.

One of the most useful features of the GE process is that it tells us automatically which of these occurs, in advance of finding the solutions.

2.8 Illustration of the various possibilities arising from the GE process and the nature of the solutions indicated.

(i) $\begin{bmatrix} 1 & 2 & 1 \\ 0 & 1 & 3 \end{bmatrix}$ unique solution.

(ii) $\begin{bmatrix} 1 & -1 & 2 \\ 0 & 0 & 1 \end{bmatrix}$ inconsistent.

(iii) $\begin{bmatrix} 1 & 3 & 3 \\ 0 & 1 & 0 \end{bmatrix}$ unique solution.

(iv) $\begin{bmatrix} 1 & 1 & 2 \\ 0 & 0 & 0 \end{bmatrix}$ infinitely many solutions.

(v) $\begin{bmatrix} 1 & 2 & 1 & 4 \\ 0 & 1 & -2 & 2 \\ 0 & 0 & 1 & 3 \end{bmatrix}$ unique solution.

(vi) $\begin{bmatrix} 1 & 0 & -1 & 5 \\ 0 & 1 & 1 & -3 \\ 0 & 0 & 0 & 1 \end{bmatrix}$ inconsistent.

(vii) $\begin{bmatrix} 1 & 3 & 0 & 2 \\ 0 & 1 & 3 & -1 \\ 0 & 0 & 1 & 0 \end{bmatrix}$ unique solution.

(viii) $\begin{bmatrix} 1 & -1 & 1 & 5 \\ 0 & 1 & 7 & 2 \\ 0 & 0 & 0 & 0 \end{bmatrix}$ infinitely many solutions.

(ix) $\begin{bmatrix} 1 & 2 & -1 & 3 \\ 0 & 0 & 1 & 2 \\ 0 & 0 & 0 & 1 \end{bmatrix}$ inconsistent.

(x) $\begin{bmatrix} 1 & 2 & 2 & 5 \\ 0 & 0 & 1 & 2 \\ 0 & 0 & 0 & 0 \end{bmatrix}$ infinitely many solutions.

Rule

Given a set of (any number of) simultaneous equations in p unknowns:

(i) there is no solution if after the GE process the last non-zero row has a 1 at the right-hand end and zeros elsewhere;

(ii) there is a unique solution if after the GE process there are exactly p non-zero rows, the last of which has a 1 in the position *second* from the right-hand end;

(iii) there are infinitely many solutions if after the GE process there are fewer than p non-zero rows and (i) above does not apply.

Example 2.8 gives various arrays resulting from the GE process, to illustrate the three possibilities above. Note that the number of unknowns is always one fewer than the number of columns in the array.

2.9 Find all values of c for which the equations

$$\begin{aligned} x+\ y&=c \\ 3x-cy&=2 \end{aligned}$$

have a solution.

GE process:

$$\begin{array}{ccc} 1 & 1 & c \\ 3 & -c & 2 \\ \hline 1 & 1 & c \\ 0 & -c-3 & 2-3c \\ \hline 1 & 1 & c \\ 0 & 1 & \dfrac{2-3c}{-c-3} \\ \hline \end{array}$$

$(2)-3\times(1)$

$(2)\div(-c-3)$

Now this last step is legitimate only if $-c-3$ is not zero. Thus, provided that $c+3\neq 0$, we can say

$$y=\frac{2-3c}{-c-3} \quad \text{and} \quad x=c-\frac{2-3c}{-c-3}.$$

If $c+3=0$ then $c=-3$, and the equations are

$$\begin{aligned} x+\ y&=-3 \\ 3x+3y&=\ \ 2. \end{aligned}$$

These are easily seen to be inconsistent. Thus the given equations have a solution if and only if $c\neq -3$.

2.10 Find all values of c for which the following equations have

(a) a unique solution,

(b) no solution,

(c) infinitely many solutions.

$$\begin{aligned} x_1+\ x_2+\ x_3&=c \\ cx_1+\ x_2+2x_3&=2 \\ x_1+cx_2+\ x_3&=4. \end{aligned}$$

GE process:

$$\begin{array}{cccc} 1 & 1 & 1 & c \\ c & 1 & 2 & 2 \\ 1 & c & 1 & 4 \\ \hline 1 & 1 & 1 & c \\ 0 & 1-c & 2-c & 2-c^2 \\ 0 & c-1 & 0 & 4-c \quad * \\ \hline \end{array}$$

Finally, Examples 2.9 and 2.10 show how the GE process can be applied even when the coefficients in the given simultaneous equations involve a parameter (or parameters) which may be unknown or unspecified. As naturally expected, the solution values for the unknowns depend on the parameter(s), but, importantly, the nature of the solution, that is to say, whether there are no solutions, a unique solution or infinitely many solutions, also depends on the value(s) of the parameter(s).

Summary

This chapter should enable the reader to apply the GE process to any given set of simultaneous linear equations to find whether solutions exist, and if they do to determine whether there is a unique solution or infinitely many solutions, and to find them.

Exercises

1. Show that the following sets of equations are inconsistent.

(i)
$$\begin{aligned} x-2y&=1\\ 2x-y&=-8\\ -x+y&=6. \end{aligned}$$

(ii)
$$\begin{aligned} 3x+y&=3\\ 2x-y&=7\\ 5x+4y&=4. \end{aligned}$$

(iii)
$$\begin{aligned} x_1-x_2-2x_3&=-1\\ -2x_1+x_2+x_3&=2\\ 3x_1+2x_2+9x_3&=4. \end{aligned}$$

(iv)
$$\begin{aligned} 2x_1-x_2+4x_3&=4\\ x_1+2x_2-3x_3&=1\\ 3x_1\qquad+3x_3&=6. \end{aligned}$$

2. Show that the following sets of equations have infinitely many solutions, and express the solutions in terms of parameters.

(i)
$$\begin{aligned} x-3y&=2\\ 2x-6y&=4. \end{aligned}$$

(ii)
$$\begin{aligned} 2x+3y&=-1\\ 8x+12y&=-4. \end{aligned}$$

(iii)
$$\begin{aligned} x_1+x_2+x_3&=5\\ -x_1+2x_2-7x_3&=-2\\ 2x_1+x_2+4x_3&=9. \end{aligned}$$

(iv)
$$\begin{aligned} x_1\qquad+2x_3&=1\\ 2x_1+x_2+3x_3&=1\\ 2x_1-x_2+5x_3&=3. \end{aligned}$$

(v)
$$\begin{aligned} x_1-x_2+3x_3&=4\\ 2x_1-2x_2+x_3&=3\\ -x_1+x_2+x_3&=0. \end{aligned}$$

(vi)
$$\begin{aligned} x_1+2x_2+x_3&=2\\ 2x_1-x_2+7x_3&=-6\\ -x_1+x_2-4x_3&=4\\ x_1-2x_2+5x_3&=-6. \end{aligned}$$

3. Show that the following sets of equations have unique solutions, and find them.

(i)
$$\begin{aligned} 2x-5y&=-1\\ 3x+y&=7. \end{aligned}$$

(ii)
$$\begin{aligned} x-2y&=-1\\ 4x+y&=14\\ 3x-4y&=1. \end{aligned}$$

(iii)
$$\begin{aligned} x_1-x_2-2x_3&=-6\\ 3x_1+x_2-2x_3&=-6\\ -2x_1-2x_2+x_3&=2. \end{aligned}$$

(iv)
$$\begin{aligned} 3x_2+x_3&=-3\\ x_1-2x_2-2x_3&=4\\ 2x_1+x_2-3x_3&=3. \end{aligned}$$

$$\begin{array}{cccc} 1 & 1 & 1 & c \\ 0 & 1 & \dfrac{2-c}{1-c} & \dfrac{2-c^2}{1-c} \\ 0 & 0 & 2-c & 4-c+2-c^2 \quad ** \end{array} \qquad \text{(provided that } c\neq 1\text{)}$$

$$\begin{array}{cccc} 1 & 1 & 1 & c \\ 0 & 1 & \dfrac{2-c}{1-c} & \dfrac{2-c^2}{1-c} \\ 0 & 0 & 1 & 3+c \end{array} \qquad \text{(provided that } c\neq 2\text{)}$$

If $c=1$ then the row marked * is 0 0 0 3, showing the equations to be inconsistent. If $c=2$ then the row marked ** is 0 0 0 0, and the equations have infinitely many solutions: $x_3=t, x_2=t, x_1=-t$ ($t\in\mathbb{R}$). Last, if $c\neq 1$ and $c\neq 2$ then there is a unique solution, given by the last array above:

$$x_3=3+c,$$

$$x_2=\frac{2-c^2}{1-c}-\frac{(2-c)(3+c)}{1-c},$$

and

$$x_1=c-\frac{2-c^2}{1-c}+\frac{(2-c)(3+c)}{1-c}-(3+c).$$

(v)
$$\begin{aligned} x_1+2x_2+3x_3&=3\\ 2x_1+3x_2+4x_3&=3\\ 3x_1+4x_2+5x_3&=3\\ x_1+x_2+x_3&=0. \end{aligned}$$

(vi)
$$\begin{aligned} -x_1+x_2+x_3+x_4&=2\\ x_1-x_2+x_3+x_4&=4\\ x_1+x_2-x_3+x_4&=6\\ x_1+x_2+x_3-x_4&=8. \end{aligned}$$

4. Find whether the following sets of equations are inconsistent, have a unique solution, or have infinitely many solutions.

(i)
$$\begin{aligned} x_1+x_2+x_3&=1\\ 2x_1+x_2-3x_3&=1\\ 3x_2-x_3&=1. \end{aligned}$$

(ii)
$$\begin{aligned} x_1+2x_2-x_3&=2\\ 2x_1+2x_2-4x_3&=0\\ -x_1+3x_3&=2. \end{aligned}$$

(iii)
$$\begin{aligned} x_1-x_2+x_3-2x_4&=-6\\ 2x_2+x_3-3x_4&=-5\\ 3x_1-x_2-4x_3-x_4&=9\\ -x_1-3x_2+3x_3+2x_4&=-5. \end{aligned}$$

(iv)
$$\begin{aligned} x_1+x_2+x_3&=2\\ 3x_1-2x_2+x_3&=3\\ 2x_1+x_3&=3\\ -x_1+3x_2+x_3&=3. \end{aligned}$$

(v)
$$\begin{aligned} x_1+x_2+x_3+x_4&=0\\ x_1+x_4&=0\\ x_1+2x_2+x_3&=0. \end{aligned}$$

5. (i) Examine the solutions of

$$\begin{aligned} x_1-x_2+x_3&=c\\ 2x_1-3x_2+4x_3&=0\\ 3x_1-4x_2+5x_3&=1, \end{aligned}$$

when $c=1$ and when $c\neq 1$.

(ii) Find all values of k for which the equations

$$\begin{aligned} 3x_1-7x_2-4x_3&=8\\ -2x_1+6x_2+11x_3&=21\\ 5x_1-21x_2+7x_3&=10k\\ x_1+23x_2+13x_3&=41 \end{aligned}$$

are consistent.

6. In the following sets of equations, determine all values of c for which the set of equations has (a) no solution, (b) infinitely many solutions, and (c) a unique solution.

(i)
$$\begin{aligned} x_1+x_2-x_3&=2\\ x_1+2x_2+x_3&=3\\ x_1+x_2+(c^2-5)x_3&=c. \end{aligned}$$

(ii)
$$\begin{aligned} x_1+x_2+x_3&=2\\ 2x_1+3x_2+2x_3&=5\\ 2x_1+3x_2+(c^2-1)x_3&=c+1. \end{aligned}$$

(iii)
$$\begin{aligned} x_1+x_2&=3\\ x_1+(c^2-8)x_2&=c. \end{aligned}$$

(iv)
$$\begin{aligned} cx_1+x_2-2x_3&=0\\ x_1+cx_2+3x_3&=0\\ 2x_1+3x_2+cx_3&=0. \end{aligned}$$

Examples

3.1 Examples of matrix notations.

$$A=\begin{bmatrix}1 & 2 & 3\\ -2 & -1 & 0\end{bmatrix}: \quad a_{11}=1,\ a_{12}=2,\ a_{13}=3,\ a_{21}=-2,\ a_{22}=-1,\ a_{23}=0.$$

$$B=\begin{bmatrix}5 & 6\\ 7 & 8\\ 9 & 10\end{bmatrix}. \quad b_{11}=5,\ b_{12}=6,\ b_{21}=7,\ b_{22}=8,\ b_{31}=9,\ b_{32}=10.$$

A is a 2×3 matrix, B is a 3×2 matrix.

3.2 Examples of matrix addition.

$$\begin{bmatrix}1 & 2 & 3\\ 4 & 5 & 6\end{bmatrix}+\begin{bmatrix}3 & 2 & 1\\ 2 & 3 & 4\end{bmatrix}=\begin{bmatrix}4 & 4 & 4\\ 6 & 8 & 10\end{bmatrix}.$$

$$\begin{bmatrix}-1 & 3 & 2\\ 4 & 0 & 1\\ -2 & 1 & 5\end{bmatrix}+\begin{bmatrix}4 & 1 & 1\\ 3 & 2 & -2\\ 1 & 2 & -3\end{bmatrix}=\begin{bmatrix}3 & 4 & 3\\ 7 & 2 & -1\\ -1 & 3 & 2\end{bmatrix}.$$

$$\begin{bmatrix}6 & 1\\ -1 & 2\\ 3 & 4\end{bmatrix}-\begin{bmatrix}3 & 2\\ 1 & -3\\ 0 & 1\end{bmatrix}=\begin{bmatrix}3 & -1\\ -2 & 5\\ 3 & 3\end{bmatrix}.$$

$$\begin{bmatrix}a_{11} & a_{12} & a_{13}\\ a_{21} & a_{22} & a_{23}\end{bmatrix}+\begin{bmatrix}b_{11} & b_{12} & b_{13}\\ b_{21} & b_{22} & b_{23}\end{bmatrix}=\begin{bmatrix}a_{11}+b_{11} & a_{12}+b_{12} & a_{13}+b_{13}\\ a_{21}+b_{21} & a_{22}+b_{22} & a_{23}+b_{23}\end{bmatrix}.$$

3.3 Examples of scalar multiples.

$$\begin{bmatrix}5 & 6\\ 7 & 8\\ 9 & 10\end{bmatrix}+\begin{bmatrix}5 & 6\\ 7 & 8\\ 9 & 10\end{bmatrix}=\begin{bmatrix}10 & 12\\ 14 & 16\\ 18 & 20\end{bmatrix}=2\begin{bmatrix}5 & 6\\ 7 & 8\\ 9 & 10\end{bmatrix}.$$

$$6\begin{bmatrix}1 & 2\\ 3 & 4\end{bmatrix}=\begin{bmatrix}6 & 12\\ 18 & 24\end{bmatrix}.$$

$$\tfrac{1}{2}\begin{bmatrix}3 & -2 & 1\\ 2 & 1 & -4\end{bmatrix}=\begin{bmatrix}\frac{3}{2} & -1 & \frac{1}{2}\\ 1 & \frac{1}{2} & -2\end{bmatrix}.$$

3.4 More scalar multiples.

Let $A=\begin{bmatrix}-1 & 1\\ 2 & 4\end{bmatrix}$ and $B=\begin{bmatrix}-1 & 3 & 2\\ 4 & 0 & 1\\ -2 & 1 & 5\end{bmatrix}$.

Then

$$2A=\begin{bmatrix}-2 & 2\\ 4 & 8\end{bmatrix},\quad 7A=\begin{bmatrix}-7 & 7\\ 14 & 28\end{bmatrix},\quad \tfrac{1}{2}A=\begin{bmatrix}-\frac{1}{2} & \frac{1}{2}\\ 1 & 2\end{bmatrix}$$

and $2B=\begin{bmatrix}-2 & 6 & 4\\ 8 & 0 & 2\\ -4 & 2 & 10\end{bmatrix}$, $-5B=\begin{bmatrix}5 & -15 & -10\\ -20 & 0 & -5\\ 10 & -5 & -25\end{bmatrix}$, $\tfrac{1}{5}B=\begin{bmatrix}-\frac{1}{5} & \frac{3}{5} & \frac{2}{5}\\ \frac{4}{5} & 0 & \frac{1}{5}\\ -\frac{2}{5} & \frac{1}{5} & 1\end{bmatrix}$.

3

Matrices and algebraic vectors

A *matrix* is nothing more than a rectangular array of numbers (for us this means real numbers). In fact the arrays which were part of the shorthand way of carrying out the Gaussian elimination process are matrices. The usefulness of matrices originates in precisely that process, but extends far beyond. We shall see in this chapter how the advantages of brevity gained through the use of arrays in Chapters 1 and 2 can be developed, and how out of this development the idea of a matrix begins to stand on its own.

An array of numbers with p rows and q columns is called a $p \times q$ matrix ('p by q matrix'), and the numbers themselves are called the *entries* in the matrix. The number in the ith row and jth column is called the (i, j)-entry. Sometimes suffixes are used to indicate position, so that a_{ij} (or b_{ij}, etc.) may be used for the (i, j)-entry. The first suffix denotes the row and the second suffix the column. See Examples 3.1. A further notation which is sometimes used is $[a_{ij}]_{p \times q}$. This denotes the $p \times q$ matrix whose (i, j)-entry is a_{ij}, for each i and j.

Immediately we can see that there are extremes allowed under this definition, namely when either p or q is 1. When p is 1 the matrix has only one row, and is called a *row vector*, and when q is 1 the matrix has only one column, and is called a *column vector*. The case when both p and q are 1 is rather trivial and need not concern us here. A column vector with p entries we shall call a *p-vector*, so a p-vector is a $p \times 1$ matrix.

Addition of matrices (including addition of row or column vectors) is very straightforward. We just add the corresponding entries. See Examples 3.2. The only point to note is that, in order for the sum of two matrices (or vectors) to make sense, they must be of the same size. To put this precisely, they must both be $p \times q$ matrices, for the same p and q. In formal terms, if A is the $p \times q$ matrix whose (i, j)-entry is a_{ij} and B is the $p \times q$ matrix whose (i, j)-entry is b_{ij} then $A + B$ is the $p \times q$ matrix whose (i, j)-entry is $a_{ij} + b_{ij}$. Likewise subtraction: $A - B$ is the $p \times q$ matrix whose (i, j)-entry is $a_{ij} - b_{ij}$.

3.5 Multiplication of a matrix with a column vector. Consider the equations:

$$\begin{aligned} 2x_1 - x_2 + 4x_3 &= 1 \\ x_1 + 3x_2 - 2x_3 &= 0 \\ -2x_1 + x_2 - 3x_3 &= 2. \end{aligned}$$

These may be represented as an equation connecting two column vectors:

$$\begin{bmatrix} 2x_1 - x_2 + 4x_3 \\ x_1 + 3x_2 - 2x_3 \\ -2x_1 + x_2 - 3x_3 \end{bmatrix} = \begin{bmatrix} 1 \\ 0 \\ 2 \end{bmatrix}.$$

The idea of multiplication of a matrix with a vector is defined so that the left-hand vector is the result of multiplying the vector of unknowns by the matrix of coefficients, thus:

$$\begin{bmatrix} 2 & -1 & 4 \\ 1 & 3 & -2 \\ -2 & 1 & -3 \end{bmatrix} \begin{bmatrix} x_1 \\ x_2 \\ x_3 \end{bmatrix} = \begin{bmatrix} 2x_1 - x_2 + 4x_3 \\ x_1 + 3x_2 - 2x_3 \\ -2x_1 + x_2 - 3x_3 \end{bmatrix}.$$

In this way the original set of simultaneous equations may be written as a matrix equation:

$$\begin{bmatrix} 2 & -1 & 4 \\ 1 & 3 & -2 \\ -2 & 1 & -3 \end{bmatrix} \begin{bmatrix} x_1 \\ x_2 \\ x_3 \end{bmatrix} = \begin{bmatrix} 1 \\ 0 \\ 2 \end{bmatrix}.$$

3.6 Examples of simultaneous equations written as matrix equations.

$$\begin{aligned} 3x_1 - 2x_2 &= 1 \\ 4x_1 + x_2 &= -2 \end{aligned} \qquad \begin{bmatrix} 3 & -2 \\ 4 & 1 \end{bmatrix} \begin{bmatrix} x_1 \\ x_2 \end{bmatrix} = \begin{bmatrix} 1 \\ -2 \end{bmatrix}.$$

$$\begin{aligned} x_1 + x_2 + x_3 &= 6 \\ x_1 - x_2 - x_3 &= 0 \end{aligned} \qquad \begin{bmatrix} 1 & 1 & 1 \\ 1 & -1 & -1 \end{bmatrix} \begin{bmatrix} x_1 \\ x_2 \\ x_3 \end{bmatrix} = \begin{bmatrix} 6 \\ 0 \end{bmatrix}.$$

$$\begin{aligned} 3x_1 - 2x_2 &= 0 \\ x_1 + x_2 &= 5 \\ -x_1 + 2x_2 &= 4 \end{aligned} \qquad \begin{bmatrix} 3 & -2 \\ 1 & 1 \\ -1 & 2 \end{bmatrix} \begin{bmatrix} x_1 \\ x_2 \end{bmatrix} = \begin{bmatrix} 0 \\ 5 \\ 4 \end{bmatrix}.$$

3.7 Multiplication of a column vector by a matrix.

$$\begin{bmatrix} 1 & 2 \\ 3 & 4 \end{bmatrix} \begin{bmatrix} x_1 \\ x_2 \end{bmatrix} = \begin{bmatrix} x_1 + 2x_2 \\ 3x_1 + 4x_2 \end{bmatrix}.$$

$$\begin{bmatrix} 1 & 2 \\ 3 & 4 \end{bmatrix} \begin{bmatrix} 5 \\ 6 \end{bmatrix} = \begin{bmatrix} 5+12 \\ 15+24 \end{bmatrix} = \begin{bmatrix} 17 \\ 39 \end{bmatrix}.$$

$$\begin{bmatrix} 1 & 2 & 1 \\ -1 & -3 & 2 \end{bmatrix} \begin{bmatrix} x_1 \\ x_2 \\ x_3 \end{bmatrix} = \begin{bmatrix} x_1 + 2x_2 + x_3 \\ -x_1 - 3x_2 + 2x_3 \end{bmatrix}.$$

$$\begin{bmatrix} 1 & 1 & 1 \\ -1 & 2 & 1 \\ 3 & 1 & 3 \end{bmatrix} \begin{bmatrix} 2 \\ -1 \\ -2 \end{bmatrix} = \begin{bmatrix} 2-1-2 \\ -2-2-2 \\ 6-1-6 \end{bmatrix} = \begin{bmatrix} -1 \\ -6 \\ -1 \end{bmatrix}.$$

In Examples 3.3 we see what happens when we add a matrix to itself. Each entry is added to itself. In other words, each entry is multiplied by 2. This obviously extends to the case where we add a matrix to itself three times or four times or any number of times. It is convenient, therefore, to introduce the idea of multiplication of a matrix (or a vector) by a number. Notice that the definition applies for any real number, not just for integers. To multiply a matrix by a number, just multiply each entry by the number. In formal terms, if A is the $p \times q$ matrix whose (i, j)-entry is a_{ij} and if k is any number, then kA is the $p \times q$ matrix whose (i, j)-entry is ka_{ij}. See Examples 3.4.

Multiplication of a matrix with a vector or with another matrix is more complicated. Example 3.5 provides some motivation. The three left-hand sides are taken as a column vector, and this column vector is the result of multiplying the 3×3 matrix of coefficients with the 3×1 matrix (3-vector) of the unknowns. In general:

$$\begin{bmatrix} a_{11} & a_{12} & a_{13} \\ a_{21} & a_{22} & a_{23} \\ a_{31} & a_{32} & a_{33} \end{bmatrix} \begin{bmatrix} x_1 \\ x_2 \\ x_3 \end{bmatrix} = \begin{bmatrix} a_{11}x_1 + a_{12}x_2 + a_{13}x_3 \\ a_{21}x_1 + a_{22}x_2 + a_{23}x_3 \\ a_{31}x_1 + a_{32}x_2 + a_{33}x_3 \end{bmatrix}.$$

Note that the right-hand side is a column vector. Further illustrations are given in Examples 3.6. This idea can be applied to any set of simultaneous equations, no matter how many unknowns or how many equations. The left-hand side can be represented as a product of a matrix with a column vector. A set of p equations in q unknowns involves a $p \times q$ matrix multiplied to a q-vector.

Now let us abstract the idea. Can we multiply any matrix with any column vector? Not by the above process. To make that work, there must be as many columns in the matrix as there are entries in the column vector. A $p \times q$ matrix can be multiplied on the right by a column vector only if it has q entries. The result of the multiplication is then a column vector with p entries. We just reverse the above process. See Examples 3.7.

3.8 Evaluate the product

$$\begin{bmatrix} 1 & 2 & 3 \\ 2 & 3 & 4 \\ 4 & 5 & 6 \end{bmatrix} \begin{bmatrix} 1 & -1 \\ 3 & -2 \\ -1 & 1 \end{bmatrix}.$$

The product is a 3×2 matrix. The first column of the product is

$$\begin{bmatrix} 1 & 2 & 3 \\ 2 & 3 & 4 \\ 4 & 5 & 6 \end{bmatrix} \begin{bmatrix} 1 \\ 3 \\ -1 \end{bmatrix}, \quad \text{i.e.} \quad \begin{bmatrix} 1+6-3 \\ 2+9-4 \\ 4+15-6 \end{bmatrix}, \quad \text{i.e.} \quad \begin{bmatrix} 4 \\ 7 \\ 13 \end{bmatrix}.$$

The second column of the product is

$$\begin{bmatrix} 1 & 2 & 3 \\ 2 & 3 & 4 \\ 4 & 5 & 6 \end{bmatrix} \begin{bmatrix} -1 \\ -2 \\ 1 \end{bmatrix}, \quad \text{i.e.} \quad \begin{bmatrix} -1-4+3 \\ -2-6+4 \\ -4-10+6 \end{bmatrix}, \quad \text{i.e.} \quad \begin{bmatrix} -2 \\ -4 \\ -8 \end{bmatrix}.$$

Hence the product matrix is

$$\begin{bmatrix} 4 & -2 \\ 7 & -4 \\ 13 & -8 \end{bmatrix}.$$

3.9 Evaluation of matrix products.

(i)
$$\begin{bmatrix} 1 & 2 \\ 3 & 4 \end{bmatrix} \begin{bmatrix} 1 & 0 & -1 \\ 0 & 1 & -1 \end{bmatrix} = \begin{bmatrix} 1+0 & 0+2 & -1-2 \\ 3+0 & 0+4 & -3-4 \end{bmatrix} = \begin{bmatrix} 1 & 2 & -3 \\ 3 & 4 & -7 \end{bmatrix}.$$

(ii)
$$[1 \quad -1] \begin{bmatrix} 2 & 1 \\ 1 & 1 \end{bmatrix} = [2-1 \quad 1+1] = [1 \quad 2].$$

(iii)
$$\begin{bmatrix} 1 & 0 & 1 \\ 0 & 1 & 1 \\ 1 & 1 & 0 \end{bmatrix} \begin{bmatrix} 0 & 0 & 1 \\ 0 & 1 & 0 \\ 1 & 0 & 0 \end{bmatrix} = \begin{bmatrix} 0+0+1 & 0+0+0 & 1+0+0 \\ 0+0+1 & 0+1+0 & 0+0+0 \\ 0+0+0 & 0+1+0 & 1+0+0 \end{bmatrix} = \begin{bmatrix} 1 & 0 & 1 \\ 1 & 1 & 0 \\ 0 & 1 & 1 \end{bmatrix}.$$

(iv)
$$\begin{bmatrix} 1 & 0 & 1 & 1 \\ 0 & 2 & -1 & 3 \end{bmatrix} \begin{bmatrix} 0 & 0 \\ 1 & 2 \\ -1 & 1 \\ 3 & -2 \end{bmatrix} = \begin{bmatrix} 0+0-1+3 & 0+0+1-2 \\ 0+2+1+9 & 0+4-1-6 \end{bmatrix} = \begin{bmatrix} 2 & -1 \\ 12 & -3 \end{bmatrix}.$$

Next we take this further, and say what is meant by the product of two matrices. The process is illustrated by Example 3.8. The columns of the product matrix are calculated in turn by finding the products of the left-hand matrix with, separately, each of the columns of the right-hand matrix. Let A be a $p \times q$ matrix whose (i, j)-entry is a_{ij}, and let B be a $q \times r$ matrix whose (i, j)-entry is b_{ij}. Then the product AB is a $p \times r$ matrix whose (i, j)-entry is $\sum_{k=1}^{q} a_{ik}b_{kj}$, i.e. the sum of all the products of the entries in the ith row of A with the respective entries in the jth column of B.

Rule

A $p \times q$ matrix can be multiplied on the right only by a matrix with q rows. If A is a $p \times q$ matrix and B is a $q \times r$ matrix, then the product AB is a $p \times r$ matrix.

There is a useful mnemonic here. We can think of matrices as dominoes. A p, q domino can be laid next to a q, r domino, and the resulting 'free' numbers are p and r.

Examples 3.9 illustrate the procedures in calculating products. It is important to notice that given matrices can be multiplied only if they have appropriate sizes, and that it may be possible to multiply matrices in one order but not in the reverse order.

The most important case of matrix multiplication is multiplication of a matrix by a column vector, so before we move on to consider properties of the general multiplication, let us recap the application to simultaneous equations. A set of simultaneous equations containing p equations in q unknowns can always be represented as a matrix equation of the form

$$A\boldsymbol{x} = \boldsymbol{h},$$

where A is a $p \times q$ matrix, $\boldsymbol{x}$ is a q-vector whose entries are the unknowns, and $\boldsymbol{h}$ is the p-vector whose entries are the right-hand sides of the given equations.

Rules

(i) $A + B = B + A$

(ii) $(A + B) + C = A + (B + C)$

(iii) $k(A + B) = kA + kB$

(iv) $(kA)B = k(AB)$

(v) $(AB)C = A(BC)$

(vi) $A(B + C) = AB + AC$

(vii) $(A + B)C = AC + BC$,

where A, B and C are any matrices whose sizes permit the formation of these sums and products, and k is any real number.

3.10 Show that for any $p\times q$ matrix A and any $q\times r$ matrices B and C,

$$A(B+C)=AB+AC.$$

Let a_{ij} denote the (i,j)-entry in A, for $1\leqslant i\leqslant p$ and $1\leqslant j\leqslant q$, let b_{ij} denote the (i,j)-entry in B, for $1\leqslant i\leqslant q$ and $1\leqslant j\leqslant r$, and let c_{ij} denote the (i,j)-entry in C, for $1\leqslant i\leqslant q$ and $1\leqslant j\leqslant r$. The (k,j)-entry in $B+C$ is then $b_{kj}+c_{kj}$. By the definition, then, the (i,j)-entry in $A(B+C)$ is

$$\sum_{k=1}^{q} a_{ik}(b_{kj}+c_{kj}),$$

i.e.

$$\sum_{k=1}^{q} a_{ik}b_{kj}+\sum_{k=1}^{q} a_{ik}c_{kj},$$

which is just the sum of the (i,j)-entries in AB and in AC. Hence

$$A(B+C)=AB+AC.$$

3.11 The commutative law fails for matrix multiplication. Let

$$A=\begin{bmatrix}1 & 2\\ 3 & 4\end{bmatrix},$$

and let

$$B=\begin{bmatrix}1 & 1\\ 0 & 1\end{bmatrix}.$$

Certainly both products AB and BA exist. Their values are different, however, as we can verify by direct calculation.

$$AB=\begin{bmatrix}1 & 2\\ 3 & 4\end{bmatrix}\begin{bmatrix}1 & 1\\ 0 & 1\end{bmatrix}=\begin{bmatrix}1 & 3\\ 3 & 7\end{bmatrix},$$

and

$$BA=\begin{bmatrix}1 & 1\\ 0 & 1\end{bmatrix}\begin{bmatrix}1 & 2\\ 3 & 4\end{bmatrix}=\begin{bmatrix}4 & 6\\ 3 & 4\end{bmatrix}.$$

Rules (i), (ii), (iii) and (iv) are easy to verify. They reflect corresponding properties of numbers, since the operations involved correspond to simple operations on the entries of the matrices. Rules (v), (vi) and (vii), while being convenient and familiar, are by no means obviously true. Proofs of them are intricate, but require no advanced methods. To illustrate the ideas, the proof of (vi) is given as Example 3.10.

There is one algebraic rule which is conspicuously absent from the above list. Multiplication of matrices does not satisfy the commutative law. The products AB and BA, even if they can both be formed, in general are not the same. See Example 3.11. This can lead to difficulties unless we are careful, particularly when multiplying out bracketed expressions. Consider the following:

$$(A+B)(A+B)=AA+AB+BA+BB,$$

so

$$(A+B)^2=A^2+AB+BA+B^2,$$

and the result must be left in this form, different from the usual expression for the square of a sum.

Finally a word about notation. Matrices we denote by upper case letters $A, B, C, \ldots, X, Y, Z, \ldots$. Column vectors we denote by bold-face lower case letters $\mathbf{a}, \mathbf{b}, \mathbf{c}, \ldots, \mathbf{x}, \mathbf{y}, \mathbf{z}, \ldots$. Thankfully, this is one situation where there is a notation which is almost universal.

Summary

Procedures for adding and multiplying vectors and matrices are given, together with rules for when sums and products can be formed. The algebraic laws satisfied by these operations are listed. It is shown how to write a set of simultaneous linear equations as a matrix equation.

Exercises

1. In each case below, evaluate the matrices $A+B$, $A\boldsymbol{x}$, $B\boldsymbol{x}$, $3A$, $\frac{1}{2}B$, where A, B and $\boldsymbol{x}$ are as given.

(i) $A=\begin{bmatrix}1 & 2\\ 3 & 4\end{bmatrix}$, $B=\begin{bmatrix}1 & -1\\ -1 & 1\end{bmatrix}$, $\boldsymbol{x}=\begin{bmatrix}x_1\\ x_2\end{bmatrix}$.

(ii) $A=\begin{bmatrix}3 & 0\\ 1 & 1\end{bmatrix}$, $B=\begin{bmatrix}-2 & 1\\ 0 & 2\end{bmatrix}$, $\boldsymbol{x}=\begin{bmatrix}1\\ 4\end{bmatrix}$.

(iii) $A=\begin{bmatrix}1 & -1 & 2\\ 0 & 1 & 1\\ 2 & 3 & -3\end{bmatrix}$, $B=\begin{bmatrix}0 & 0 & -1\\ 1 & 2 & 2\\ -3 & 1 & 0\end{bmatrix}$, $\boldsymbol{x}=\begin{bmatrix}x_1\\ x_2\\ x_3\end{bmatrix}$.

(iv) $A=\begin{bmatrix}-2 & -1 & 1\\ 0 & 1 & 4\\ 6 & -2 & 1\end{bmatrix}$, $B=\begin{bmatrix}1 & 1 & 1\\ 1 & 1 & 1\\ 1 & 1 & 1\end{bmatrix}$, $\boldsymbol{x}=\begin{bmatrix}2\\ 1\\ -1\end{bmatrix}$.

2. Evaluate all the following products of a matrix with a vector.

(i) $\begin{bmatrix}1 & 2 & -2\\ 3 & -1 & -1\\ -2 & 2 & 0\end{bmatrix}\begin{bmatrix}1\\ 2\\ 3\end{bmatrix}$.

(ii) $\begin{bmatrix}2 & 2\\ 1 & -2\\ 0 & 3\end{bmatrix}\begin{bmatrix}1\\ 1\end{bmatrix}$.

(iii) $\begin{bmatrix}1 & 1 & -1 & 3\\ 2 & -2 & 0 & 1\end{bmatrix}\begin{bmatrix}0\\ 1\\ 1\\ -1\end{bmatrix}$.

(iv) $\begin{bmatrix}2 & 2\\ 1 & -2\\ 0 & 3\\ 3 & 0\\ 0 & 0\end{bmatrix}\begin{bmatrix}1\\ 1\end{bmatrix}$.

(v) $\begin{bmatrix}1 & 1 & 1 & 1 & 1\\ 1 & 1 & 1 & 1 & 1\\ 1 & 1 & 1 & 1 & 1\end{bmatrix}\begin{bmatrix}1\\ 2\\ 3\\ 4\\ 5\end{bmatrix}$.

(vi) $\begin{bmatrix}1 & 0 & 0\\ 0 & 1 & 0\\ 0 & 0 & 1\end{bmatrix}\begin{bmatrix}-2\\ 1\\ 3\end{bmatrix}$.

3. Let

$$A=\begin{bmatrix}5 & 2 & 3\\ 2 & -3 & 4\end{bmatrix},\quad B=\begin{bmatrix}2 & -1 & 1 & 0\\ 0 & 2 & 2 & 2\\ 3 & 0 & -1 & 3\end{bmatrix},$$

$$C=\begin{bmatrix}1 & 0 & 2\\ 2 & -3 & 0\\ 0 & 0 & 3\\ 2 & 1 & 0\end{bmatrix},\quad D=\begin{bmatrix}2 & -1\\ 1 & 2\\ 3 & -2\end{bmatrix}.$$

Evaluate the products AB, AD, BC, CB and CD. Is there any other product of two of these matrices which exists? Evaluate any such. Evaluate the products $A(BC)$ and $(AB)C$.

4. Evaluate the following matrix products.

(i) $\begin{bmatrix}2 & 1\\ -1 & 1\\ 3 & 2\end{bmatrix}\begin{bmatrix}3 & 4\\ 5 & 2\end{bmatrix}$.

(ii) $$\begin{bmatrix} 1 & 0 & 1 & 0 & 1 \\ 2 & 3 & -1 & 1 & 0 \\ 2 & 1 & -2 & -1 & 1 \\ 1 & 4 & 3 & -3 & 1 \end{bmatrix} \times \begin{bmatrix} 2 & 2 & 1 \\ -1 & -1 & 2 \\ 0 & 1 & 0 \\ -3 & -2 & 2 \\ 1 & 1 & 1 \end{bmatrix}.$$

(iii) $$\begin{bmatrix} 0 & 1 \\ 1 & 0 \end{bmatrix}\begin{bmatrix} 1 & 2 \\ 3 & 4 \end{bmatrix}.$$

(iv) $$\begin{bmatrix} 2 & -1 & 0 \\ 1 & -1 & 4 \end{bmatrix}\begin{bmatrix} 0 & 1 & 1 \\ -2 & 3 & 3 \\ 1 & 2 & -1 \end{bmatrix}\begin{bmatrix} 1 \\ -1 \\ 1 \end{bmatrix}.$$

5. Obtain A^3-2A^2+A-I, when

$$A=\begin{bmatrix} 1 & 1 & 2 \\ 1 & 1 & 1 \\ 2 & 1 & 1 \end{bmatrix}.$$

6. How must the sizes of matrices A and B be related in order for both of the products AB and BA to exist?

Examples

4.1 Properties of a zero matrix.

$$\begin{bmatrix}0&0\\0&0\end{bmatrix}+\begin{bmatrix}a&b\\c&d\end{bmatrix}=\begin{bmatrix}a&b\\c&d\end{bmatrix},$$

$$\begin{bmatrix}0&0\\0&0\end{bmatrix}\begin{bmatrix}a&b\\c&d\end{bmatrix}=\begin{bmatrix}0&0\\0&0\end{bmatrix},$$

$$\begin{bmatrix}a&b\\c&d\end{bmatrix}\begin{bmatrix}0&0\\0&0\end{bmatrix}=\begin{bmatrix}0&0\\0&0\end{bmatrix},$$

$$\begin{bmatrix}0&0\\0&0\end{bmatrix}\begin{bmatrix}a&b&c\\d&e&f\end{bmatrix}=\begin{bmatrix}0&0&0\\0&0&0\end{bmatrix}.$$

4.2 Properties of an identity matrix.

$$\begin{bmatrix}1&0\\0&1\end{bmatrix}\begin{bmatrix}a&b\\c&d\end{bmatrix}=\begin{bmatrix}a&b\\c&d\end{bmatrix},$$

$$\begin{bmatrix}a&b\\c&d\end{bmatrix}\begin{bmatrix}1&0\\0&1\end{bmatrix}=\begin{bmatrix}a&b\\c&d\end{bmatrix},$$

$$\begin{bmatrix}1&0&0\\0&1&0\\0&0&1\end{bmatrix}\begin{bmatrix}a&b&c\\d&e&f\\g&h&k\end{bmatrix}=\begin{bmatrix}a&b&c\\d&e&f\\g&h&k\end{bmatrix},$$

$$\begin{bmatrix}a&b&c\\d&e&f\\g&h&k\end{bmatrix}\begin{bmatrix}1&0&0\\0&1&0\\0&0&1\end{bmatrix}=\begin{bmatrix}a&b&c\\d&e&f\\g&h&k\end{bmatrix}.$$

4.3 Examples of diagonal matrices.

$$\begin{bmatrix}3&0\\0&2\end{bmatrix},\quad\begin{bmatrix}3&0\\0&0\end{bmatrix},\quad\begin{bmatrix}-1&0\\0&-1\end{bmatrix},$$

$$\begin{bmatrix}6&0&0\\0&-2&0\\0&0&1\end{bmatrix},\quad\begin{bmatrix}2&0&0\\0&2&0\\0&0&2\end{bmatrix},\quad\begin{bmatrix}0&0&0\\0&1&0\\0&0&0\end{bmatrix},$$

$$\begin{bmatrix}1&0&0&0\\0&2&0&0\\0&0&3&0\\0&0&0&4\end{bmatrix}.$$

4.4 The following matrices are upper triangular.

$$\begin{bmatrix}1&2\\0&1\end{bmatrix},\quad\begin{bmatrix}1&1&3\\0&-2&2\\0&0&-1\end{bmatrix},\quad\begin{bmatrix}2&1&0\\0&1&1\\0&0&1\end{bmatrix},\quad\begin{bmatrix}0&1&1\\0&0&2\\0&0&1\end{bmatrix}.$$

The following matrices are lower triangular.

$$\begin{bmatrix}1&0\\2&1\end{bmatrix},\quad\begin{bmatrix}1&0&0\\2&-2&0\\3&1&-1\end{bmatrix},\quad\begin{bmatrix}2&0&0\\1&1&0\\0&1&1\end{bmatrix},\quad\begin{bmatrix}0&0&0\\1&0&0\\1&2&1\end{bmatrix}.$$

4

Special matrices

Example 4.1 shows the properties of a *zero* matrix. This form of special matrix does not raise any problems. A matrix which consists entirely of 0s (a zero matrix) behaves just as we would expect. We normally use 0 (zero) to denote a zero matrix, and **0** to denote a zero column vector. Of course we should bear in mind that there are many zero matrices having different sizes.

From matrices which act like zero we turn to matrices which act like 1. A square matrix which has 1s down the diagonal from top left to bottom right (this diagonal is called the main diagonal) and has 0s elsewhere is called an *identity matrix*. Example 4.2 shows the property which such matrices have, namely

$$AI = IA = A,$$

where I is an identity matrix and A is a square matrix of the same size. Notice that identity matrices are square and that there is one $p \times p$ identity matrix for each number p. We denote it by I_p, or just I if the size is not important.

There are other sorts of special matrix which are distinctive because of their algebraic properties or because of their appearance (or both). We describe some types here, although their significance will not be clear till later.

A *diagonal matrix* is a square matrix which has zero entries at all points off the main diagonal. One particular sort of diagonal matrix is an identity matrix. Other examples are given in Examples 4.3. Of course we do not insist that all the entries on the main diagonal are non-zero. We might even consider a zero matrix to be a diagonal matrix. The sum and product of two $p \times p$ diagonal matrices are $p \times p$ diagonal matrices.

The main diagonal divides a square matrix into two triangles. A square matrix which has zeros at all positions below the main diagonal is called an

4.5 Sums and products of triangular matrices.

(i) A sum of upper triangular matrices is upper triangular.

$$\begin{bmatrix} 1 & 1 & 3 \\ 0 & -2 & 2 \\ 0 & 0 & -1 \end{bmatrix} + \begin{bmatrix} 2 & 1 & 0 \\ 0 & 1 & 1 \\ 0 & 0 & 1 \end{bmatrix} = \begin{bmatrix} 3 & 2 & 3 \\ 0 & -1 & 3 \\ 0 & 0 & 0 \end{bmatrix}.$$

(ii) A product of upper triangular matrices is upper triangular.

$$\begin{bmatrix} 1 & 1 & 3 \\ 0 & -2 & 2 \\ 0 & 0 & -1 \end{bmatrix} \begin{bmatrix} 2 & 1 & 0 \\ 0 & 1 & 1 \\ 0 & 0 & 1 \end{bmatrix} = \begin{bmatrix} 2 & 2 & 4 \\ 0 & -2 & 0 \\ 0 & 0 & -1 \end{bmatrix}.$$

(iii) A product of lower triangular matrices is lower triangular.

$$\begin{bmatrix} 1 & 0 & 0 \\ 2 & -2 & 0 \\ 3 & 1 & 1 \end{bmatrix} \begin{bmatrix} 2 & 0 & 0 \\ 1 & 1 & 0 \\ 0 & 1 & 1 \end{bmatrix} = \begin{bmatrix} 2 & 0 & 0 \\ 2 & -2 & 0 \\ 7 & 2 & 1 \end{bmatrix}.$$

4.6 Examples of transposed matrices.

$$\begin{bmatrix} 1 & 2 \\ 3 & 4 \end{bmatrix}^{\mathrm{T}} = \begin{bmatrix} 1 & 3 \\ 2 & 4 \end{bmatrix}, \quad \begin{bmatrix} 1 & 2 & 3 \\ 4 & 5 & 6 \end{bmatrix}^{\mathrm{T}} = \begin{bmatrix} 1 & 4 \\ 2 & 5 \\ 3 & 6 \end{bmatrix},$$

$$[1 \;\; 2 \;\; 3]^{\mathrm{T}} = \begin{bmatrix} 1 \\ 2 \\ 3 \end{bmatrix}, \quad \text{and} \quad \begin{bmatrix} 1 \\ 2 \\ 3 \end{bmatrix}^{\mathrm{T}} = [1 \;\; 2 \;\; 3],$$

$$\begin{bmatrix} 3 & 1 & 2 \\ 1 & 1 & -4 \\ 2 & -4 & -1 \end{bmatrix}^{\mathrm{T}} = \begin{bmatrix} 3 & 1 & 2 \\ 1 & 1 & -4 \\ 2 & -4 & -1 \end{bmatrix},$$

so this matrix is symmetric.

4.7 A sum of symmetric matrices is symmetric. Let A and B be symmetric matrices with the same size. Then $A^{\mathrm{T}} = A$ and $B^{\mathrm{T}} = B$.

$$(A+B)^{\mathrm{T}} = A^{\mathrm{T}} + B^{\mathrm{T}} = A + B,$$

and so $A+B$ is symmetric. Here is a particular case:

$$A = \begin{bmatrix} 3 & 1 & 2 \\ 1 & 1 & -4 \\ 2 & -4 & -1 \end{bmatrix}, \quad B = \begin{bmatrix} 1 & 2 & -1 \\ 2 & 2 & 0 \\ -1 & 0 & 3 \end{bmatrix}.$$

Then A and B are both symmetric, and

$$A+B = \begin{bmatrix} 4 & 3 & 1 \\ 3 & 3 & -4 \\ 1 & -4 & 2 \end{bmatrix},$$

which is symmetric.

A product of two symmetric matrices is generally *not* symmetric. With A and B as above,

$$AB = \begin{bmatrix} 3 & 8 & 3 \\ 7 & 4 & -13 \\ -5 & -4 & -5 \end{bmatrix},$$

which is not symmetric.

upper triangular matrix. A square matrix which has zeros at all positions above the main diagonal is called a *lower triangular* matrix. A matrix of one or other of these kinds is called a triangular matrix. Such matrices have convenient properties which make them useful in some applications. But we can see now, as in Example 4.5, that sums and products of upper (lower) triangular matrices are upper (lower) triangular. Notice that when the GE process is applied to a square matrix the result is always an upper triangular matrix.

The main diagonal also plays a part in our next kind of special matrix. A square matrix is *symmetric* if reflection in the main diagonal leaves the matrix unchanged. In formal terms, if A is any matrix whose (i,j)-entry is a_{ij}, the *transpose* of A (denoted by A^{T}) is the matrix whose (i,j)-entry is a_{ji}, i.e. the matrix obtained by reflecting in the main diagonal. A is symmetric if $A^{\mathrm{T}}=A$. Notice that the rows of A^{T} are the columns of A, and vice versa. See Example 4.6. Such matrices figure prominently in more advanced work, but we can see now (Example 4.7) that sums of symmetric matrices are symmetric, but products in general are not. There are three important rules about transposes.

4.8 Let A and B be any $p \times p$ matrices. Then $(AB)^T = B^T A^T$. To see this, let the (i, j)-entries of A and B be denoted by a_{ij} and b_{ij} respectively. The (i, j)-entry in $(AB)^T$ is the (j, i)-entry in AB, which is

$$\sum_{k=1}^{p} a_{jk} b_{ki}.$$

The (i, j)-entry in $B^T A^T$ is

$$\sum_{k=1}^{p} b^T_{ik} a^T_{kj},$$

where b^T_{ik} is the (i, k)-entry in B^T and a^T_{kj} is the (k, j)-entry in A^T. Now from the definition of the transpose, we have

$$b^T_{ik} = b_{ki} \quad \text{and} \quad a^T_{kj} = a_{jk}.$$

Hence the (i, j)-entry in $B^T A^T$ is

$$\sum_{k=1}^{p} b_{ki} a_{jk}, \quad \text{i.e.} \quad \sum_{k=1}^{p} a_{jk} b_{ki},$$

which is the same as the (i, j)-entry in $(AB)^T$. This proves the result.

4.9 Examples of skew-symmetric matrices.

(i) $$\begin{bmatrix} 0 & 2 \\ -2 & 0 \end{bmatrix}^T = \begin{bmatrix} 0 & -2 \\ 2 & 0 \end{bmatrix} = -\begin{bmatrix} 0 & 2 \\ -2 & 0 \end{bmatrix}.$$

(ii) $$\begin{bmatrix} 0 & 1 & 2 \\ -1 & 0 & -3 \\ -2 & 3 & 0 \end{bmatrix}^T = \begin{bmatrix} 0 & -1 & -2 \\ 1 & 0 & 3 \\ 2 & -3 & 0 \end{bmatrix} = -\begin{bmatrix} 0 & 1 & 2 \\ -1 & 0 & -3 \\ -2 & 3 & 0 \end{bmatrix}.$$

4.10 Examples of orthogonal matrices.

(i) Let $A = \begin{bmatrix} \frac{1}{\sqrt{2}} & \frac{1}{\sqrt{2}} \\ -\frac{1}{\sqrt{2}} & \frac{1}{\sqrt{2}} \end{bmatrix}$. Then $A^T = \begin{bmatrix} \frac{1}{\sqrt{2}} & -\frac{1}{\sqrt{2}} \\ \frac{1}{\sqrt{2}} & \frac{1}{\sqrt{2}} \end{bmatrix}$,

so

$$A^T A = \begin{bmatrix} \frac{1}{2}+\frac{1}{2} & \frac{1}{2}-\frac{1}{2} \\ \frac{1}{2}-\frac{1}{2} & \frac{1}{2}+\frac{1}{2} \end{bmatrix} = \begin{bmatrix} 1 & 0 \\ 0 & 1 \end{bmatrix},$$

and

$$AA^T = \begin{bmatrix} \frac{1}{2}+\frac{1}{2} & -\frac{1}{2}+\frac{1}{2} \\ -\frac{1}{2}+\frac{1}{2} & \frac{1}{2}+\frac{1}{2} \end{bmatrix} = \begin{bmatrix} 1 & 0 \\ 0 & 1 \end{bmatrix}.$$

Hence A is orthogonal.

(ii) Let $B = \begin{bmatrix} \frac{2}{3} & -\frac{1}{3} & \frac{2}{3} \\ \frac{2}{3} & \frac{2}{3} & -\frac{1}{3} \\ -\frac{1}{3} & \frac{2}{3} & \frac{2}{3} \end{bmatrix}$. Then $B^T = \begin{bmatrix} \frac{2}{3} & \frac{2}{3} & -\frac{1}{3} \\ -\frac{1}{3} & \frac{2}{3} & \frac{2}{3} \\ \frac{2}{3} & -\frac{1}{3} & \frac{2}{3} \end{bmatrix}$.

Then by direct evaluation we verify that $B^T B = I$ and $BB^T = I$.

Rules

(i) $(A^{\mathrm{T}})^{\mathrm{T}} = A$.
(ii) $(A+B)^{\mathrm{T}} = A^{\mathrm{T}} + B^{\mathrm{T}}$,
(iii) $(AB)^{\mathrm{T}} = B^{\mathrm{T}}A^{\mathrm{T}}$.

The last of these is important because of the change of the order of the multiplication. Remember this! The first two are quite easy to justify. The third is rather intricate, though not essentially difficult. A proof is given in Example 4.8.

The transpose of a matrix A may be related to A in other ways. A *skew-symmetric* matrix is a matrix for which $A^{\mathrm{T}} = -A$. See Example 4.9. An *orthogonal* matrix is a square matrix for which $A^{\mathrm{T}}A = I$ and $AA^{\mathrm{T}} = I$. See Example 4.10.

4.11 Examples of elementary matrices.

$$E_1=\begin{bmatrix}0&1&0\\1&0&0\\0&0&1\end{bmatrix},$$

obtained by interchanging the first two rows of an identity matrix.

$$E_2=\begin{bmatrix}1&0&0\\0&1&0\\0&0&5\end{bmatrix},$$

obtained by multiplying the third row of an identity matrix by 5.

$$E_3=\begin{bmatrix}1&3&0\\0&1&0\\0&0&1\end{bmatrix},$$

obtained by adding three times the second row to the first row in an identity matrix.

4.12 Let

$$A=\begin{bmatrix}1&2&3\\4&5&6\\7&8&9\end{bmatrix}.$$

Check the effects of premultiplying A by E_1, E_2 and E_3 above.

$$E_1A=\begin{bmatrix}0&1&0\\1&0&0\\0&0&1\end{bmatrix}\begin{bmatrix}1&2&3\\4&5&6\\7&8&9\end{bmatrix}=\begin{bmatrix}4&5&6\\1&2&3\\7&8&9\end{bmatrix}.$$

(The first two rows are interchanged.)

$$E_2A=\begin{bmatrix}1&0&0\\0&1&0\\0&0&5\end{bmatrix}\begin{bmatrix}1&2&3\\4&5&6\\7&8&9\end{bmatrix}=\begin{bmatrix}1&2&3\\4&5&6\\35&40&45\end{bmatrix}.$$

(The third row is multiplied by 5.)

$$E_3A=\begin{bmatrix}1&3&0\\0&1&0\\0&0&1\end{bmatrix}\begin{bmatrix}1&2&3\\4&5&6\\7&8&9\end{bmatrix}=\begin{bmatrix}13&17&21\\4&5&6\\7&8&9\end{bmatrix}.$$

(Three times the second row is added to the first row.)

Examples 4.11 illustrate the notion of *elementary* matrix. An elementary matrix is a square matrix which is obtained from an identity matrix by the application of a single elementary row operation (see Chapter 1). The significance of such matrices lies in the following. Let E be obtained from a $p \times p$ identity matrix by application of a single elementary row operation, and let A be any $p \times q$ matrix. Then the product matrix EA is the same as the matrix obtained from A by applying the same elementary row operation directly to it. Examples 4.12 illustrate this. Our knowledge of the GE process enables us to say: given any square matrix A, there exists a sequence $E_1, E_2, \ldots, E_r$ of elementary matrices such that the product $E_rE_{r-1} \ldots E_2E_1A$ is an upper triangular matrix. These elementary matrices correspond to the elementary row operations carried out in the course of the GE process. For an explicit case of this, see Example 4.13.

Another important property of elementary matrices arises from the preceding discussion. Let E be an elementary matrix, obtained from an identity matrix by application of a single elementary row operation. Certainly E can be converted back into the identity matrix by application of another elementary row operation. Let F be the elementary matrix corresponding (as above) to this elementary row operation. Then $FE=I$. The two row operations cancel each other out, and the two elementary matrices correspondingly combine to give the identity matrix. It is not hard to see that $EF=I$ here also. Such matrices are called inverses of each other. We shall discuss that idea at length later. Examples 4.14 show some elementary matrices and their inverses. The reader should check that their products are identity matrices. Also, from the definition of an orthogonal matrix it is apparent that an orthogonal matrix and its transpose are inverses of each other.

Summary

Various special kinds of matrices are described: zero matrices, identity, diagonal, triangular, symmetric, skew-symmetric, orthogonal and elementary matrices. Some algebraic properties of these are discussed. The transpose of a square matrix is defined, and rules for transposition of sums and products are given. The correspondence between elementary matrices and elementary row operations is pointed out.

4.13 Find a sequence $E_1, E_2, \ldots, E_r$ of elementary matrices such that the product $E_r E_{r-1} \ldots E_1 A$ is an upper triangular matrix, where

$$A=\begin{bmatrix} 0 & 1 & -3 & 2 \\ 1 & 2 & 1 & 1 \\ 1 & 1 & 4 & 2 \end{bmatrix}.$$

We proceed with the standard GE process, noting the elementary matrix which corresponds to each row operation.

$$\begin{bmatrix} 0 & 1 & -3 & 2 \\ 1 & 2 & 1 & 1 \\ 1 & 1 & 4 & 2 \end{bmatrix}$$

$$\begin{bmatrix} 1 & 2 & 1 & 1 \\ 0 & 1 & -3 & 2 \\ 1 & 1 & 4 & 2 \end{bmatrix} \text{ interchange rows} \qquad E_1=\begin{bmatrix} 0 & 1 & 0 \\ 1 & 0 & 0 \\ 0 & 0 & 1 \end{bmatrix}$$

$$\begin{bmatrix} 1 & 2 & 1 & 1 \\ 0 & 1 & -3 & 2 \\ 0 & -1 & 3 & 1 \end{bmatrix} \;(3)-(1) \qquad E_2=\begin{bmatrix} 1 & 0 & 0 \\ 0 & 1 & 1 \\ -1 & 0 & 1 \end{bmatrix}$$

$$\begin{bmatrix} 1 & 2 & 1 & 1 \\ 0 & 1 & -3 & 2 \\ 0 & 0 & 0 & 3 \end{bmatrix} \;(3)+(2) \qquad E_3=\begin{bmatrix} 1 & 0 & 0 \\ 0 & 1 & 0 \\ 0 & 1 & 1 \end{bmatrix}$$

$$\begin{bmatrix} 1 & 2 & 1 & 1 \\ 0 & 1 & -3 & 2 \\ 0 & 0 & 0 & 1 \end{bmatrix} \;(3)\div 3 \qquad E_4=\begin{bmatrix} 1 & 0 & 0 \\ 0 & 1 & 0 \\ 0 & 0 & \frac{1}{3} \end{bmatrix}.$$

Hence

$$E_4E_3E_2E_1A=\begin{bmatrix} 1 & 2 & 1 & 1 \\ 0 & 1 & -3 & 2 \\ 0 & 0 & 0 & 1 \end{bmatrix}.$$

4.14 Elementary matrices and their inverses.

$$\begin{bmatrix} 0 & 1 & 0 \\ 1 & 0 & 0 \\ 0 & 0 & 1 \end{bmatrix} \text{ has inverse } \begin{bmatrix} 0 & 1 & 0 \\ 1 & 0 & 0 \\ 0 & 0 & 1 \end{bmatrix}.$$

$$\begin{bmatrix} 1 & 0 & 0 \\ 0 & 1 & 0 \\ 0 & 0 & 5 \end{bmatrix} \text{ has inverse } \begin{bmatrix} 1 & 0 & 0 \\ 0 & 1 & 0 \\ 0 & 0 & \frac{1}{5} \end{bmatrix}.$$

$$\begin{bmatrix} 1 & 3 & 0 \\ 0 & 1 & 0 \\ 0 & 0 & 1 \end{bmatrix} \text{ has inverse } \begin{bmatrix} 1 & -3 & 0 \\ 0 & 1 & 0 \\ 0 & 0 & 0 \end{bmatrix}.$$

The way to see these is to consider the effect of premultiplying by first one and then the other of each given pair. The second 'undoes' the effect of the first.

Exercises

1. Evaluate A^2, A^3, and A^4, where

$$A=\begin{bmatrix}1&1&1\\0&1&1\\0&0&1\end{bmatrix}.$$

Carry out the same calculations for the matrix

$$B=\begin{bmatrix}1&0&0\\1&1&0\\1&1&1\end{bmatrix}.$$

2. Let I be the 3×3 identity matrix. Show that $AI=A$ whenever A is a 2×3 matrix. Does this hold for any $p\times 3$ matrix A, irrespective of the value of p? Likewise, is it the case that $IB=B$ for every $3\times q$ matrix B?

3. In each case below, evaluate AB, where A and B are as given.

(i) $A=\begin{bmatrix}1&-1&2\\0&2&1\\0&0&-1\end{bmatrix}$, $B=\begin{bmatrix}0&1&1\\0&2&-2\\0&0&1\end{bmatrix}$.

(ii) $A=\begin{bmatrix}1&2&1\\0&1&-2\\0&0&1\end{bmatrix}$, $B=\begin{bmatrix}x_1\\x_2\\x_3\end{bmatrix}$.

(iii) $A=\begin{bmatrix}1&0&0\\2&-1&0\\-2&1&3\end{bmatrix}$, $B=\begin{bmatrix}-1&0&0\\1&2&0\\2&1&1\end{bmatrix}$.

(iv) $A=[1\quad 2\quad 3]$, $B=\begin{bmatrix}1&0&0\\1&1&0\\1&1&1\end{bmatrix}$.

4. Evaluate the product

$$[x_1\quad x_2\quad x_3]\begin{bmatrix}1&0&0\\-1&1&0\\3&-2&1\end{bmatrix}.$$

Hence find values of x_1, x_2 and x_3 for which the product is equal to $[-1\quad 4\quad 1]$.

5. Let A be any square matrix. Show that $A+A^{\mathrm{T}}$ is a symmetric matrix. Show also that the products $A^{\mathrm{T}}A$ and AA^{T} are symmetric matrices.

6. Which of the following matrices are symmetric, and which are skew-symmetric (and which are neither)?

$$\begin{bmatrix}1&2\\2&3\end{bmatrix},\quad \begin{bmatrix}1&2\\-2&3\end{bmatrix},\quad \begin{bmatrix}0&2\\-2&0\end{bmatrix},\quad \begin{bmatrix}1&2\\3&1\end{bmatrix},\quad \begin{bmatrix}1&0\\0&-1\end{bmatrix},$$

$$\begin{bmatrix}-1&0&1\\0&2&2\\2&1&-1\end{bmatrix},\quad \begin{bmatrix}0&1&-2\\-1&0&3\\2&-3&0\end{bmatrix},$$

$$\begin{bmatrix}1&2&0\\-2&0&-1\\0&1&1\end{bmatrix},\quad \begin{bmatrix}2&3&1\\3&0&-1\\1&-1&2\end{bmatrix},$$

$$\begin{bmatrix} 1 & 0 & 1 \\ 0 & 1 & 0 \\ 1 & 0 & 1 \end{bmatrix}, \quad \begin{bmatrix} 1 & 0 & -1 \\ 0 & 1 & 0 \\ 1 & 0 & 1 \end{bmatrix}, \quad \begin{bmatrix} 1 & 1 & 0 \\ 1 & 0 & -1 \\ 0 & -1 & -1 \end{bmatrix}.$$

7. Show that the following matrices are orthogonal.

$$\begin{bmatrix} \frac{1}{\sqrt{5}} & -\frac{2}{\sqrt{5}} \\ \frac{2}{\sqrt{5}} & \frac{1}{\sqrt{5}} \end{bmatrix}, \quad \begin{bmatrix} -\frac{3}{5} & \frac{4}{5} \\ \frac{4}{5} & \frac{3}{5} \end{bmatrix}, \quad \begin{bmatrix} 0 & \frac{2}{\sqrt{6}} & -\frac{1}{\sqrt{3}} \\ \frac{1}{\sqrt{2}} & \frac{1}{\sqrt{6}} & \frac{1}{\sqrt{3}} \\ -\frac{1}{\sqrt{2}} & \frac{1}{\sqrt{6}} & \frac{1}{\sqrt{3}} \end{bmatrix}.$$

8. Show that a product of two orthogonal matrices of the same size is an orthogonal matrix.

9. Describe in words the effect of premultiplying a 4×4 matrix by each of the elementary matrices below. Also in each case write down the elementary matrix which has the reverse effect.

(i) $\begin{bmatrix} 1 & 0 & 0 & 0 \\ 0 & 0 & 1 & 0 \\ 0 & 1 & 0 & 0 \\ 0 & 0 & 0 & 1 \end{bmatrix}$. (ii) $\begin{bmatrix} 1 & 0 & 2 & 0 \\ 0 & 1 & 0 & 0 \\ 0 & 0 & 1 & 0 \\ 0 & 0 & 0 & 1 \end{bmatrix}$.

(iii) $\begin{bmatrix} 1 & 0 & 0 & 0 \\ 0 & 1 & 0 & 0 \\ -3 & 0 & 1 & 0 \\ 0 & 0 & 0 & 1 \end{bmatrix}$. (iv) $\begin{bmatrix} 1 & 0 & 0 & 0 \\ 0 & -2 & 0 & 0 \\ 0 & 0 & 1 & 0 \\ 0 & 0 & 0 & 1 \end{bmatrix}$.

10. Apply the Gaussian elimination process to the matrix

$$A = \begin{bmatrix} 0 & 1 & 3 \\ 1 & 2 & -1 \\ 2 & 3 & 1 \end{bmatrix},$$

noting at each stage the elementary matrix corresponding to the row operation applied. Evaluate the product T of these elementary matrices and check your answer by evaluating the product TA (which should be the same as the result of the GE process).

11. Repeat Exercise 10, with the matrix

$$A = \begin{bmatrix} 1 & -1 & 2 & 1 \\ -1 & 3 & 0 & 1 \\ 2 & 1 & 1 & -1 \end{bmatrix}.$$

Examples

5.1 Show that the inverse of a matrix (if it exists) is unique.

Let $AB=BA=I$ (so that B satisfies the requirements for the inverse of A). Now suppose that $AX=XA=I$. Then

$$BAX=(BA)X=IX=X.$$

Also

$$BAX=B(AX)=BI=B.$$

Hence $X=B$. Consequently B is the only matrix with the properties of the inverse of A.

5.2 An example of a matrix which does not have an inverse is

$$\begin{bmatrix} 1 & -1 \\ -1 & 1 \end{bmatrix}.$$

There is no matrix B such that

$$\begin{bmatrix} 1 & -1 \\ -1 & 1 \end{bmatrix} B=I.$$

To see this, let

$$B=\begin{bmatrix} a & b \\ c & d \end{bmatrix}.$$

Then

$$\begin{bmatrix} 1 & -1 \\ -1 & 1 \end{bmatrix}\begin{bmatrix} a & b \\ c & d \end{bmatrix}=\begin{bmatrix} a-c & b-d \\ -a+c & -b+d \end{bmatrix}.$$

This cannot equal

$$\begin{bmatrix} 1 & 0 \\ 0 & 1 \end{bmatrix},$$

for if $a-c=1$ then $-a+c=-1$, not 0.

5.3 Let A be a diagonal matrix, say $A=[a_{ij}]_{p\times p}$, with $a_{ij}=0$ when $i\neq j$. Suppose also that for $1\leqslant i\leqslant p$ we have $a_{ii}\neq 0$ (there are no 0s on the main diagonal). Then A is invertible.

To see this, we show that B is the inverse of A, where $B=[b_{ij}]_{p\times p}$ is the diagonal matrix with $b_{ii}=1/a_{ii}$, for $1\leqslant i\leqslant p$. Calculate the product AB. The (i, i)-entry in AB is

$$\sum_{k=1}^{p} a_{ik}b_{ki},$$

which is equal to $a_{ii}b_{ii}$, since for $k\neq i$ we have $a_{ik}=b_{ki}=0$. By the choice of b_{ii}, then, $a_{ii}b_{ii}=1$ for each i, and so $AB=I$. Similarly $BA=I$. We are assuming the result that a product of diagonal matrices is a diagonal matrix (see Example 4.4).

5

Matrix inverses

At the end of Chapter 4 we discovered matrices E and F with the property that $EF=I$ and $FE=I$, and we said that they were inverses of each other. Generally, if A is a square matrix and B is a matrix of the same size with $AB=I$ and $BA=I$, then B is said to be the *inverse* of A. The inverse of A is denoted by A^{-1}. Example 5.1 is a proof that the inverse of a matrix (if it exists at all) is unique. Example 5.2 gives a matrix which does not have an inverse. So we must take care: not every matrix has an inverse. A matrix which does have an inverse is said to be *invertible* (or non-singular). Note that an invertible matrix must be square. A square matrix which is not invertible is said to be *singular*.

Following our discussion in Chapter 4 we can say that every elementary matrix is invertible and every orthogonal matrix is invertible. Example 5.3 shows that every diagonal matrix with no zeros on the main diagonal is invertible. There is, however, a standard procedure for testing whether a given matrix is invertible, and, if it is, of finding its inverse. This process is described in this chapter. It is an extension of the GE process.

5.4 Let $A=\begin{bmatrix}1 & 1 & 1\\ 1 & 2 & 3\\ 0 & 1 & 1\end{bmatrix}$.

Find whether A is invertible and, if it is, find A^{-1}.

First carry out the standard GE process on A, at the same time performing the same operations on an identity matrix.

$$\begin{array}{rrr|rrr} 1 & 1 & 1 & 1 & 0 & 0\\ 1 & 2 & 3 & 0 & 1 & 0\\ 0 & 1 & 1 & 0 & 0 & 1 \end{array}$$

$$\begin{array}{rrr|rrrl} 1 & 1 & 1 & 1 & 0 & 0 & \\ 0 & 1 & 2 & -1 & 1 & 0 & (2)-(1)\\ 0 & 1 & 1 & 0 & 0 & 1 & \end{array} \qquad E_1=\begin{bmatrix}1 & 0 & 0\\ -1 & 1 & 0\\ 0 & 0 & 1\end{bmatrix}$$

$$\begin{array}{rrr|rrrl} 1 & 1 & 1 & 1 & 0 & 0 & \\ 0 & 1 & 2 & -1 & 1 & 0 & \\ 0 & 0 & -1 & 1 & -1 & 1 & (3)-(2) \end{array} \qquad E_2=\begin{bmatrix}1 & 0 & 0\\ 0 & 1 & 0\\ 0 & -1 & 1\end{bmatrix}$$

$$\begin{array}{rrr|rrrl} 1 & 1 & 1 & 1 & 0 & 0 & \\ 0 & 1 & 2 & -1 & 1 & 0 & \\ 0 & 0 & 1 & -1 & 1 & -1 & (3)\times -1 \end{array} \qquad E_3=\begin{bmatrix}1 & 0 & 0\\ 0 & 1 & 0\\ 0 & 0 & -1\end{bmatrix}.$$

This is where the standard process ends. The matrix A' referred to in the text is

$$\begin{bmatrix}1 & 1 & 1\\ 0 & 1 & 2\\ 0 & 0 & 1\end{bmatrix}.$$

The process of finding the inverse continues with further row operations, with the objective of transforming it into an identity matrix.

$$\begin{array}{rrr|rrrl} 1 & 1 & 0 & 2 & -1 & 1 & (1)-(3)\\ 0 & 1 & 2 & -1 & 1 & 0 & \\ 0 & 0 & 1 & -1 & 1 & -1 & \end{array} \qquad E_4=\begin{bmatrix}1 & 0 & -1\\ 0 & 1 & 0\\ 0 & 0 & 1\end{bmatrix}$$

$$\begin{array}{rrr|rrrl} 1 & 1 & 0 & 2 & -1 & 1 & \\ 0 & 1 & 0 & 1 & -1 & 2 & (2)-2\times(3)\\ 0 & 0 & 1 & -1 & 1 & -1 & \end{array} \qquad E_5=\begin{bmatrix}1 & 0 & 0\\ 0 & 1 & -2\\ 0 & 0 & 1\end{bmatrix}$$

$$\begin{array}{rrr|rrrl} 1 & 0 & 0 & 1 & 0 & -1 & (1)-(2)\\ 0 & 1 & 0 & 1 & -1 & 2 & \\ 0 & 0 & 1 & -1 & 1 & -1 & \end{array} \qquad E_6=\begin{bmatrix}1 & -1 & 0\\ 0 & 1 & 0\\ 0 & 0 & 1\end{bmatrix}.$$

The process has been successful, so A is invertible, and

$$A^{-1}=\begin{bmatrix}1 & 0 & -1\\ 1 & -1 & 2\\ -1 & 1 & -1\end{bmatrix}.$$

Now check (just this once) that this is equal to the product of the elementary matrices $E_6E_5E_4E_3E_2E_1$. In normal applications of this process there is no need to keep a note of the elementary matrices used.

Example 5.4 illustrates the basis of the procedure. Starting with a square matrix A, the GE process leads to an upper triangular matrix, say A'. In the example, continue as follows. Subtract a multiple of the third row of A' from the second row in order to get 0 in the (2, 3)-place. Next, subtract a multiple of the third row from the first row in order to get 0 in the (1, 3)-place. Last, subtract a multiple of the second row from the first row in order to get 0 in the (1, 2)-place. By the GE process followed by this procedure we convert A into an identity matrix by elementary row operations. There exist, therefore, elementary matrices $E_1, E_2, \ldots, E_s$ such that

$$I = E_s E_{s-1} \ldots E_2 E_1 A.$$

Now if we set $B = E_s E_{s-1} \ldots E_2 E_1$, then we have $BA = I$. We shall show that $AB = I$ also. Let $F_1, F_2, \ldots, F_s$ be the inverses of $E_1, E_2, \ldots, E_s$ respectively. Then

$$\begin{aligned} F_1 F_2 \ldots F_s &= F_1 F_2 \ldots F_s I \\ &= F_1 F_2 \ldots F_s E_s E_{s-1} \ldots E_2 E_1 A \\ &= IA = A, \end{aligned}$$

since $F_s E_s = I$, $F_{s-1} E_{s-1} = I, \ldots, F_1 E_1 = I$. Consequently,

$$AB = F_1 F_2 \ldots F_s E_s E_{s-1} \ldots E_2 E_1 = I.$$

Hence B is the inverse of A. Our procedure for finding the inverse of A must therefore calculate for us the product $E_s E_{s-1} \ldots E_2 E_1$. This product can be written as $E_s E_{s-1} \ldots E_2 E_1 I$, and this gives the hint. We convert A to I by certain elementary row operations. The *same* row operations convert I into A^{-1} (if it exists). Explicitly,

$$\text{if} \quad I = E_s \ldots E_1 A \quad \text{then} \quad A^{-1} = E_s \ldots E_1 I.$$

5.5 Find the inverse of the matrix $\begin{bmatrix} 1 & 0 & 2 \\ 0 & 1 & 2 \\ 1 & 2 & 0 \end{bmatrix}$.

$$\begin{array}{ccc|cccl}
1 & 0 & 2 & 1 & 0 & 0 & \\
0 & 1 & 2 & 0 & 1 & 0 & \\
1 & 2 & 0 & 0 & 0 & 1 & \\
\hline
1 & 0 & 2 & 1 & 0 & 0 & \\
0 & 1 & 2 & 0 & 1 & 0 & \\
0 & 2 & -2 & -1 & 0 & 1 & (3)-(1) \\
\hline
1 & 0 & 2 & 1 & 0 & 0 & \\
0 & 1 & 2 & 0 & 1 & 0 & \\
0 & 0 & -6 & -1 & -2 & 1 & (3)-2\times(2) \\
\hline
1 & 0 & 2 & 1 & 0 & 0 & \\
0 & 1 & 2 & 0 & 1 & 0 & \\
0 & 0 & 1 & \frac{1}{6} & \frac{1}{3} & -\frac{1}{6} & (3)\div -6 \\
\hline
\end{array}$$

(At this stage we can be sure that the given matrix *is* invertible, and that the process will succeed in finding the inverse.)

$$\begin{array}{ccc|cccl}
1 & 0 & 0 & \frac{2}{3} & -\frac{2}{3} & \frac{1}{3} & (1)-2\times(3) \\
0 & 1 & 0 & -\frac{1}{3} & \frac{1}{3} & \frac{1}{3} & (2)-2\times(3) \\
0 & 0 & 1 & \frac{1}{6} & \frac{1}{3} & -\frac{1}{6} & \\
\hline
\end{array}$$

This is the end of the process, since the left-hand matrix is an identity matrix. We have shown that

$$\begin{bmatrix} 1 & 0 & 2 \\ 0 & 1 & 2 \\ 1 & 2 & 0 \end{bmatrix}^{-1} = \begin{bmatrix} \frac{2}{3} & -\frac{2}{3} & \frac{1}{3} \\ -\frac{1}{3} & \frac{1}{3} & \frac{1}{3} \\ \frac{1}{6} & \frac{1}{3} & -\frac{1}{6} \end{bmatrix}.$$

5.6 Find (if possible) the inverse of the matrix $\begin{bmatrix} 1 & 2 & 3 \\ 1 & 1 & 2 \\ 0 & 1 & 1 \end{bmatrix}$.

$$\begin{array}{ccc|cccl}
1 & 2 & 3 & 1 & 0 & 0 & \\
1 & 1 & 2 & 0 & 1 & 0 & \\
0 & 1 & 1 & 0 & 0 & 1 & \\
\hline
1 & 2 & 3 & 1 & 0 & 0 & \\
0 & -1 & -1 & -1 & 1 & 0 & (2)-(1) \\
0 & 1 & 1 & 0 & 0 & 1 & \\
\hline
1 & 2 & 3 & 1 & 0 & 0 & \\
0 & 1 & 1 & 1 & -1 & 0 & (2)\div -1 \\
0 & 1 & 1 & 0 & 0 & 1 & \\
\hline
1 & 2 & 3 & 1 & 0 & 0 & \\
0 & 1 & 1 & 1 & -1 & 0 & \\
0 & 0 & 0 & -1 & 1 & 1 & (3)-(2) \\
\hline
\end{array}$$

The practical process for finding inverses is illustrated by Example 5.5. Apply elementary row operations to the given matrix A to convert it to I. At the same time, apply the same elementary row operations to I, thus converting it into A^{-1} (provided A^{-1} exists, as it does in this example). This shows how to find the inverse of a 3×3 matrix, but the method extends to any size of square matrix. Apply elementary row operations to obtain zeros below the main diagonal, as in the GE process, and, once this is complete, carry on with the procedure for obtaining zeros above the main diagonal as well. Remember that there is a simple way to check the answer when finding the inverse of a given matrix A. If your answer is B, calculate the product AB. It should be I. If it is not, then you have made a mistake.

What happens to our process for finding inverses if the original matrix A is not invertible? The method depended on obtaining, during the process, the matrix A' which had 1s on the main diagonal and 0s below it. As we saw in Chapter 2, this need not always be possible. It could happen that the last row (after the GE process) consists entirely of 0s. In such a case the process for finding the inverse breaks down at this point. There is no way to obtain 0s in the other places in the last column. Example 5.6 illustrates this. It is precisely in these cases that the original matrix is not invertible. We can see this quite easily. Suppose that the matrix A' has last row all 0s. There exist elementary matrices $E_1, E_2 \ldots, E_r$ such that

$$A' = E_r E_{r-1} \ldots E_2 E_1 A.$$

Now suppose (by way of contradiction) that A is invertible. Then $AA^{-1} = I$. Let $F_1, F_2, \ldots, F_r$ be the inverses of $E_1, E_2, \ldots, E_r$ respectively. Then

$$\begin{aligned} A'(A^{-1}F_1F_2 \ldots F_r) &= (E_rE_{r-1} \ldots E_2E_1A)A^{-1}F_1 \ldots F_r \\ &= E_rE_{r-1} \ldots E_2E_1IF_1 \ldots F_r \\ &= I. \end{aligned}$$

Here the matrix A' is

$$\begin{bmatrix} 1 & 2 & 3 \\ 0 & 1 & 1 \\ 0 & 0 & 0 \end{bmatrix},$$

and the process for finding the inverse cannot be continued, because of the zero in the (3, 3)-place. The conclusion that we draw is that the given matrix is not invertible.

5.7 Let A be a $p \times p$ matrix whose pth (last) row consists entirely of 0s, and let X be any $p \times p$ matrix. Show the product AX has pth row consisting entirely of 0s.

Let $A = [a_{ij}]_{p \times p}$, with $a_{pj} = 0$ for $1 \leqslant j \leqslant p$. Let $X = [x_{ij}]_{p \times p}$. In AX the (p, j)-entry is $\sum_{k=1}^{p} a_{pk} x_{kj}$. But $a_{pk} = 0$ for all k, so this sum is zero. Hence the pth row of AX consists entirely of 0s.

5.8 A formula for the inverse of a 2×2 matrix.

$$\begin{bmatrix} a & b \\ c & d \end{bmatrix}^{-1} = \frac{1}{ad - bc} \begin{bmatrix} d & -b \\ -c & a \end{bmatrix},$$

provided that $ad - bc \neq 0$.

This is easily verified by multiplying out.

5.9 Show that if A and B are square matrices with $AB = I$, then A is invertible and $A^{-1} = B$.

Suppose that $AB = I$ and A is singular. Then, by the discussion in the text, there is an invertible matrix X such that XA has last row consisting of 0s. It follows that XAB has last row consisting of 0s (see Example 5.7). But $XAB = X$, since $AB = I$. But X cannot have its last row all 0s, because it is invertible (think about the process for finding the inverse). From this contradiction we may deduce that A is invertible. It remains to show that $A^{-1} = B$.

We have $AB = I$, and so

$$A^{-1}(AB) = A^{-1}I = A^{-1},$$

i.e. $$(A^{-1}A)B = A^{-1},$$

i.e. $$B = A^{-1}.$$

Notice that from $BA = I$ we can conclude by the same argument that B is invertible and $B^{-1} = A$. From this it follows that A is invertible and $A^{-1} = (B^{-1})^{-1} = B$.

Example 5.7 shows that such a product $A'X$, for any matrix X, has last row all 0s, and so $A'A^{-1}F_1 \ldots F_r$ has last row all 0s. But I does not. Hence the supposition that A is invertible is false.

Example 5.8 gives a formula for the inverse of a 2×2 matrix, if it exists. Example 5.10 is another calculation of an inverse.

Rule

A square matrix A is invertible if and only if the procedure given above reaches an intermediate stage with matrix A' having 1s on the main diagonal and 0s below it.

The definition of the matrix inverse required two conditions: B is the inverse of A if $AB = I$ *and* $BA = I$. It can be shown, however, that either one of these conditions is sufficient. Each condition implies the other. For a part proof of this, see Example 5.9. In practice it is very useful to use only one condition rather than two.

Next, a rule for inverses of products. Suppose that A and B are invertible $p \times p$ matrices. Must AB be invertible, and if so what is its inverse? Here is the trick:

$$\begin{aligned}(AB)(B^{-1}A^{-1}) &= A(BB^{-1})A^{-1} \quad \text{(rule (v) on page 27)}\\ &= AIA^{-1}\\ &= AA^{-1} = I,\end{aligned}$$

and

$$\begin{aligned}(B^{-1}A^{-1})(AB) &= B^{-1}(A^{-1}A)B\\ &= B^{-1}IB = I.\end{aligned}$$

Thus the matrix $B^{-1}A^{-1}$ has the required properties and, since inverses are unique if they exist, we have:

Rule

If A and B are both invertible $p \times p$ matrices, then so is AB, and $(AB)^{-1} = B^{-1}A^{-1}$.

(Take note of the change in the order of the multiplication.)

This rule extends to products of any number of matrices. The order reverses. Indeed, we have come across an example of this already. The elementary matrices $E_1, E_2, \ldots, E_r$ had inverses $F_1, F_2, \ldots, F_r$ respectively, and

$$(E_rE_{r-1} \ldots E_2E_1)(F_1F_2 \ldots F_{r-1}F_r) = I,$$

and

$$(F_1F_2 \ldots F_{r-1}F_r)(E_rE_{r-1} \ldots E_2E_1) = I,$$

so the inverse of $E_rE_{r-1} \ldots E_2E_1$ is $F_1F_2 \ldots F_{r-1}F_r$, and vice versa.

5.10 Find the inverse (if it exists) of the matrix

$$\begin{bmatrix} 2 & 1 & 1 \\ 1 & 0 & -1 \\ 1 & 3 & 2 \end{bmatrix}.$$

$$\begin{array}{ccc|ccc l}
2 & 1 & 1 & 1 & 0 & 0 & \\
1 & 0 & -1 & 0 & 1 & 0 & \\
1 & 3 & 2 & 0 & 0 & 1 & \\
\hline
1 & \frac{1}{2} & \frac{1}{2} & \frac{1}{2} & 0 & 0 & (1)\div 2 \\
1 & 0 & -1 & 0 & 1 & 0 & \\
1 & 3 & 2 & 0 & 0 & 1 & \\
\hline
1 & \frac{1}{2} & \frac{1}{2} & \frac{1}{2} & 0 & 0 & \\
0 & -\frac{1}{2} & -\frac{3}{2} & -\frac{1}{2} & 1 & 0 & (2)-(1) \\
0 & \frac{5}{2} & \frac{3}{2} & -\frac{1}{2} & 0 & 1 & (3)-(1) \\
\hline
1 & \frac{1}{2} & \frac{1}{2} & \frac{1}{2} & 0 & 0 & \\
0 & 1 & 3 & 1 & -2 & 0 & (2)\div -\frac{1}{2} \\
0 & \frac{5}{2} & \frac{3}{2} & -\frac{1}{2} & 0 & 1 & \\
\hline
1 & \frac{1}{2} & \frac{1}{2} & \frac{1}{2} & 0 & 0 & \\
0 & 1 & 3 & 1 & -2 & 0 & \\
0 & 0 & -6 & -3 & 5 & 1 & (3)-\frac{5}{2}\times(2) \\
\hline
1 & \frac{1}{2} & \frac{1}{2} & \frac{1}{2} & 0 & 0 & \\
0 & 1 & 3 & 1 & -2 & 0 & \\
0 & 0 & 1 & \frac{1}{2} & -\frac{5}{6} & -\frac{1}{6} & (3)\div -6 \\
\hline
\end{array}$$

(And here we can say that our matrix is invertible.)

$$\begin{array}{ccc|ccc l}
1 & \frac{1}{2} & 0 & \frac{1}{4} & \frac{5}{12} & \frac{1}{12} & (1)-\frac{1}{2}\times(3) \\
0 & 1 & 1 & -\frac{1}{2} & \frac{1}{2} & \frac{1}{2} & (2)-3\times(3) \\
0 & 0 & 1 & \frac{1}{2} & -\frac{5}{6} & -\frac{1}{6} & \\
\hline
1 & 0 & 0 & \frac{1}{2} & \frac{1}{6} & -\frac{1}{6} & (1)-\frac{1}{2}\times(2) \\
0 & 1 & 0 & -\frac{1}{2} & \frac{1}{2} & \frac{1}{2} & \\
0 & 0 & 1 & \frac{1}{2} & -\frac{5}{6} & -\frac{1}{6} & \\
\hline
\end{array}$$

Hence

$$\begin{bmatrix} 2 & 1 & 1 \\ 1 & 0 & -1 \\ 1 & 3 & 2 \end{bmatrix}^{-1} = \begin{bmatrix} \frac{1}{2} & \frac{1}{6} & -\frac{1}{6} \\ -\frac{1}{2} & \frac{1}{2} & \frac{1}{2} \\ \frac{1}{2} & -\frac{5}{6} & -\frac{1}{6} \end{bmatrix}.$$

Summary
The definitions are given of invertible and singular matrices. A procedure is given for deciding whether a given matrix is invertible and, if it is, finding the inverse. The validity of the procedure is established. Also a rule is given for writing down inverses of products of invertible matrices.

Exercises

1. Find which of the following matrices are invertible and which are singular. Find the inverses of those which are invertible. Verify your answers.

(i) $\begin{bmatrix} 1 & 2 \\ 1 & 3 \end{bmatrix}$. (ii) $\begin{bmatrix} 1 & 0 \\ 3 & 1 \end{bmatrix}$, (iii) $\begin{bmatrix} -2 & -3 \\ 4 & 6 \end{bmatrix}$.

(iv) $\begin{bmatrix} 1 & 2 & 1 \\ 0 & 1 & 2 \\ 0 & 0 & 1 \end{bmatrix}$. (v) $\begin{bmatrix} 1 & -1 & 2 \\ -1 & 2 & -1 \\ 1 & -3 & 1 \end{bmatrix}$.

(vi) $\begin{bmatrix} 0 & 1 & 1 \\ 1 & 0 & 1 \\ 1 & 1 & 0 \end{bmatrix}$. (vii) $\begin{bmatrix} 1 & 2 & -1 \\ 2 & 2 & -4 \\ -1 & 0 & 3 \end{bmatrix}$.

(viii) $\begin{bmatrix} 1 & -2 & -1 \\ 0 & 3 & 4 \\ -3 & 1 & 1 \end{bmatrix}$. (ix) $\begin{bmatrix} 1 & -1 & 4 \\ 2 & 3 & 3 \\ 3 & 1 & 8 \end{bmatrix}$.

(x) $\begin{bmatrix} 2 & 3 & -2 & 3 \\ 1 & 0 & 2 & 1 \\ -1 & 1 & 4 & -2 \\ 3 & 0 & 0 & 4 \end{bmatrix}$. (xi) $\begin{bmatrix} 1 & 1 & 1 & 1 \\ -2 & 1 & 0 & 3 \\ 3 & 0 & -2 & 5 \\ 1 & -1 & -1 & -3 \end{bmatrix}$.

2. Find the inverse of the diagonal matrix

$$\begin{bmatrix} a & 0 & 0 \\ 0 & b & 0 \\ 0 & 0 & c \end{bmatrix},$$

where a, b and c are non-zero.

3. Let A, B and C be invertible matrices of the same size. Show that the inverse of the product ABC is $C^{-1}B^{-1}A^{-1}$.

4. Let $\boldsymbol{x}$ and $\boldsymbol{y}$ be p-vectors. Show that $\boldsymbol{x}\boldsymbol{y}^{\mathrm{T}}$ is a $p \times p$ matrix and is singular. Pick some vectors $\boldsymbol{x}$ and $\boldsymbol{y}$ at random and verify that $\boldsymbol{x}\boldsymbol{y}^{\mathrm{T}}$ is singular.

5. Show that, for any invertible matrix A,

$(A^{-1})^{\mathrm{T}}A^{\mathrm{T}} = I$ and $A^{\mathrm{T}}(A^{-1})^{\mathrm{T}} = I$.

Deduce that A^{T} is invertible and that its inverse is the transpose of A^{-1}. Deduce also that if A is symmetric then A^{-1} is also symmetric.

6. Let X and A be $p \times p$ matrices such that X is singular and A is invertible. Show that the products XA and AX are both singular. (Hint: suppose that an inverse matrix exists and derive a contradiction to the fact that X does not have an inverse.)

Examples

6.1 Illustrations of linear dependence.

(i) $\left(\begin{bmatrix}3\\-3\end{bmatrix},\begin{bmatrix}1\\-1\end{bmatrix}\right)$ is LD: $2\begin{bmatrix}3\\-3\end{bmatrix}-6\begin{bmatrix}1\\-1\end{bmatrix}=\begin{bmatrix}0\\0\end{bmatrix}$.

(ii) $\left(\begin{bmatrix}1\\2\end{bmatrix},\begin{bmatrix}-1\\3\end{bmatrix},\begin{bmatrix}2\\0\end{bmatrix}\right)$ is LD: $6\begin{bmatrix}1\\2\end{bmatrix}-4\begin{bmatrix}-1\\3\end{bmatrix}-5\begin{bmatrix}2\\0\end{bmatrix}=\begin{bmatrix}0\\0\end{bmatrix}$.

(iii) $\left(\begin{bmatrix}3\\2\\1\end{bmatrix},\begin{bmatrix}0\\1\\-1\end{bmatrix},\begin{bmatrix}5\\1\\4\end{bmatrix}\right)$ is LD: $-5\begin{bmatrix}3\\2\\1\end{bmatrix}+7\begin{bmatrix}0\\1\\-1\end{bmatrix}+3\begin{bmatrix}5\\1\\4\end{bmatrix}=\begin{bmatrix}0\\0\\0\end{bmatrix}$.

6.2 A list of two non-zero 2-vectors is LD if and only if each is a multiple of the other. Let $\begin{bmatrix}a_1\\b_1\end{bmatrix},\begin{bmatrix}a_2\\b_2\end{bmatrix}$ be two non-zero 2-vectors.

First, suppose that they constitute a LD list. Then there exist numbers x_1 and x_2, not both zero, such that

$$x_1\begin{bmatrix}a_1\\b_1\end{bmatrix}+x_2\begin{bmatrix}a_2\\b_2\end{bmatrix}=\begin{bmatrix}0\\0\end{bmatrix}.$$

Without loss of generality, say $x_1\neq 0$. Then since $\begin{bmatrix}a_1\\b_1\end{bmatrix}\neq\mathbf{0}$, we must have $x_2\neq 0$ also. Consequently,

$$\begin{bmatrix}a_1\\b_1\end{bmatrix}=-(x_2/x_1)\begin{bmatrix}a_2\\b_2\end{bmatrix}\quad\text{and}\quad\begin{bmatrix}a_2\\b_2\end{bmatrix}=-(x_1/x_2)\begin{bmatrix}a_1\\b_1\end{bmatrix},$$

i.e. each is a multiple of the other.

Conversely, suppose that each is a multiple of the other, say

$$\begin{bmatrix}a_1\\b_1\end{bmatrix}=k\begin{bmatrix}a_2\\b_2\end{bmatrix}\quad\text{and}\quad\begin{bmatrix}a_2\\b_2\end{bmatrix}=(1/k)\begin{bmatrix}a_1\\b_1\end{bmatrix}.$$

Then

$$\begin{bmatrix}a_1\\b_1\end{bmatrix}-k\begin{bmatrix}a_2\\b_2\end{bmatrix}=\begin{bmatrix}0\\0\end{bmatrix},\ \text{and so}\ \left(\begin{bmatrix}a_1\\b_1\end{bmatrix},\begin{bmatrix}a_2\\b_2\end{bmatrix}\right)\ \text{is LD.}$$

6.3 Show that any list of three 2-vectors is LD. To show that

$$\left(\begin{bmatrix}a_1\\b_1\end{bmatrix},\begin{bmatrix}a_2\\b_2\end{bmatrix},\begin{bmatrix}a_3\\b_3\end{bmatrix}\right)\ \text{is LD,}$$

we seek numbers x_1, x_2 and x_3, not all zero, such that

$$x_1\begin{bmatrix}a_1\\b_1\end{bmatrix}+x_2\begin{bmatrix}a_2\\b_2\end{bmatrix}+x_3\begin{bmatrix}a_3\\b_3\end{bmatrix}=\begin{bmatrix}0\\0\end{bmatrix},$$

i.e.

$$\left.\begin{aligned}a_1x_1+a_2x_2+a_3x_3&=0\\b_1x_1+b_2x_2+b_3x_3&=0\end{aligned}\right\}.$$

In other words, we seek solutions other than $x_1=x_2=x_3=0$ to this set of simultaneous equations. These equations are consistent (because $x_1=x_2=x_3=0$ do satisfy them), so by the rules in Chapter 2 there are infinitely many solutions. Thus there do exist non-trivial solutions, and so the given list of vectors is LD.

6

Linear independence and rank

Examples 6.1 illustrate what is meant by linear dependence of a list of vectors. More formally: given a list of vectors $v_1, \ldots, v_k$ of the same size (i.e. all are $p \times 1$ matrices for the same p), a *linear combination* of these vectors is a sum of multiples of them, i.e. $x_1v_1 + x_2v_2 + \cdots + x_kv_k$, where $x_1, \ldots, x_k$ are any numbers. A list of vectors is said to be *linearly dependent* (abbreviated to LD) if there is some non-trivial linear combination of them which is equal to the zero vector. Of course, in a trivial way, we can always obtain the zero vector by taking all of the coefficients $x_1, \ldots, x_k$ to be 0. A *non-trivial* linear combination is one in which at least one of the coefficients is non-zero.

A list of vectors of the same size which is not linearly dependent is said to be *linearly independent* (abbreviated to LI).

Example 6.2 deals with the case of a list of two 2-vectors. A list of two non-zero 2-vectors is LD if and only if each is a multiple of the other. Example 6.3 deals with a list of three 2-vectors. Such a list is *always* LD. Why? Because a certain set of simultaneous equations *must* have a solution of a certain kind.

6.4 Find whether the list

$$\left(\begin{bmatrix}1\\2\\5\end{bmatrix}, \begin{bmatrix}2\\-2\\4\end{bmatrix}, \begin{bmatrix}1\\1\\4\end{bmatrix}\right)$$

is LI or LD.

Seek numbers x_1, x_2 and x_3, not all zero, such that

$$x_1\begin{bmatrix}1\\2\\5\end{bmatrix}+x_2\begin{bmatrix}2\\-2\\4\end{bmatrix}+x_3\begin{bmatrix}1\\1\\4\end{bmatrix}=\begin{bmatrix}0\\0\\0\end{bmatrix},$$

i.e.

$$\left.\begin{aligned}x_1+2x_2+\ x_3&=0\\2x_1-2x_2+\ x_3&=0\\5x_1+4x_2+4x_3&=0\end{aligned}\right\}.$$

Apply the standard GE process (details omitted):

$$\begin{bmatrix}1&2&1&0\\2&-2&1&0\\5&4&4&0\end{bmatrix}\rightarrow\begin{bmatrix}1&2&1&0\\0&1&\frac{1}{6}&0\\0&0&0&0\end{bmatrix}.$$

From this we conclude that the set of equations has infinitely many solutions, and so the given list of vectors is LD.

6.5 Find whether the list

$$\left(\begin{bmatrix}1\\2\\-1\end{bmatrix}, \begin{bmatrix}2\\2\\0\end{bmatrix}, \begin{bmatrix}1\\4\\3\end{bmatrix}\right)$$

is LI or LD.

Following the same procedure as in Example 6.4, we seek a non-trivial solution to

$$\begin{bmatrix}1&2&1\\2&2&4\\-1&0&3\end{bmatrix}\begin{bmatrix}x_1\\x_2\\x_3\end{bmatrix}=\begin{bmatrix}0\\0\\0\end{bmatrix}$$

(here writing the three simultaneous equations as a matrix equation). Apply the standard GE process:

$$\begin{bmatrix}1&2&1&0\\2&2&4&0\\-1&0&3&0\end{bmatrix}\rightarrow\begin{bmatrix}1&2&1&0\\0&1&-1&0\\0&0&1&0\end{bmatrix}.$$

From this we conclude that there is a unique solution to the equation, namely $x_1=x_2=x_3=0$. Consequently there does not exist a non-trivial linear combination of the given vectors which is equal to the zero vector. The given list is therefore LI.

6.6 Find whether the list

$$\left(\begin{bmatrix}1\\-1\\5\end{bmatrix}, \begin{bmatrix}1\\2\\-1\end{bmatrix}, \begin{bmatrix}2\\1\\2\end{bmatrix}, \begin{bmatrix}1\\2\\7\end{bmatrix}\right)$$

is LI or LD.

For the moment, however, let us see precisely how linear dependence and simultaneous equations are connected. Consider three 3-vectors, as in Example 6.4. Let us seek to show that these vectors constitute a list which is LD (even though they may not). So we seek coefficients x_1, x_2, x_3 (not all zero) to make the linear combination equal to the zero vector. Now the vector equation which x_1, x_2 and x_3 must satisfy, if we separate out the corresponding entries on each side, becomes a set of three simultaneous equations in the unknowns x_1, x_2 and x_3. We can use our standard procedure (the GE process) to solve these equations. But there is a particular feature of these equations. The right-hand sides are all 0s, so, as we noted earlier, there certainly is one solution (at least), namely $x_1 = x_2 = x_3 = 0$. What we seek is another solution (any other solution), and from our earlier work we know that if there is to be another solution then there must be infinitely many solutions, since the only possibilities are: no solutions, a unique solution, and infinitely many solutions. What is more, we know what form the result of the GE process must take if there are to be infinitely many solutions. The last row must consist entirely of 0s. In Example 6.4 it does, so the given vectors are LD. In Example 6.5 it does not, so the given vectors are LI.

Because the right-hand sides of the equations are all 0s in calculations of this kind, we can neglect this column (or omit it, as we customarily shall). Referring to Chapter 2 we can see:

Rule

Let $v_1, \ldots, v_q$ be a list of p-vectors. To test whether this set is LD or LI, form a matrix A with the vectors $v_1, \ldots, v_q$ as columns (so that A is a $p \times q$ matrix) and carry out the standard GE process on A. If the resulting matrix has fewer than q non-zero rows then the given list of vectors is LD. Otherwise it is LI.

Example 6.6 shows what happens with a list of four 3-vectors. It will always turn out to be LD. The matrix after the GE process is bound to have fewer than four non-zero rows. This illustrates a general rule.

Here we reduce the working to the bare essentials. Apply the GE process to the matrix

$$\begin{bmatrix} 1 & 1 & 2 & 1 \\ -1 & 2 & 1 & 2 \\ 5 & -1 & 2 & 7 \end{bmatrix}.$$

We obtain (details omitted) the matrix

$$\begin{bmatrix} 1 & 1 & 2 & 1 \\ 0 & 1 & 1 & 1 \\ 0 & 0 & 1 & -4 \end{bmatrix}.$$

This matrix has fewer than four non-zero rows, so if we were proceeding as in the previous examples and seeking solutions to equations we would conclude that there were infinitely many solutions. Consequently the given list of vectors is LD.

6.7 Illustrations of calculations of ranks of matrices.

(i) $\begin{bmatrix} 1 & 2 & -1 \\ 2 & 2 & -4 \\ -1 & 0 & 3 \end{bmatrix}$ has rank 2.

GE process:

$$\begin{bmatrix} 1 & 2 & -1 \\ 2 & 2 & -4 \\ -1 & 0 & 3 \end{bmatrix} \rightarrow \begin{bmatrix} 1 & 2 & -1 \\ 0 & 1 & 1 \\ 0 & 0 & 0 \end{bmatrix},$$

a matrix with *two* non-zero rows.

(ii) $\begin{bmatrix} 1 & 2 \\ -2 & 1 \end{bmatrix}$ has rank 2.

GE process:

$$\begin{bmatrix} 1 & 2 \\ -2 & 1 \end{bmatrix} \rightarrow \begin{bmatrix} 1 & 2 \\ 0 & 1 \end{bmatrix} \quad \text{(two non-zero rows).}$$

(iii) $\begin{bmatrix} 1 & 1 & 1 & 1 \\ 2 & 2 & -1 & 1 \\ -1 & 7 & 5 & 2 \end{bmatrix}$ has rank 3.

GE process:

$$\begin{bmatrix} 1 & 1 & 1 & 2 \\ 2 & 2 & -1 & 1 \\ -1 & 7 & 5 & 2 \end{bmatrix} \rightarrow \begin{bmatrix} 1 & 1 & 1 & 2 \\ 0 & 1 & \frac{3}{4} & \frac{1}{2} \\ 0 & 0 & 1 & 1 \end{bmatrix} \quad \text{(three non-zero rows).}$$

(iv) $\begin{bmatrix} 1 & 1 & 1 \\ 1 & 2 & 3 \\ 0 & 1 & 1 \end{bmatrix}$ has rank 3.

GE process:

$$\begin{bmatrix} 1 & 1 & 1 \\ 1 & 2 & 3 \\ 0 & 1 & 1 \end{bmatrix} \rightarrow \begin{bmatrix} 1 & 1 & 1 \\ 0 & 1 & 2 \\ 0 & 0 & 1 \end{bmatrix} \quad \text{(three non-zero rows).}$$

(v) $\begin{bmatrix} 1 & 2 & -1 \\ 2 & 4 & -2 \\ -1 & -2 & 1 \end{bmatrix}$ has rank 1.

Rule
Any list of p-vectors which contains more than p distinct vectors is LD.

Following Chapter 5, we have another rule.

Rule
If a matrix is invertible then its columns form a LI list of vectors.

(Recall that a $p \times p$ matrix is invertible if and only if the standard GE process leads to a matrix with p non-zero rows.)

Another important idea is already implicit in the above. The *rank* of a matrix is the number of non-zero rows remaining after the standard GE process. Examples 6.7 show how ranks are calculated. It is obvious that the rank of a $p \times q$ matrix is necessarily less than or equal to p. It is also less than or equal to q. To see this, think about the shape of the matrix remaining after the GE process. It has 0s everywhere below the main diagonal, which starts at the top left. The largest possible number of non-zero rows occurs when the main diagonal itself contains no 0s, and in that case the first q rows are non-zero and the remaining rows are all 0s.

Consideration of rank is useful when stating criteria for equations to have particular sorts of solutions. We shall pursue this in Chapter 8.

In the meantime let us consider a first version of what we shall call the *Equivalence Theorem*, which brings together, through the GE process, all the ideas covered so far.

GE process:

$$\begin{bmatrix} 1 & 2 & -1 \\ 2 & 4 & -2 \\ -1 & -2 & 1 \end{bmatrix} \rightarrow \begin{bmatrix} 1 & 2 & -1 \\ 0 & 0 & 0 \\ 0 & 0 & 0 \end{bmatrix} \quad \text{(one non-zero row).}$$

6.8 Illustration of the Equivalence Theorem.

(i) $$A = \begin{bmatrix} 1 & 2 & 3 \\ 1 & 1 & 2 \\ 1 & -1 & 1 \end{bmatrix}.$$

The GE process leads to

$$\begin{bmatrix} 1 & 2 & 3 \\ 0 & 1 & 1 \\ 0 & 0 & 1 \end{bmatrix}.$$

From this we can tell that the rank of A is 3, that the columns of A form a LI list, and that the process for finding the inverse of A will succeed, so A is invertible.

(ii) $$A = \begin{bmatrix} 1 & 1 & 1 & 0 \\ 1 & 1 & 0 & 1 \\ 1 & 0 & 1 & 1 \\ 0 & 1 & 1 & 1 \end{bmatrix}.$$

The GE process leads to

$$\begin{bmatrix} 1 & 1 & 1 & 0 \\ 0 & 1 & 0 & -1 \\ 0 & 0 & 1 & -1 \\ 0 & 0 & 0 & 1 \end{bmatrix}.$$

Consequently the rank of A is 4, the columns of A form a LI list, and A is invertible.

(iii) $$A = \begin{bmatrix} 1 & 3 & -1 \\ -2 & 1 & -5 \\ 4 & 5 & 3 \end{bmatrix}.$$

The GE process leads to

$$\begin{bmatrix} 1 & 3 & -1 \\ 0 & 1 & -1 \\ 0 & 0 & 0 \end{bmatrix}.$$

Consequently the rank of A is 2 (not equal to 3), the columns of A form a list which is LD (not LI), and the process for finding the inverse of A will fail, so A is not invertible. Also the equation $Ax = 0$ has infinitely many solutions, so all conditions of the Equivalence Theorem fail.

Theorem

Let A be a $p \times p$ matrix. The following are equivalent.

(i) A is invertible.

(ii) The rank of A is equal to p.

(iii) The columns of A form a LI list.

(iv) The set of simultaneous equations which can be written $A\boldsymbol{x}=\mathbf{0}$ has no solution other than $\boldsymbol{x}=\mathbf{0}$.

The justification for this theorem is that in each of these situations the GE process leads to a matrix with p non-zero rows. See Examples 6.8.

In Chapter 7 we shall introduce another equivalent condition, involving determinants.

Summary

The definitions are given of linear dependence and linear independence of lists of vectors. A method is given for testing linear dependence and independence. The idea of rank is introduced, and the equivalence of invertibility with conditions involving rank, linear independence and solutions to equations is demonstrated, via the GE process.

Exercises

1. Find numbers x and y such that

$$x\begin{bmatrix}3\\1\\1\end{bmatrix}+y\begin{bmatrix}2\\-1\\1\end{bmatrix}=\begin{bmatrix}1\\-3\\1\end{bmatrix}.$$

2. For each given list of vectors, find whether the third vector is a linear combination of the first two.

(i) $\left(\begin{bmatrix}1\\2\end{bmatrix},\begin{bmatrix}0\\1\end{bmatrix},\begin{bmatrix}-1\\1\end{bmatrix}\right)$ (ii) $\left(\begin{bmatrix}0\\0\end{bmatrix},\begin{bmatrix}1\\2\end{bmatrix},\begin{bmatrix}2\\3\end{bmatrix}\right)$.

(iii) $\left(\begin{bmatrix}1\\1\\2\end{bmatrix},\begin{bmatrix}0\\1\\0\end{bmatrix},\begin{bmatrix}-1\\5\\-2\end{bmatrix}\right)$.

(iv) $\left(\begin{bmatrix}2\\-1\\-3\end{bmatrix},\begin{bmatrix}1\\5\\-1\end{bmatrix},\begin{bmatrix}3\\0\\-5\end{bmatrix}\right)$.

3. Show that each of the following lists is linearly dependent, and in each case find a linear combination of the given vectors which is equal to the zero vector.

(i) $\left(\begin{bmatrix}3\\-1\end{bmatrix},\begin{bmatrix}6\\-2\end{bmatrix}\right)$. (ii) $\left(\begin{bmatrix}-1\\1\end{bmatrix},\begin{bmatrix}0\\2\end{bmatrix},\begin{bmatrix}3\\4\end{bmatrix}\right)$.

(iii) $\left(\begin{bmatrix}-2\\1\\1\end{bmatrix},\begin{bmatrix}1\\1\\4\end{bmatrix},\begin{bmatrix}3\\-1\\0\end{bmatrix}\right)$.

(iv) $\left(\begin{bmatrix}1\\5\\9\end{bmatrix},\begin{bmatrix}-3\\1\\5\end{bmatrix},\begin{bmatrix}2\\3\\4\end{bmatrix}\right)$.

(v) $\left(\begin{bmatrix}0\\2\\6\end{bmatrix},\begin{bmatrix}3\\1\\6\end{bmatrix},\begin{bmatrix}4\\-2\\-2\end{bmatrix}\right)$.

(vi) $\left(\begin{bmatrix}1\\-2\\3\\8\end{bmatrix},\begin{bmatrix}2\\1\\1\\1\end{bmatrix},\begin{bmatrix}5\\3\\2\\1\end{bmatrix}\right)$.

4. Find, in each case below, whether the given list of vectors is linearly dependent or independent.

(i) $\left(\begin{bmatrix}2\\3\end{bmatrix},\begin{bmatrix}1\\-1\end{bmatrix}\right)$, (ii) $\left(\begin{bmatrix}1\\2\end{bmatrix},\begin{bmatrix}0\\0\end{bmatrix}\right)$.

(iii) $\left(\begin{bmatrix}3\\-1\\2\end{bmatrix},\begin{bmatrix}0\\0\\0\end{bmatrix},\begin{bmatrix}1\\0\\0\end{bmatrix}\right)$.

(iv) $\left(\begin{bmatrix}1\\1\\1\end{bmatrix},\begin{bmatrix}-2\\1\\2\end{bmatrix},\begin{bmatrix}1\\3\\4\end{bmatrix}\right)$.

(v) $\left(\begin{bmatrix}0\\1\\-1\end{bmatrix}, \begin{bmatrix}3\\4\\1\end{bmatrix}, \begin{bmatrix}-2\\2\\0\end{bmatrix}\right)$.

(vi) $\left(\begin{bmatrix}1\\-2\\-1\end{bmatrix}, \begin{bmatrix}3\\-1\\7\end{bmatrix}, \begin{bmatrix}1\\1\\5\end{bmatrix}\right)$.

(vii) $\left(\begin{bmatrix}1\\1\\2\\1\end{bmatrix}, \begin{bmatrix}-1\\0\\1\\-1\end{bmatrix}, \begin{bmatrix}2\\1\\1\\-2\end{bmatrix}, \begin{bmatrix}0\\1\\0\\0\end{bmatrix}\right)$.

5. Calculate the rank of each of the matrices given below.

$$\begin{bmatrix}3 & 6\\-1 & -2\end{bmatrix}, \begin{bmatrix}-1 & 0 & 3\\1 & 2 & 4\end{bmatrix}, \begin{bmatrix}2 & 1\\3 & -1\end{bmatrix}, \begin{bmatrix}1 & 0\\2 & 0\end{bmatrix},$$

$$\begin{bmatrix}-2 & 1 & 3\\1 & 1 & -1\\1 & 4 & 0\end{bmatrix}, \begin{bmatrix}1 & -3 & 2\\5 & 1 & 3\\9 & 5 & 4\end{bmatrix}, \begin{bmatrix}1 & -2 & 1\\1 & 1 & 3\\1 & 2 & 4\end{bmatrix},$$

$$\begin{bmatrix}0 & 3 & -2\\1 & 4 & 2\\-1 & 1 & 0\end{bmatrix}, \begin{bmatrix}1 & 2\\1 & -1\\0 & 2\end{bmatrix}, \begin{bmatrix}1 & 2\\-1 & -2\\3 & 6\end{bmatrix},$$

$$\begin{bmatrix}1 & 0\\2 & 0\\3 & 0\end{bmatrix}, \begin{bmatrix}2 & 2 & 3\\1 & -1 & 1\end{bmatrix}, \begin{bmatrix}1 & 0 & -2\\-2 & 0 & 4\end{bmatrix},$$

$$\begin{bmatrix}1 & -1 & 1 & 2\\3 & 1 & -2 & 0\\1 & 3 & -4 & -4\end{bmatrix}, \begin{bmatrix}0 & 1 & -1 & 1\\2 & 0 & 1 & 1\\-1 & 1 & 2 & 3\end{bmatrix},$$

$$\begin{bmatrix}1 & -1 & 2 & 0\\1 & 0 & 1 & 1\\2 & 1 & 1 & 0\\1 & -1 & -2 & 0\end{bmatrix}, \begin{bmatrix}1 & -1 & 2 & 0 & 3\\1 & 0 & 1 & 1 & 4\\2 & 1 & 1 & 0 & 2\\1 & -1 & -2 & 0 & -2\end{bmatrix},$$

$$\begin{bmatrix}-1 & 2 & 1 & 1\\0 & 1 & 2 & -1\\-1 & 1 & -1 & 2\\-1 & 3 & 3 & 0\end{bmatrix}, \begin{bmatrix}2 & 1 & -1 & 2\\1 & 2 & 1 & 2\\3 & 1 & -1 & 0\\0 & 2 & 1 & 4\end{bmatrix}.$$

6. Let $\boldsymbol{x}$ and $\boldsymbol{y}$ be 3-vectors. Then $\boldsymbol{xy}^{\mathrm{T}}$ is a 3×3 matrix which is singular (see Chapter 5, Exercise 4). What is the rank of $\boldsymbol{xy}^{\mathrm{T}}$? Try out some particular examples to see what happens. Does this result hold for p-vectors, for every p?

Examples

7.1 Evaluation of 2×2 determinants.

(i) $\begin{vmatrix} 1 & 2 \\ 3 & 4 \end{vmatrix} = 4 - 6 = -2.$

(ii) $\begin{vmatrix} 3 & 1 \\ -2 & 5 \end{vmatrix} = 15 - (-2) = 17.$

(iii) $\begin{vmatrix} 0 & 1 \\ 4 & 5 \end{vmatrix} = 0 - 4 = -4.$

(iv) $\begin{vmatrix} 3 & 0 \\ 2 & 7 \end{vmatrix} = 21 - 0 = 21.$

(v) $\begin{vmatrix} 0 & 1 \\ 1 & 0 \end{vmatrix} = 0 - 1 = -1.$

(vi) $\begin{vmatrix} 0 & 0 \\ 2 & 5 \end{vmatrix} = 0 - 0 = 0.$

7

Determinants

A 2×2 determinant is written

$$\begin{vmatrix} a & b \\ c & d \end{vmatrix}.$$

What this means is just the number $ad-bc$. Examples 7.1 show some simple 2×2 determinants and their values. Determinants are not the same thing as matrices. A determinant has a numerical (or algebraic) value. A matrix is an array. However, it makes sense, given any 2×2 matrix A, to talk of *the determinant of* A, written det A.

$$\text{If} \quad A=\begin{bmatrix} a & b \\ c & d \end{bmatrix} \quad \text{then} \quad \det A=\begin{vmatrix} a & b \\ c & d \end{vmatrix}=ad-bc.$$

The significance and usefulness of determinants will be more apparent when we deal with 3×3 and larger determinants, but the reader will find this expression $ad-bc$ occurring previously in Example 5.8. Also it may be instructive (as an exercise) to go through the details of the solution (for x and y) of the simultaneous equations

$$\left.\begin{aligned} ax+by&=h \\ cx+dy&=k \end{aligned}\right\}.$$

A 3×3 determinant is written

$$\begin{vmatrix} a_1 & a_2 & a_3 \\ b_1 & b_2 & b_3 \\ c_1 & c_2 & c_3 \end{vmatrix}.$$

What this means is

$$a_1b_2c_3+a_2b_3c_1+a_3b_1c_2-a_1b_3c_2-a_2b_1c_3-a_3b_2c_1.$$

Again, this is a number, calculated from the numbers in the given array. It makes sense, therefore, to talk of the determinant of A where A is a 3×3 matrix, with the obvious meaning.

7.2 (i) Evaluate the determinant

$$\begin{vmatrix} 1 & 2 & 3 \\ 0 & 1 & 2 \\ 1 & 0 & 3 \end{vmatrix}.$$

Here let us use the first method, with an extended array.

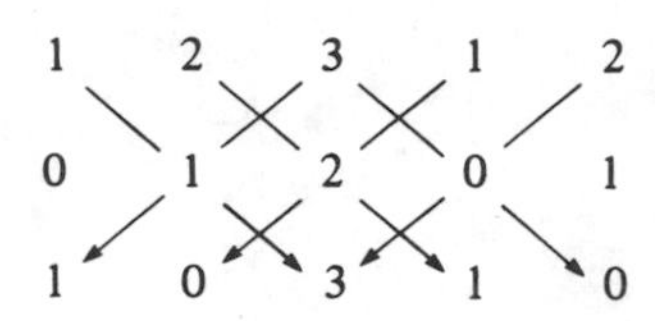

The value is $3+4+0-3-0-0$, i.e. 4.

(ii) Evaluate the determinant

$$\begin{vmatrix} 1 & -2 & -1 \\ 1 & -1 & -3 \\ 2 & -1 & 9 \end{vmatrix}.$$

Array:

$$\begin{matrix} 1 & -2 & -1 & 1 & -2 \\ 1 & -1 & -3 & 1 & -1 \\ 2 & -1 & 9 & 2 & -1 \end{matrix}$$

Value is: $-9+12+1-2-3-(-18)$, i.e. 17.

7.3 Evaluation of determinants using expansion by the first row.

(i) $$\begin{vmatrix} 1 & -2 & -1 \\ 1 & -1 & -3 \\ 2 & -1 & 9 \end{vmatrix} = 1\begin{vmatrix} -1 & -3 \\ -1 & 9 \end{vmatrix} - (-2)\begin{vmatrix} 1 & -3 \\ 2 & 9 \end{vmatrix} + (-1)\begin{vmatrix} 1 & -1 \\ 2 & -1 \end{vmatrix}$$

$$= (-12) + 2(15) - (1) = 17.$$

(ii) $$\begin{vmatrix} 1 & 2 & 3 \\ 0 & 1 & 2 \\ 1 & 0 & 3 \end{vmatrix} = 1\begin{vmatrix} 1 & 2 \\ 0 & 3 \end{vmatrix} - 2\begin{vmatrix} 0 & 2 \\ 1 & 3 \end{vmatrix} + 3\begin{vmatrix} 0 & 1 \\ 1 & 0 \end{vmatrix}$$

$$= 3 - 2(-2) + 3(-1) = 4.$$

(iii) $$\begin{vmatrix} 0 & 1 & 0 \\ 1 & 2 & 1 \\ 1 & 1 & 0 \end{vmatrix} = 0\begin{vmatrix} 2 & 1 \\ 1 & 0 \end{vmatrix} - 1\begin{vmatrix} 1 & 1 \\ 1 & 0 \end{vmatrix} + 0\begin{vmatrix} 1 & 2 \\ 1 & 1 \end{vmatrix}$$

$$= 0 - 1(-1) + 0 = 1.$$

(iv) $$\begin{vmatrix} 1 & 2 & 1 \\ 0 & 1 & 0 \\ 1 & 1 & 0 \end{vmatrix} = 1\begin{vmatrix} 1 & 0 \\ 1 & 0 \end{vmatrix} - 2\begin{vmatrix} 0 & 0 \\ 1 & 0 \end{vmatrix} + 1\begin{vmatrix} 0 & 1 \\ 1 & 1 \end{vmatrix}$$

$$= 0 - 2(0) + 1(-1) = -1.$$

How are we to cope with this imposing formula? There are several ways. Here is one way. Write down the given array, and then write the first and second columns again, on the right.

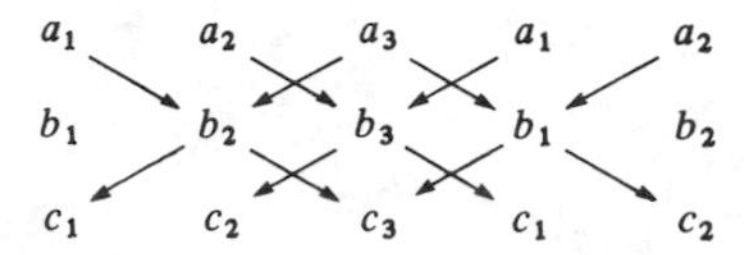

Add together the products of the numbers on the left-to-right arrows, and subtract the products of the numbers on the right-to-left arrows. A quick check with the definition above shows that we obtain the correct expression. Example 7.2 shows this method in operation.

The second method is called *expansion by a row or column*, and it is the key to the development of 4×4 and larger determinants so, although this method may seem more complicated, it is important and should be understood.

$$\begin{vmatrix} a_1 & a_2 & a_3 \\ b_1 & b_2 & b_3 \\ c_1 & c_2 & c_3 \end{vmatrix} = a_1\begin{vmatrix} b_2 & b_3 \\ c_2 & c_3 \end{vmatrix} - a_2\begin{vmatrix} b_1 & b_3 \\ c_1 & c_3 \end{vmatrix} + a_3\begin{vmatrix} b_1 & b_2 \\ c_1 & c_2 \end{vmatrix}.$$

Notice the form of the right-hand side. Each entry in the first row is multiplied with the determinant obtained by deleting the first row and the column containing that particular entry. To see that this gives the correct value for the determinant, we just have to multiply out the right-hand side as

$$a_1(b_2c_3-b_3c_2)-a_2(b_1c_3-b_3c_1)+a_3(b_1c_2-b_2c_1)$$

and compare with the original definition. When using this method, we must be careful to remember the negative sign which appears in front of the middle term. See Example 7.3.

There are similar expressions for a 3×3 determinant in which the coefficients are the entries in any of the rows or columns. We list these explicitly below.

$$-b_1\begin{vmatrix} a_2 & a_3 \\ c_2 & c_3 \end{vmatrix} + b_2\begin{vmatrix} a_1 & a_3 \\ c_1 & c_3 \end{vmatrix} - b_3\begin{vmatrix} a_1 & a_2 \\ c_1 & c_2 \end{vmatrix} \quad \text{(second row).}$$

$$c_1\begin{vmatrix} a_2 & a_3 \\ b_2 & b_3 \end{vmatrix} - c_2\begin{vmatrix} a_1 & a_3 \\ b_1 & b_3 \end{vmatrix} + c_3\begin{vmatrix} a_1 & a_2 \\ b_1 & b_2 \end{vmatrix} \quad \text{(third row).}$$

$$a_1\begin{vmatrix} b_2 & b_3 \\ c_2 & c_3 \end{vmatrix} - b_1\begin{vmatrix} a_2 & a_3 \\ c_2 & c_3 \end{vmatrix} + c_1\begin{vmatrix} a_2 & a_3 \\ b_2 & b_3 \end{vmatrix} \quad \text{(first column).}$$

$$-a_2\begin{vmatrix} b_1 & b_3 \\ c_1 & c_3 \end{vmatrix} + b_2\begin{vmatrix} a_1 & a_3 \\ c_1 & c_3 \end{vmatrix} - c_2\begin{vmatrix} a_1 & a_3 \\ b_1 & b_3 \end{vmatrix} \quad \text{(second column).}$$

7.4 Evaluation of determinants using expansion by a row or column.

(i)
$$\begin{vmatrix} 1 & 1 & 1 \\ 2 & 4 & -1 \\ 1 & -1 & 0 \end{vmatrix} = 1\begin{vmatrix} 1 & 1 \\ 4 & -1 \end{vmatrix} - (-1)\begin{vmatrix} 1 & 1 \\ 2 & -1 \end{vmatrix} + 0\begin{vmatrix} 1 & 1 \\ 2 & 4 \end{vmatrix}$$
$$= -5 - 3 = -8.$$

(Expansion by the third row.)

(ii)
$$\begin{vmatrix} 2 & 1 & 1 \\ 1 & 0 & -1 \\ 1 & 3 & 2 \end{vmatrix} = (-1)\begin{vmatrix} 1 & 1 \\ 3 & 2 \end{vmatrix} + 0\begin{vmatrix} 2 & 1 \\ 1 & 2 \end{vmatrix} - (-1)\begin{vmatrix} 2 & 1 \\ 1 & 3 \end{vmatrix}$$
$$= -(-1) + 5 = 6.$$

(Expansion by the second row.)

(iii)
$$\begin{vmatrix} 3 & 0 & 1 \\ -2 & -1 & 1 \\ 2 & 2 & -4 \end{vmatrix} = 0\begin{vmatrix} -2 & 1 \\ 2 & -4 \end{vmatrix} + (-1)\begin{vmatrix} 3 & 1 \\ 2 & -4 \end{vmatrix} - 2\begin{vmatrix} 3 & 1 \\ -2 & 1 \end{vmatrix}$$
$$= -(-14) - 2(5) = 4.$$

(Expansion by the second column.)

(iv)
$$\begin{vmatrix} 1 & 2 & 3 \\ 0 & 1 & 2 \\ 1 & 0 & 3 \end{vmatrix} = 1\begin{vmatrix} 1 & 2 \\ 0 & 3 \end{vmatrix} - 0\begin{vmatrix} 2 & 3 \\ 0 & 3 \end{vmatrix} + 1\begin{vmatrix} 2 & 3 \\ 1 & 2 \end{vmatrix}$$
$$= 3 + 1$$
$$= 4.$$

(Expansion by the first column.)

7.5 (i) Show that interchanging rows in a 3×3 determinant changes the sign of the value of the determinant.

We verify this by straight algebraic computation.

$$\begin{vmatrix} a_1 & a_2 & a_3 \\ b_1 & b_2 & b_3 \\ c_1 & c_2 & c_3 \end{vmatrix} = a_1\begin{vmatrix} b_2 & b_3 \\ c_2 & c_3 \end{vmatrix} - a_2\begin{vmatrix} b_1 & b_3 \\ c_1 & c_3 \end{vmatrix} + a_3\begin{vmatrix} b_1 & b_2 \\ c_1 & c_2 \end{vmatrix},$$

$$\begin{vmatrix} b_1 & b_2 & b_3 \\ a_1 & a_2 & a_3 \\ c_1 & c_2 & c_3 \end{vmatrix} = -a_1\begin{vmatrix} b_2 & b_3 \\ c_2 & c_3 \end{vmatrix} + a_2\begin{vmatrix} b_1 & b_3 \\ c_1 & c_3 \end{vmatrix} - a_3\begin{vmatrix} b_1 & b_2 \\ c_1 & c_2 \end{vmatrix}.$$

(The second determinant is evaluated by the second row.)

Similar calculations demonstrate the result for other possible single interchanges of rows.

(ii) Show that multiplying one row of a 3×3 determinant by a number k has the effect of multiplying the value of the determinant by k.

Again, straight computation gives this result.

$$\begin{vmatrix} a_1 & a_2 & a_3 \\ kb_1 & kb_2 & kb_3 \\ c_1 & c_2 & c_3 \end{vmatrix} = -kb_1\begin{vmatrix} a_2 & a_3 \\ c_2 & c_3 \end{vmatrix} + kb_2\begin{vmatrix} a_1 & a_3 \\ c_1 & c_3 \end{vmatrix} - kb_3\begin{vmatrix} a_1 & a_2 \\ c_1 & c_2 \end{vmatrix}$$
$$= k\left(-b_1\begin{vmatrix} a_2 & a_3 \\ c_2 & c_3 \end{vmatrix} + b_2\begin{vmatrix} a_1 & a_3 \\ c_1 & c_3 \end{vmatrix} - b_3\begin{vmatrix} a_1 & a_2 \\ c_1 & c_2 \end{vmatrix}\right)$$

$$a_3\begin{vmatrix} b_1 & b_2 \\ c_1 & c_2 \end{vmatrix} - b_3\begin{vmatrix} a_1 & a_2 \\ c_1 & c_2 \end{vmatrix} + c_3\begin{vmatrix} a_1 & a_2 \\ b_1 & b_2 \end{vmatrix} \quad \text{(third column).}$$

Again, justification of these is by multiplying out and comparing with the definition. It is important to notice the pattern of signs. Each entry always appears (as a coefficient) associated with a positive or negative sign, the same in each of the above expressions. This pattern is easy to remember from the array

$$\begin{bmatrix} + & - & + \\ - & + & - \\ + & - & + \end{bmatrix}.$$

Examples 7.4 give some further illustrations of evaluation of 3×3 determinants. Expansion by certain rows or columns can make the calculation easier in particular cases, especially when some of the entries are zeros.

Now let us try to connect these ideas with previous ideas. Recall the three kinds of elementary row operation given in Chapter 1, which form the basis of the GE process. How are determinants affected by these operations? We find the answers for 3×3 determinants.

Rule

(i) Interchanging two rows in a determinant changes the sign of the determinant.

(ii) Multiplying one row of a determinant by a number k has the effect of multiplying the determinant by k.

(iii) Adding a multiple of one row in a determinant to another row does not change the value of the determinant.

Proofs: (i) and (ii) are quite straightforward. See Examples 7.5. The third requires a little more discussion. Consider a particular case, as follows.

$$\begin{vmatrix} a_1 & a_2 & a_3 \\ b_1+kc_1 & b_2+kc_2 & b_3+kc_3 \\ c_1 & c_2 & c_3 \end{vmatrix}$$

$$= -(b_1+kc_1)\begin{vmatrix} a_2 & a_3 \\ c_2 & c_3 \end{vmatrix} + (b_2+kc_2)\begin{vmatrix} a_1 & a_3 \\ c_1 & c_3 \end{vmatrix}$$

$$-(b_3+kc_3)\begin{vmatrix} a_1 & a_2 \\ c_1 & c_2 \end{vmatrix}$$

$$=k\begin{vmatrix} a_1 & a_2 & a_3 \\ b_1 & b_2 & b_3 \\ c_1 & c_2 & c_3 \end{vmatrix}.$$

Similar calculations demonstrate the result when a multiplier is applied to another row.

7.6 Show that a 3×3 determinant in which two rows are identical has value zero.

Expansion by the other row gives the result:

$$\begin{vmatrix} a_1 & a_2 & a_3 \\ a_1 & a_2 & a_3 \\ c_1 & c_2 & c_3 \end{vmatrix} = c_1\begin{vmatrix} a_2 & a_3 \\ a_2 & a_3 \end{vmatrix} - c_2\begin{vmatrix} a_1 & a_3 \\ a_1 & a_3 \end{vmatrix} + c_3\begin{vmatrix} a_1 & a_2 \\ a_1 & a_2 \end{vmatrix}$$

$$= c_1(0) - c_2(0) + c_3(0)$$

$$= 0.$$

$$\begin{vmatrix} a_1 & a_2 & a_3 \\ b_1 & b_2 & b_3 \\ a_1 & a_2 & a_3 \end{vmatrix} = -b_1\begin{vmatrix} a_2 & a_3 \\ a_2 & a_3 \end{vmatrix} + b_2\begin{vmatrix} a_1 & a_3 \\ a_1 & a_3 \end{vmatrix} - b_3\begin{vmatrix} a_1 & a_2 \\ a_1 & a_2 \end{vmatrix}$$

$$= -b_1(0) + b_2(0) - b_3(0)$$

$$= 0.$$

(And similarly when the second and third rows are identical.)

7.7 Evaluate the 4×4 determinant

$$\begin{vmatrix} 2 & 0 & 1 & -1 \\ 1 & 1 & -1 & 0 \\ 0 & 3 & 1 & 3 \\ -2 & 1 & -1 & 1 \end{vmatrix}.$$

Expand by the first row, obtaining

$$2\begin{vmatrix} 1 & -1 & 0 \\ 3 & 1 & 3 \\ 1 & -1 & 1 \end{vmatrix} - 0\begin{vmatrix} 1 & -1 & 0 \\ 0 & 1 & 3 \\ -2 & -1 & 1 \end{vmatrix}$$

$$+1\begin{vmatrix} 1 & 1 & 0 \\ 0 & 3 & 3 \\ -2 & 1 & 1 \end{vmatrix} - (-1)\begin{vmatrix} 1 & 1 & -1 \\ 0 & 3 & 1 \\ -2 & 1 & -1 \end{vmatrix}$$

$$= 2\left(\begin{vmatrix} 1 & 3 \\ -1 & 1 \end{vmatrix} + \begin{vmatrix} 3 & 3 \\ 1 & 1 \end{vmatrix}\right) + \left(\begin{vmatrix} 3 & 3 \\ 1 & 1 \end{vmatrix} - \begin{vmatrix} 0 & 3 \\ -2 & 1 \end{vmatrix}\right)$$

$$+\left(3\begin{vmatrix} 1 & -1 \\ -2 & -1 \end{vmatrix} - \begin{vmatrix} 1 & 1 \\ -2 & 1 \end{vmatrix}\right)$$

(the determinants being evaluated by the first row, the first row and the second row respectively)

$$= 2(4+0) + (0-6) + (-9-3)$$

$$= -10.$$

$$= -b_1\begin{vmatrix} a_2 & a_3 \\ c_2 & c_3 \end{vmatrix} + b_2\begin{vmatrix} a_1 & a_3 \\ c_1 & c_3 \end{vmatrix} - b_3\begin{vmatrix} a_1 & a_2 \\ c_1 & c_2 \end{vmatrix}$$

$$+k\left(-c_1\begin{vmatrix} a_2 & a_3 \\ c_2 & c_3 \end{vmatrix} + c_2\begin{vmatrix} a_1 & a_3 \\ c_1 & c_3 \end{vmatrix} - c_3\begin{vmatrix} a_1 & a_2 \\ c_1 & c_2 \end{vmatrix}\right)$$

$$= \begin{vmatrix} a_1 & a_2 & a_3 \\ b_1 & b_2 & b_3 \\ c_1 & c_2 & c_3 \end{vmatrix} + k(-c_1(a_2c_3 - a_3c_2) + c_2(a_1c_3 - a_3c_1) - c_3(a_1c_2 - a_2c_1))$$

$$= \begin{vmatrix} a_1 & a_2 & a_3 \\ b_1 & b_2 & b_3 \\ c_1 & c_2 & c_3 \end{vmatrix},$$

since the expression in brackets is identically zero (work it out!). This process works in exactly the same way for the other possible cases. Incidentally, we have come across another result of interest in the course of the above.

Rule

A determinant in which two rows are the same has value zero.

Proof: Notice that in the previous proof the expression in brackets is in fact the expansion by the second row of the determinant

$$\begin{vmatrix} a_1 & a_2 & a_3 \\ c_1 & c_2 & c_3 \\ c_1 & c_2 & c_3 \end{vmatrix},$$

and this is seen above to equal zero. Other possible cases work similarly. See Examples 7.6.

We shall not pursue the detailed discussion of larger determinants. A 4×4 determinant is defined most easily by means of expansion by the first row:

$$\begin{vmatrix} a_1 & a_2 & a_3 & a_4 \\ b_1 & b_2 & b_3 & b_4 \\ c_1 & c_2 & c_3 & c_4 \\ d_1 & d_2 & d_3 & d_4 \end{vmatrix} = a_1\begin{vmatrix} b_2 & b_3 & b_4 \\ c_2 & c_3 & c_4 \\ d_2 & d_3 & d_4 \end{vmatrix} - a_2\begin{vmatrix} b_1 & b_3 & b_4 \\ c_1 & c_3 & c_4 \\ d_1 & d_3 & d_4 \end{vmatrix}$$

$$+ a_3\begin{vmatrix} b_1 & b_2 & b_4 \\ c_1 & c_2 & c_4 \\ d_1 & d_2 & d_4 \end{vmatrix} - a_4\begin{vmatrix} b_1 & b_2 & b_3 \\ c_1 & c_2 & c_3 \\ d_1 & d_2 & d_3 \end{vmatrix}.$$

See Example 7.7. Expansions by the other rows and by the columns produce the same value, provided we remember the alternating pattern of signs

7.8 Evaluation of determinants of triangular matrices.

(i) $$\begin{vmatrix} a & b & c \\ 0 & d & e \\ 0 & 0 & f \end{vmatrix} = a\begin{vmatrix} d & e \\ 0 & f \end{vmatrix} = adf.$$ (Expanding by the first column.)

(ii) $$\begin{vmatrix} a_1 & a_2 & a_3 & a_4 \\ 0 & b_2 & b_3 & b_4 \\ 0 & 0 & c_3 & c_4 \\ 0 & 0 & 0 & d_4 \end{vmatrix} = a_1\begin{vmatrix} b_2 & b_3 & b_4 \\ 0 & c_3 & c_4 \\ 0 & 0 & d_4 \end{vmatrix}$$ (by first column)

$$= a_1(b_2c_3d_4) \quad \text{(by part (i))}$$
$$= a_1b_2c_3d_4.$$

(iii) Larger determinants yield corresponding results. The determinant of *any* triangular matrix is equal to the product of the entries on the main diagonal. You should be able to see why this happens. A proper proof would use the Principle of Mathematical Induction. (We have dealt here with upper triangular matrices. Similar arguments apply in the case of lower triangular matrices.)

7.9 Evaluation of determinants using the GE process.

(i) $$\begin{vmatrix} 1 & 3 & -1 \\ 2 & 0 & 1 \\ 1 & 1 & 4 \end{vmatrix}.$$

Proceed with the GE process, noting the effect on the determinant of each row operation performed.

1	3	-1		Leaves the determinant unchanged.
0	-6	3	$(2)-2\times(1)$	
0	-2	5	$(3)-(1)$	
1	3	-1		Divides the determinant by -6.
0	1	$-\frac{1}{2}$	$(2)\div -6$	
0	-2	5		
1	3	-1		Leaves the determinant unchanged.
0	1	$-\frac{1}{2}$		
0	0	4	$(3)+2\times(2)$	

This last matrix is upper triangular, and has determinant equal to 4. Hence the original determinant has value $4\times(-6)$, i.e. -24.

(ii) $$\begin{vmatrix} 0 & 1 & 1 & 1 \\ 1 & 0 & 1 & 1 \\ 1 & 1 & 0 & 1 \\ 1 & 1 & 1 & 0 \end{vmatrix}.$$

GE process as above:

1	0	1	1	interchange rows	Changes the sign of the determinant.
0	1	1	1		
1	1	0	1		
1	1	1	0		

$$\begin{bmatrix} + & - & + & - \\ - & + & - & + \\ + & - & + & - \\ - & + & - & + \end{bmatrix}.$$

Larger determinants are defined similarly.

The above rules hold for determinants of all sizes. Indeed, corresponding results also hold for columns and column operations in all determinants, but we shall not go into the details of these. (Elementary column operations are exactly analogous to elementary row operations.)

It is apparent that evaluation of large determinants will be a lengthy business. The results of this chapter can be used to provide a short cut, however. If we apply the standard GE process, keeping track of all the row operations used, we end up with a triangular matrix, whose determinant is a multiple of the given determinant. Evaluation of determinants of triangular matrices is a simple matter (see Example 7.8), so here is another use for our GE process. Some determinants are evaluated by this procedure in Examples 7.9.

Now recall the Equivalence Theorem from Chapter 6. Four different sets of circumstances led to the GE process applied to a $p \times p$ matrix ending with a matrix with p non-zero rows. The argument above demonstrates that in such a case the $p \times p$ matrix concerned must have a non-zero determinant. The upper triangular matrix resulting from the GE process applied to a $p \times p$ matrix has determinant zero if and only if its last row consists entirely of 0s, in which case it has fewer than p non-zero rows.

Here then is the complete Equivalence Theorem.

Theorem

Let A be a $p \times p$ matrix. The following are equivalent.

(i) A is invertible.

(ii) The rank of A is equal to p.

(iii) The columns of A form a LI list.

(iv) The set of simultaneous equations which can be written $A\boldsymbol{x}=\mathbf{0}$ has no solution other than $\boldsymbol{x}=0$.

(v) (This is the new part.) The determinant of A is non-zero.

We end this chapter with some new notions, which are developed in further study of linear algebra (but not in this book).

In the expansion of a determinant by a row or column, each entry of the chosen row or column is multiplied by a *signed* determinant of the next smaller size. This signed determinant is called the *cofactor* of that particular

1	0	1	1		Determinant unchanged
0	1	1	1		
0	1	−1	0	(3)−(1)	
0	1	0	−1	(4)−(1)	

1	0	1	1		Determinant unchanged.
0	1	1	1		
0	0	−2	−1	(3)−(2)	
0	0	−1	−2	(4)−(2)	

1	0	1	1		Divides the determinant by −2.
0	1	1	1		
0	0	1	$\frac{1}{2}$	(3)÷ −2	
0	0	−1	−2		

1	0	1	1		Determinant unchanged.
0	1	1	1		
0	0	1	$\frac{1}{2}$		
0	0	0	$-\frac{3}{2}$	(4)+(3)	

This last matrix has determinant equal to $-\frac{3}{2}$. Hence the original determinant has value $(-\frac{3}{2})\times(-2)$, i.e. 3.

7.10 Find adj A, where

$$A=\begin{bmatrix} 2 & 1 & 1 \\ -1 & 1 & 0 \\ 1 & -1 & 1 \end{bmatrix}.$$

$$A_{11}=\begin{vmatrix} 1 & 0 \\ -1 & 1 \end{vmatrix}=1, \qquad A_{12}=-\begin{vmatrix} -1 & 0 \\ 1 & 1 \end{vmatrix}=1,$$

$$A_{13}=\begin{vmatrix} -1 & 1 \\ 1 & -1 \end{vmatrix}=0, \qquad A_{21}=-\begin{vmatrix} 1 & 1 \\ -1 & 1 \end{vmatrix}=-2,$$

$$A_{22}=\begin{vmatrix} 2 & 1 \\ 1 & 1 \end{vmatrix}=1, \qquad A_{23}=-\begin{vmatrix} 2 & 1 \\ 1 & -1 \end{vmatrix}=3,$$

$$A_{31}=\begin{vmatrix} 1 & 1 \\ 1 & 0 \end{vmatrix}=-1, \qquad A_{32}=-\begin{vmatrix} 2 & 1 \\ -1 & 0 \end{vmatrix}=-1,$$

$$A_{33}=\begin{vmatrix} 2 & 1 \\ -1 & 1 \end{vmatrix}=3.$$

So

$$\text{adj}\, A=\begin{bmatrix} A_{11} & A_{21} & A_{31} \\ A_{12} & A_{22} & A_{32} \\ A_{13} & A_{23} & A_{33} \end{bmatrix}=\begin{bmatrix} 1 & -2 & -1 \\ 1 & 1 & -1 \\ 0 & 3 & 3 \end{bmatrix}.$$

entry. We illustrate this in the 3×3 case, but the following ideas apply to determinants of any size. In the determinant

$$\begin{vmatrix} a_1 & a_2 & a_3 \\ b_1 & b_2 & b_3 \\ c_1 & c_2 & c_3 \end{vmatrix},$$

the cofactor of a_1 is

$$\begin{vmatrix} b_2 & b_3 \\ c_2 & c_3 \end{vmatrix},$$

the cofactor of a_2 is

$$-\begin{vmatrix} b_1 & b_3 \\ c_1 & c_3 \end{vmatrix},$$

and so on. If we change notation to a double suffix and let

$$A = \begin{bmatrix} a_{11} & a_{12} & a_{13} \\ a_{21} & a_{22} & a_{23} \\ a_{31} & a_{32} & a_{33} \end{bmatrix},$$

then there is a convenient notation for cofactors. The cofactor of a_{ij} is denoted by A_{ij}. For example:

$$A_{11} = \begin{vmatrix} a_{22} & a_{23} \\ a_{32} & a_{33} \end{vmatrix} \quad \text{and} \quad A_{12} = -\begin{vmatrix} a_{21} & a_{23} \\ a_{31} & a_{33} \end{vmatrix}.$$

In this notation, then, we can write

$$\det A = a_{11}A_{11} + a_{12}A_{12} + a_{13}A_{13},$$

which is the expansion by the first row, and similarly for expansions by the other rows and the columns. Note that the negative signs are incorporated in the cofactors, so the determinant is represented in this way as a *sum* of terms.

The *adjoint* matrix of A (written adj A) is defined as follows. The (i, j)-entry of adj A is A_{ji}. Note the order of the suffixes. To obtain adj A from A, replace each entry a_{ij} of A by its own cofactor A_{ij}, and then transpose the resulting matrix. This yields adj A. See Example 7.10. This process is impossibly long in practice, even for matrices as small as 3×3, so the significance of the adjoint matrix is mainly theoretical. We can write down one (perhaps surprising) result, however.

Theorem

If A is an invertible matrix, then

$$A^{-1} = \frac{1}{\det A} \operatorname{adj} A.$$

7.11 Show that if A is an invertible 3×3 matrix then

$$A^{-1} = 1/(\det A)\,\text{adj}\,A.$$

Let $A = [a_{ij}]_{3 \times 3}$. The adjoint of A is the *transposed* matrix of cofactors, so the (k, j)-entry in adj A is A_{jk}. Hence the (i, j)-entry in the product $A(\text{adj}\,A)$ is

$$\sum_{k=1}^{3} a_{ik}A_{jk}, \quad \text{i.e } a_{i1}A_{j1} + a_{i2}A_{j2} + a_{i3}A_{j2} \quad (*).$$

Now if $j = i$ then this is equal to

$$a_{i1}A_{i1} + a_{i2}A_{i2} + a_{i3}A_{i3},$$

which is the value of det A (expanded by the ith row). So every entry of $A(\text{adj}\,A)$ on the main diagonal is det A. Moreover, if $j \neq i$, then the (i, j)-entry in $A(\text{adj}\,A)$ is zero. This is because the expression (*) then is in effect the expansion of a determinant in which two rows are identical. For example, if $i = 2$ and $j = 1$:

$$a_{21}A_{11} + a_{22}A_{12} + a_{23}A_{13} = \begin{vmatrix} a_{21} & a_{22} & a_{23} \\ a_{21} & a_{22} & a_{23} \\ a_{31} & a_{32} & a_{33} \end{vmatrix} = 0.$$

Hence all entries in $A(\text{adj}\,A)$ which are off the main diagonal are zero. So

$$A(\text{adj}\,A) = \begin{bmatrix} \det A & 0 & 0 \\ 0 & \det A & 0 \\ 0 & 0 & \det A \end{bmatrix} = (\det A)I.$$

It follows that

$$A(1/(\det A)\,\text{adj}\,A) = I.$$

(The supposition that A is invertible ensures that det $A \neq 0$.)

Consequently (see Example 5.10), we have the required result:

$$A^{-1} = 1/(\det A)\,\text{adj}\,A.$$

The above argument can be extended to deal with the case of a $p \times p$ matrix, for any value of p.

Proof: See Example 7.11. See also Example 5.8.

Finally, an important property of determinants. We omit the proof, which is not easy. The interested reader may find a proof in the book by Kolman. (A list of books for further reading is given on page 146.)

Theorem

If A and B are $p \times p$ matrices, then

$$\det(AB) = (\det A)(\det B).$$

Summary

Definitions are given of 2×2 and 3×3 determinants, and methods are described for evaluating such determinants. It is shown how larger determinants can be defined and evaluated. The effects on determinants of elementary row operations are shown. The application of the GE process to evaluating determinants is demonstrated, and it is used to show that a square matrix is invertible if and only if it has a non-zero determinant. Cofactors and the adjoint matrix are defined. The theorem on the determinant of a product of matrices is stated.

Exercises

1. Evaluate the following determinants.

$$\begin{vmatrix} 2 & 1 \\ 3 & -2 \end{vmatrix}, \quad \begin{vmatrix} -1 & -2 \\ -3 & -4 \end{vmatrix}, \quad \begin{vmatrix} 3 & -5 \\ 0 & 1 \end{vmatrix}, \quad \begin{vmatrix} 4 & -2 \\ 0 & 0 \end{vmatrix},$$

$$\begin{vmatrix} 4 & 2 \\ 6 & -4 \end{vmatrix}, \quad \begin{vmatrix} -2 & -1 \\ -3 & 2 \end{vmatrix}, \quad \begin{vmatrix} 3 & -2 \\ 2 & 1 \end{vmatrix}, \quad \begin{vmatrix} 0 & 1 \\ 3 & -5 \end{vmatrix}.$$

2. Evaluate the following determinants (by any of the procedures described in the text for 3×3 determinants).

$$\begin{vmatrix} 0 & 1 & 1 \\ 1 & 0 & 1 \\ 1 & 1 & 0 \end{vmatrix}, \quad \begin{vmatrix} 3 & 1 & 3 \\ -2 & -1 & 0 \\ 1 & 1 & 1 \end{vmatrix}, \quad \begin{vmatrix} 0 & 0 & 1 \\ -2 & 1 & 2 \\ 1 & 4 & -6 \end{vmatrix},$$

$$\begin{vmatrix} 1 & 2 & 3 \\ 4 & 5 & 6 \\ 7 & 8 & 9 \end{vmatrix}, \quad \begin{vmatrix} 4 & 5 & 1 \\ 1 & 1 & -1 \\ 3 & 2 & 2 \end{vmatrix}, \quad \begin{vmatrix} 1 & 1 & -1 \\ 4 & 5 & 1 \\ 3 & 2 & 2 \end{vmatrix},$$

$$\begin{vmatrix} 2 & 4 & -6 \\ -1 & -2 & 3 \\ 1 & 1 & 4 \end{vmatrix}, \quad \begin{vmatrix} 3 & 0 & 3 \\ -1 & 1 & 1 \\ -2 & 0 & -2 \end{vmatrix}, \quad \begin{vmatrix} 2 & -2 & 3 \\ 1 & 0 & 4 \\ 0 & 1 & 1 \end{vmatrix}.$$

3. Evaluate the following 4×4 determinants.

$$\begin{vmatrix} 2 & 0 & 1 & -1 \\ -1 & 1 & 0 & 3 \\ 0 & 2 & 1 & 1 \\ 3 & 3 & 1 & -1 \end{vmatrix}, \quad \begin{vmatrix} 0 & 3 & 1 & -2 \\ 2 & 2 & -2 & 1 \\ 1 & 0 & 1 & 0 \\ 0 & 2 & -3 & 3 \end{vmatrix}.$$

4. Let A be a 3×3 skew-symmetric matrix. Prove that $\det A = 0$. Is this true for all skew-symmetric matrices?

5. Using the fact that, for any square matrices A and B of the same size, $\det(AB) = (\det A)(\det B)$, show that if either A or B is singular (or both are) then AB is singular. Show also that if A is invertible then $\det(A^{-1}) = 1/(\det A)$.

6. Let A be a 3×3 matrix, and let k be any real number. Show that $\det(kA) = k^3(\det A)$.

7. Let A be a square matrix such that $A^3 = 0$. Show that $\det A = 0$, so that A is singular. Extend this to show that every square matrix A which satisfies $A^n = 0$, for some n, is singular.

8. Evaluate $\det A$ and adj A, where

$$A = \begin{bmatrix} 0 & 1 & 1 \\ 1 & 0 & 1 \\ 1 & 1 & 0 \end{bmatrix}.$$

Check your answers by evaluating the product $A(\text{adj } A)$.

Examples

8.1 By finding the ranks of appropriate matrices, decide whether the following set of equations has any solutions.

$$\begin{aligned} x_1+3x_2+2x_3 &= 3 \\ -x_1+x_2+2x_3 &= -3 \\ 2x_1+4x_2-2x_3 &= 10. \end{aligned}$$

$$A=\begin{bmatrix} 1 & 3 & 2 \\ -1 & 1 & 2 \\ 2 & 4 & -2 \end{bmatrix}, \quad [A\,\vdots\,\boldsymbol{h}]=\begin{bmatrix} 1 & 3 & 2 & 3 \\ -1 & 1 & 2 & -3 \\ 2 & 4 & -2 & 10 \end{bmatrix}.$$

The GE process yields (respectively)

$$\begin{bmatrix} 1 & 3 & 2 \\ 0 & 1 & 1 \\ 0 & 0 & 1 \end{bmatrix} \qquad \begin{bmatrix} 1 & 3 & 2 & 3 \\ 0 & 1 & 1 & 0 \\ 0 & 0 & 1 & -1 \end{bmatrix}$$

The rank of A is 3, the rank of $[A\,\vdots\,\boldsymbol{h}]$ is 3, so there do exist solutions.

8.2 Examples of ranks of augmented matrices.

(i) $$[A\,\vdots\,\boldsymbol{h}]=\begin{bmatrix} 0 & 1 & 2 & -3 \\ 1 & -1 & 0 & -2 \\ 3 & -2 & 2 & 1 \end{bmatrix}.$$

The GE process leads to

$$\begin{bmatrix} 1 & -1 & 0 & -2 \\ 0 & 1 & 2 & -3 \\ 0 & 0 & 0 & 1 \end{bmatrix},$$

so the rank of A is 2 and the rank of $[A\,\vdots\,\boldsymbol{h}]$ is 3.

(ii) (Arising from a set of four equations in three unknowns.)

$$[A\,\vdots\,\boldsymbol{h}]=\begin{bmatrix} 1 & -1 & 1 & 4 \\ 0 & 2 & 2 & 6 \\ 2 & 3 & -1 & 8 \\ -1 & 2 & 0 & -1 \end{bmatrix}.$$

The GE process leads to

$$\begin{bmatrix} 1 & -1 & 1 & 4 \\ 0 & 1 & 1 & 3 \\ 0 & 0 & 1 & \frac{15}{8} \\ 0 & 0 & 0 & 0 \end{bmatrix},$$

so the rank of A is 3 and the rank of $[A\,\vdots\,\boldsymbol{h}]$ is 3.

(iii) (Arising from a set of four equations in three unknowns.)

$$[A\,\vdots\,\boldsymbol{h}]=\begin{bmatrix} 1 & 1 & 1 & 1 \\ 3 & 4 & -1 & -2 \\ -1 & 0 & 2 & 1 \\ 0 & 2 & 1 & 0 \end{bmatrix}.$$

The GE process leads to

$$\begin{bmatrix} 1 & 1 & 1 & 1 \\ 0 & 1 & -4 & -5 \\ 0 & 0 & 1 & 1 \\ 0 & 0 & 0 & 1 \end{bmatrix}.$$

8

Solutions to simultaneous equations 2

We have developed ideas since Chapter 2 which are applicable to solving simultaneous linear equations, so let us reconsider our methods in the light of these ideas. Recall that a set of p equations in q unknowns can be written in the form $A\boldsymbol{x}=\boldsymbol{h}$, where A is the $p\times q$ matrix of coefficients, $\boldsymbol{x}$ is the q-vector of unknowns and $\boldsymbol{h}$ is the p-vector of the right-hand sides of the equations. Recall also that there can be three possible situations: no solution, a unique solution or infinitely many solutions.

Example 8.1 illustrates the criterion for deciding whether there are any solutions. Let $[A\,\vdots\,\boldsymbol{h}]$ denote the *augmented matrix* obtained by adding $\boldsymbol{h}$ as an extra column to A ($[A\,\vdots\,\boldsymbol{h}]$ is the matrix on which we carry out the GE process). As we saw in Chapter 2, the equations are *in*consistent if and only if the last non-zero row (after the GE process) consists entirely of 0s except for the last entry. Consider what this means with regard to the ranks of the matrices A and $[A\,\vdots\,\boldsymbol{h}]$. The GE process applied to $[A\,\vdots\,\boldsymbol{h}]$ is identical to the GE process applied to A, as far as the first q columns are concerned. In the above situation, then, the rank of A is less than the rank of $[A\,\vdots\,\boldsymbol{h}]$, since the last non-zero row after the GE process on $[A\,\vdots\,\boldsymbol{h}]$ corresponds to a row of 0s in the matrix obtained from A by the GE process. We therefore have:

Rule

The equation $A\boldsymbol{x}=\boldsymbol{h}$ has a solution if and only if the rank of $[A\,\vdots\,\boldsymbol{h}]$ is the same as the rank of A.

Examples 8.2 provide illustration of the different cases which arise.

Notice the special case of *homogeneous* simultaneous equations, that is, the case when $\boldsymbol{h}=\boldsymbol{0}$. As we observed before, such a set of equations *must* be consistent, because a solution is obtained by taking every unknown to be zero. A moment's thought should convince the reader that here the rank of $[A\,\vdots\,\boldsymbol{h}]$ is bound to be the same as the rank of A.

Here the rank of A is 3 and the rank of $[A \mid \boldsymbol{h}]$ is 4, so the set of equations would have been inconsistent.

8.3 Illustrations of the equation $A\boldsymbol{x}=\boldsymbol{h}$, with A singular.

(i) $$[A \mid \boldsymbol{h}]=\begin{bmatrix} 1 & 0 & 3 & 5 \\ -2 & 5 & -1 & 0 \\ -1 & 4 & 1 & 4 \end{bmatrix}.$$

The GE process applied to A:

$$\begin{bmatrix} 1 & 0 & 3 \\ -2 & 5 & -1 \\ -1 & 4 & 1 \end{bmatrix} \to \begin{bmatrix} 1 & 0 & 3 \\ 0 & 1 & 1 \\ 0 & 0 & 0 \end{bmatrix} \quad (A \text{ is singular}).$$

The GE process applied to $[A \mid \boldsymbol{h}]$:

$$\begin{bmatrix} 1 & 0 & 3 & 5 \\ -2 & 5 & -1 & 0 \\ -1 & 4 & 1 & 4 \end{bmatrix} \to \begin{bmatrix} 1 & 0 & 3 & 5 \\ 0 & 1 & 1 & 2 \\ 0 & 0 & 0 & 1 \end{bmatrix}.$$

In this case there would be no solutions, since the ranks of A and $[A \mid \boldsymbol{h}]$ are unequal.

(ii) $$[A \mid \boldsymbol{h}]=\begin{bmatrix} 1 & -1 & 3 & -4 \\ 2 & 3 & 1 & 7 \\ 4 & 3 & 5 & 5 \end{bmatrix}.$$

The GE process applied to A:

$$\begin{bmatrix} 1 & -1 & 3 \\ 2 & 3 & 1 \\ 4 & 3 & 5 \end{bmatrix} \to \begin{bmatrix} 1 & -1 & 3 \\ 0 & 1 & -1 \\ 0 & 0 & 0 \end{bmatrix} \quad (A \text{ is singular}).$$

The GE process applied to $[A \mid \boldsymbol{h}]$:

$$\begin{bmatrix} 1 & -1 & 3 & -4 \\ 2 & 3 & 1 & 7 \\ 4 & 3 & 5 & 5 \end{bmatrix} \to \begin{bmatrix} 1 & -1 & 3 & -4 \\ 0 & 1 & -1 & 3 \\ 0 & 0 & 0 & 0 \end{bmatrix}.$$

In this case there would be infinitely many solutions. The ranks of A and of $[A \mid \boldsymbol{h}]$ are the same, but less than 3.

8.4 Solution involving two parameters.

$$[A \mid \boldsymbol{h}]=\begin{bmatrix} 1 & -1 & 3 & -2 \\ 2 & -2 & 6 & -4 \\ -1 & 1 & 3 & 2 \end{bmatrix}.$$

The GE process applied to $[A \mid \boldsymbol{h}]$ leads to

$$\begin{bmatrix} 1 & -1 & 3 & -2 \\ 0 & 0 & 0 & 0 \\ 0 & 0 & 0 & 0 \end{bmatrix}.$$

To solve the matrix equation $\begin{bmatrix} x_1 \\ x_2 \\ x_3 \end{bmatrix} = \begin{bmatrix} -2 \\ -4 \\ 2 \end{bmatrix}$ we have, in effect, only the single equation

$$x_1 - x_2 + 3x_3 = -2.$$

Introduce parameters $x_2 = t$ and $x_3 = u$, and substitute to obtain $x_1 = t - 3u - 2$ (the method of Chapter 2).

Next, given a set of equations which are known to have a solution, what criterion determines whether there is a unique solution or infinitely many? Part of the answer is very easy to see.

Rule

If A is an invertible matrix then the equation $A\boldsymbol{x}=\boldsymbol{h}$ has a unique solution. (The solution is $\boldsymbol{x}=A^{-1}\boldsymbol{h}$.)

The other part of the answer is the converse of this, namely:

Rule

If A is a singular matrix then the equation $A\boldsymbol{x}=\boldsymbol{h}$ (if it has solutions at all) has infinitely many solutions.

Proof: See Examples 8.3 for illustration. We must consider the GE process and the process for inverting a matrix. If A is singular, then the GE process applied to A yields a matrix whose last row consists entirely of 0s. The GE process applied to $[A \,\vdots\, \boldsymbol{h}]$ may have last row all 0s or may have last row all 0s except for the last entry. In the latter case there are no solutions, and in the former case we have to look at the last non-zero row in order to decide whether there are no solutions, or infinitely many solutions. See the rule given in Chapter 2.

Example 8.4 shows a particularly trivial way in which there can be infinitely many solutions. In that case there are two parameters in the solution. Example 8.5 (in which the matrix is 4×4) shows that two parameters can arise in the solution of non-trivial cases also. Can you suggest a 4×4 set of equations in which the set of solutions has three parameters? All these are examples of a general rule.

Rule

If A is a $p\times p$ matrix whose rank is r, and $\boldsymbol{h}$ is a p-vector, then the equation $A\boldsymbol{x}=\boldsymbol{h}$ has solutions provided that the rank of $[A \,\vdots\, \boldsymbol{h}]$ is also equal to r, and in that case the number of parameters needed to specify the solutions is $p-r$. (This covers the case when $r=p$, A is invertible and there is a unique solution which requires no parameters.)

A proof of this rule is beyond the scope of this book, but the reader should be able to see intuitively why it happens by visualising the possible outcomes of the GE process.

Rule

If A is a $p\times q$ matrix with $p\geqslant q$, and $\boldsymbol{h}$ is a p-vector, then the equation $A\boldsymbol{x}=\boldsymbol{h}$ has a unique solution if and only if the rank of A and the rank of $[A \,\vdots\, \boldsymbol{h}]$ are both equal to q, the number of columns of A.

8.5 Solution involving two parameters.

$$[A \,\vdots\, \boldsymbol{h}] = \begin{bmatrix} 1 & 2 & 1 & 0 & 1 \\ 1 & -1 & -2 & 3 & -2 \\ -2 & 1 & 3 & -5 & 3 \\ 0 & 2 & 2 & -2 & 2 \end{bmatrix}.$$

The GE process applied to $[A \,\vdots\, \boldsymbol{h}]$ leads to

$$\begin{bmatrix} 1 & 2 & 1 & 0 & 1 \\ 0 & 1 & 1 & -1 & 1 \\ 0 & 0 & 0 & 0 & 0 \\ 0 & 0 & 0 & 0 & 0 \end{bmatrix}.$$

Solving equations in this case would entail solving *two* equations in four unknowns:

$$\begin{aligned} x_1 + 2x_2 + x_3 \phantom{{}-x_4} &= 1 \\ x_2 + x_3 - x_4 &= 1. \end{aligned}$$

Introduce parameters $x_3 = t$, and $x_4 = u$, and substitute to obtain $x_2 = 1 - t - u$ and $x_1 = -1 + t - 2u$.

8.6 Uniqueness of solutions when A is not square. Listed below are four possible results of applying the GE process (to the augmented matrix) where A is a 4×3 matrix and $\boldsymbol{h}$ is a 4-vector.

(i) $$\begin{bmatrix} 1 & -1 & 2 & 1 \\ 0 & 1 & 3 & -3 \\ 0 & 0 & 1 & -1 \\ 0 & 0 & 0 & 0 \end{bmatrix}.$$

Here the ranks of A and of $[A \,\vdots\, \boldsymbol{h}]$ are both 3, and there is a unique solution.

(ii) $$\begin{bmatrix} 1 & 0 & 1 & -3 \\ 0 & 1 & 1 & 0 \\ 0 & 0 & 1 & 2 \\ 0 & 0 & 0 & 1 \end{bmatrix}.$$

Here the ranks of A and of $[A \,\vdots\, \boldsymbol{h}]$ are different, so there are no solutions.

(iii) $$\begin{bmatrix} 1 & 1 & -2 & 2 \\ 0 & 1 & 1 & 3 \\ 0 & 0 & 0 & 1 \\ 0 & 0 & 0 & 0 \end{bmatrix}.$$

Here the ranks of A and of $[A \,\vdots\, \boldsymbol{h}]$ are different, so there are no solutions.

(iv) $$\begin{bmatrix} 1 & 0 & 3 & -2 \\ 0 & 0 & 1 & 2 \\ 0 & 0 & 0 & 0 \\ 0 & 0 & 0 & 0 \end{bmatrix}.$$

Here the ranks of A and of $[A \,\vdots\, \boldsymbol{h}]$ are both 2, so there are infinitely many solutions.

To see this, consider what must be the result of the GE process if there is to be a unique solution. The first q rows of the matrix must have 1s in the $(1,1)$-, $(2,2)$-, ..., (q,q)-places and 0s below this diagonal, and all subsequent rows must consist of 0s only. This is just the situation when both A and $[A \mid \boldsymbol{h}]$ have rank equal to q. Example 8.6 illustrates this.

Notice that if A is a $p\times q$ matrix with $p<q$, then the equation $A\boldsymbol{x}=\boldsymbol{h}$ *cannot* have a unique solution.

Summary

Rules are given and discussed regarding the solution of equations of the form $A\boldsymbol{x}=\boldsymbol{h}$. These involve the rank of A and the rank of the augmented matrix $[A \mid \boldsymbol{h}]$, and whether (in the case where A is a square matrix) A is invertible or singular.

Exercises

By considering the ranks of the matrix of coefficients and the augmented matrix, decide in each case below whether the given set of equations is consistent or not and, if it is, whether there is a unique solution.

(i) $$\begin{aligned} 2x-y&=1\\ x+3y&=11. \end{aligned}$$

(ii) $$\begin{aligned} 3x-6y&=5\\ x-2y&=1. \end{aligned}$$

(iii) $$\begin{aligned} -4x+3y&=0\\ 12x-9y&=0. \end{aligned}$$

(iv) $$\begin{aligned} x+2y&=0\\ 3x-4y&=0. \end{aligned}$$

(v) $$\begin{aligned} x_1-x_2+2x_3&=3\\ 2x_1-3x_2-x_3&=-8\\ 2x_1+x_2+x_3&=3. \end{aligned}$$

(vi) $$\begin{aligned} 2x_2+x_3&=0\\ x_1-3x_2+2x_3&=0\\ 2x_1+x_2-x_3&=0. \end{aligned}$$

(vii) $$\begin{aligned} -x_1+2x_2-4x_3&=1\\ 2x_1+3x_2+x_3&=-2\\ x_1-x_2+3x_3&=2. \end{aligned}$$

(viii) $$\begin{aligned} x_1+2x_2+3x_3&=0\\ x_1-x_2-3x_3&=0\\ -3x_1+x_2-2x_3&=0. \end{aligned}$$

(ix) $$\begin{aligned} 2x_1+x_2+5x_3&=3\\ -x_1+2x_2&=1\\ x_1+2x_2+4x_3&=3. \end{aligned}$$

(x) $$\begin{aligned} x_1-2x_2-x_3&=2\\ x_2+x_3&=0\\ x_1+x_3&=2\\ x_1+3x_2+4x_3&=2. \end{aligned}$$

(xi) $$\begin{aligned} x_1-2x_2-x_3&=0\\ x_2+x_2&=0\\ x_1+x_3&=0\\ x_1+3x_2+4x_3&=0. \end{aligned}$$

(xii) $$\begin{aligned} x_1+x_2-x_3&=8\\ -x_1+x_2+x_3&=2\\ x_1-x_2+x_3&=0\\ x_1+x_2+x_3&=10. \end{aligned}$$

Examples

9.1

(i)

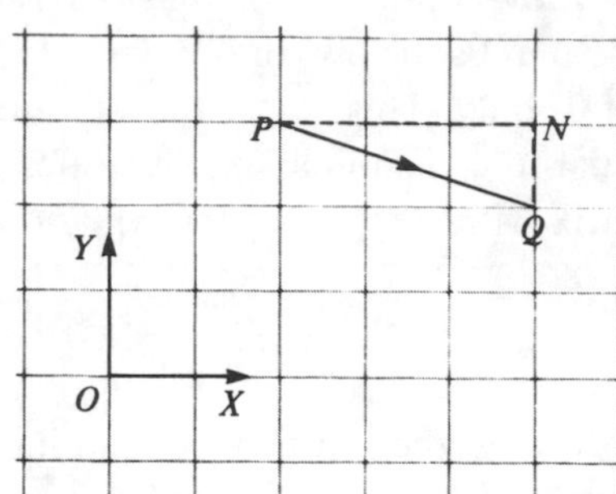

$\overrightarrow{PQ} = \begin{bmatrix} 3 \\ -1 \end{bmatrix}$

(ii)

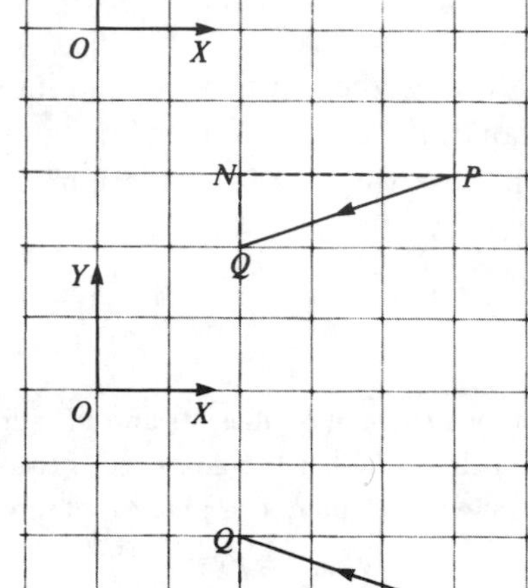

$\overrightarrow{PQ} = \begin{bmatrix} -3 \\ -1 \end{bmatrix}$

(iii)

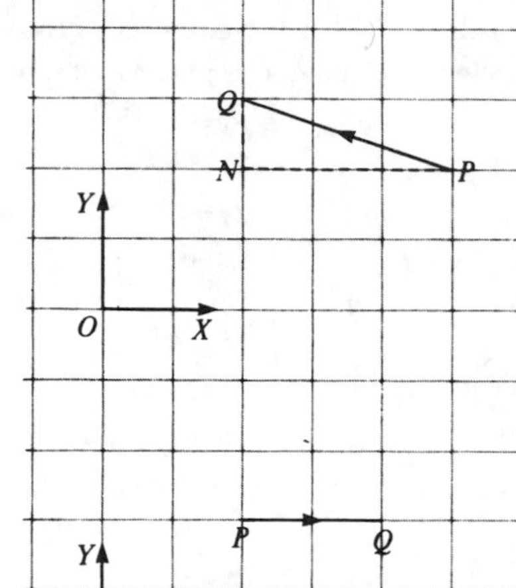

$\overrightarrow{PQ} = \begin{bmatrix} -3 \\ 1 \end{bmatrix}$

(iv)

$\overrightarrow{PQ} = \begin{bmatrix} 2 \\ 0 \end{bmatrix}$

(v)

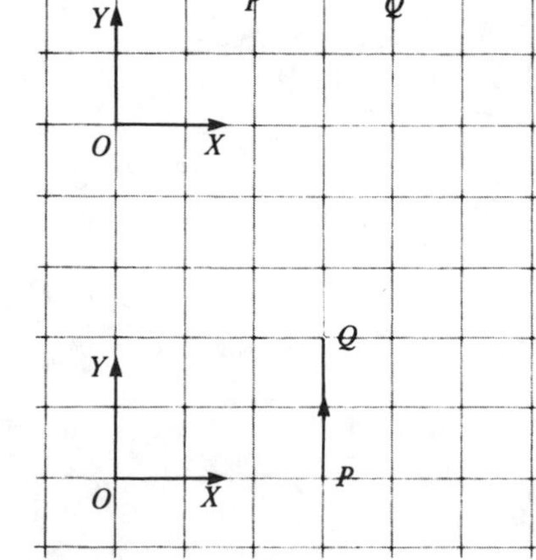

$\overrightarrow{PQ} = \begin{bmatrix} 0 \\ 2 \end{bmatrix}$

9

Vectors in geometry

Linear algebra and geometry are fundamentally related in a way which can be useful in the study of either topic. Ideas from each can provide helpful insights in the other.

The basic idea is that a column vector may be used to represent the position of a point in relation to another point, when coordinate axes are given. This applies in both two-dimensional geometry and three-dimensional geometry, but to start with it will be easier to think of the two-dimensional case. Let P and Q be two (distinct) points, and let OX and OY be given coordinate axes. Draw through P a straight line parallel to OX and through Q a straight line parallel to OY, and let these lines meet at N, as shown.

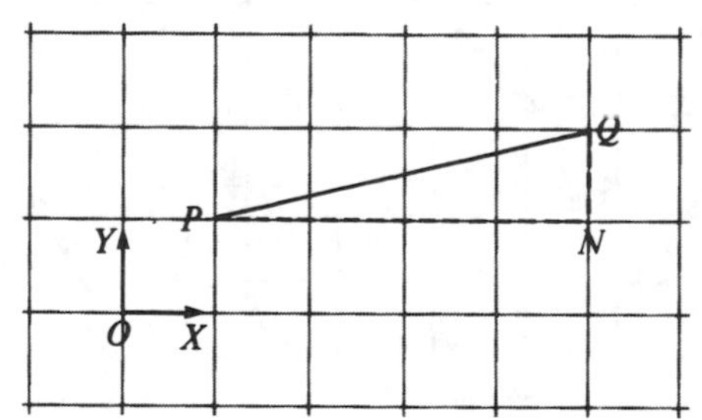

The position of Q relative to P can now be specified by a pair of numbers determined by the lengths and directions of the lines PN and NQ. The sizes of the numbers are the lengths of the lines. The signs of the numbers depend on whether the directions (P to N and N to Q) are the same as or opposite to the directions of the coordinate axes OX and OY respectively.

We can thus associate with the (ordered) pair of points P, Q a column vector $\begin{bmatrix} a \\ b \end{bmatrix}$, where a and b are the two numbers determined by the above

9.2

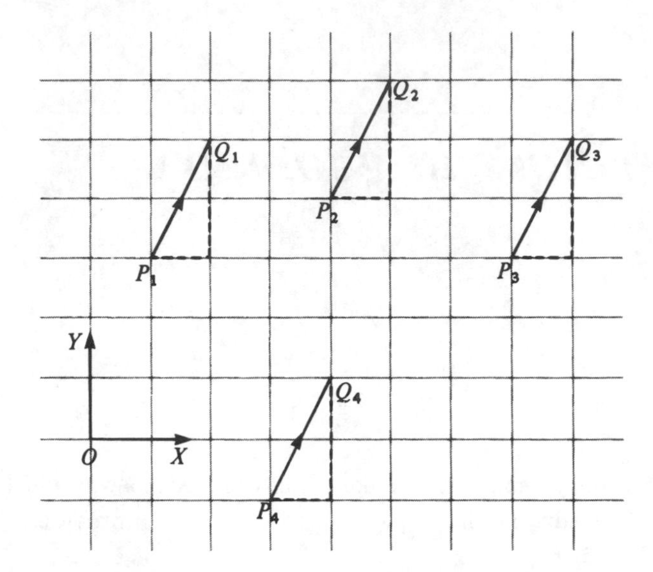

$$\overrightarrow{P_1Q_1}=\begin{bmatrix}2\\1\end{bmatrix},\quad \overrightarrow{P_2Q_2}=\begin{bmatrix}2\\1\end{bmatrix},\quad \overrightarrow{P_3Q_3}=\begin{bmatrix}2\\1\end{bmatrix},\quad \overrightarrow{P_4Q_4}=\begin{bmatrix}2\\1\end{bmatrix}.$$

9.3

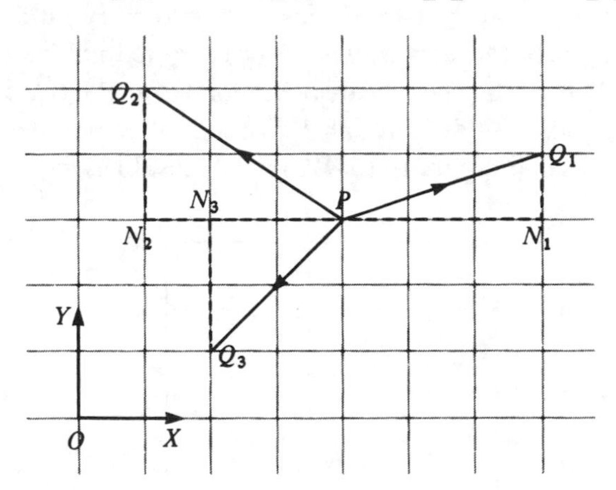

Let $\mathfrak{v}_1=\begin{bmatrix}3\\1\end{bmatrix}$.

Then Q_1 is the only point such that $\overrightarrow{PQ_1}=\mathfrak{v}_1$.

Let $\mathfrak{v}_2=\begin{bmatrix}-3\\2\end{bmatrix}$.

Then Q_2 is the only point such that $\overrightarrow{PQ_2}=\mathfrak{v}_2$.

Let $\mathfrak{v}_3=\begin{bmatrix}-2\\-2\end{bmatrix}$.

Then Q_3 is the only point such that $\overrightarrow{PQ_3}=\mathfrak{v}_3$.

process. The notation $\overrightarrow{PQ}$ is used for this. Examples 9.1 give several pairs of points and their associated column vectors, illustrating the way in which negative numbers can arise. Note the special cases given in Examples 9.1(iv) and (v).

Example 9.2 shows clearly that for any given column vector, many different pairs of points will be associated with it in this way. The diagram shows the properties that the lines P_1Q_1, P_2Q_2, P_3Q_3, and P_4Q_4 have which cause this. They are parallel, they have the same (not opposite) directions, and they have equal lengths.

A column vector is associated with a direction and a length, as we have just seen. Thus, given a (non-zero) column vector $\mathfrak{v}$ and a point P, there will always be one and only one point Q such that $\overrightarrow{PQ} = \mathfrak{v}$. Example 9.3 illustrates this.

To summarise:

1. Given any points P and Q, there is a unique column vector $\overrightarrow{PQ}$ which represents the position of Q relative to P.
2. Given any non-zero column vector $\mathfrak{v}$, there are infinitely many pairs of points P, Q such that $\overrightarrow{PQ} = \mathfrak{v}$.
3. Given any point P and any non-zero column vector $\mathfrak{v}$, there is a unique point Q such that $\overrightarrow{PQ} = \mathfrak{v}$.

9.4

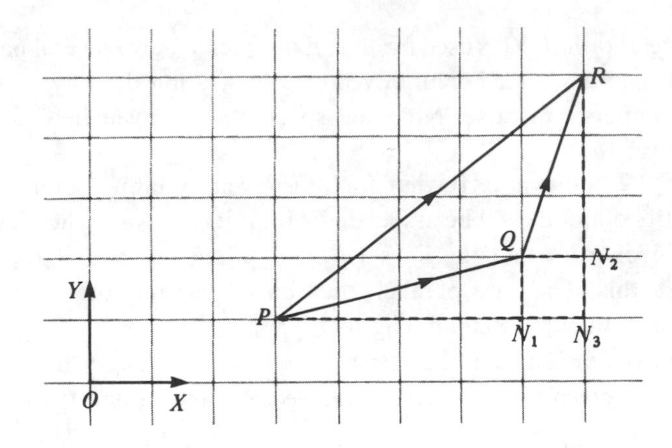

Here we treat only the case when the components of the vectors $\overrightarrow{PQ}$ and $\overrightarrow{QR}$ are all positive. If any are negative, the diagrams will be different and the argument will have to be modified slightly. Let

$$\overrightarrow{PQ}=\begin{bmatrix} a_1 \\ a_2 \end{bmatrix} \quad \text{and} \quad \overrightarrow{QR}=\begin{bmatrix} b_1 \\ b_2 \end{bmatrix}.$$

Then $|PN_1|=a_1$, $|N_1Q|=a_2$, $|QN_2|=b_1$, $|N_2R|=b_2$. So $|PN_3|=|PN_1|+|N_1N_3|=|PN_1|+|QN_2|=a_1+b_1$, and $|N_3R|=|N_3N_2|+|N_2R|=|N_1Q|+|N_2R|=a_2+b_2$.

9.5

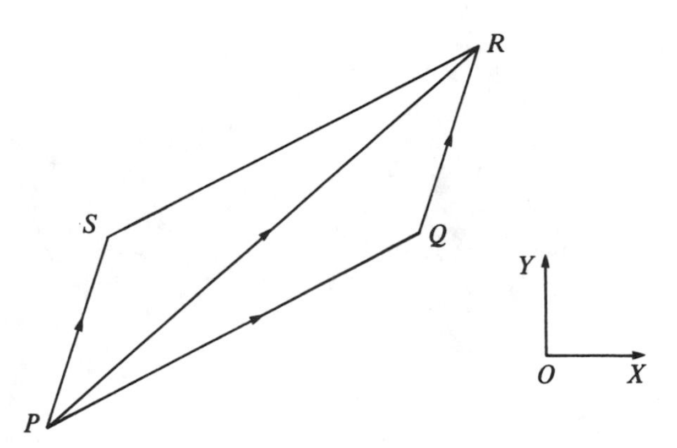

PQRS is a parallelogram. Hence *QR* and *PS* have the same direction and the same length. Consequently they are associated with the same column vector, i.e. $\overrightarrow{QR}=\overrightarrow{PS}$. By the Triangle Law, $\overrightarrow{PQ}+\overrightarrow{QR}=\overrightarrow{PR}$, so it follows immediately that $\overrightarrow{PQ}+\overrightarrow{PS}=\overrightarrow{PR}$.

Where is the advantage in this? It is in the way that addition of column vectors corresponds with a geometrical operation. Let P, Q and R be three distinct points, and let

$$\overrightarrow{PQ}=\begin{bmatrix}a_1\\a_2\end{bmatrix} \quad\text{and}\quad \overrightarrow{QR}=\begin{bmatrix}b_1\\b_2\end{bmatrix}.$$

Then

$$\overrightarrow{PR}=\begin{bmatrix}a_1+b_1\\a_2+b_2\end{bmatrix}=\begin{bmatrix}a_1\\a_2\end{bmatrix}+\begin{bmatrix}b_1\\b_2\end{bmatrix}=\overrightarrow{PQ}+\overrightarrow{QR}.$$

This rule is called the Triangle Law, and is of fundamental importance in vector geometry. See Example 9.4 for a partial justification for it.

Rule (The Triangle Law)
If PQR is a triangle then

$$\overrightarrow{PQ}+\overrightarrow{QR}=\overrightarrow{PR}.$$

The Triangle Law is sometimes expressed in a different form:

Rule (The Parallelogram Law)
If $PQRS$ is a parallelogram then

$$\overrightarrow{PQ}+\overrightarrow{PS}=\overrightarrow{PR}.$$

See Example 9.5 for a justification of this, using the Triangle Law.

9.6 Position vectors.

(i)

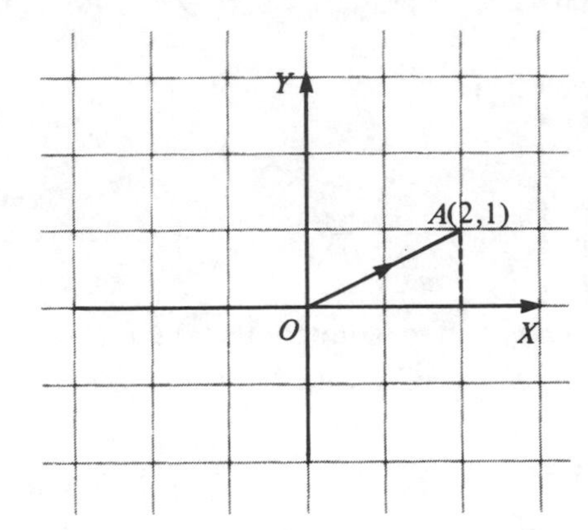

$$r_A = \begin{bmatrix} 2 \\ 1 \end{bmatrix}$$

(ii)

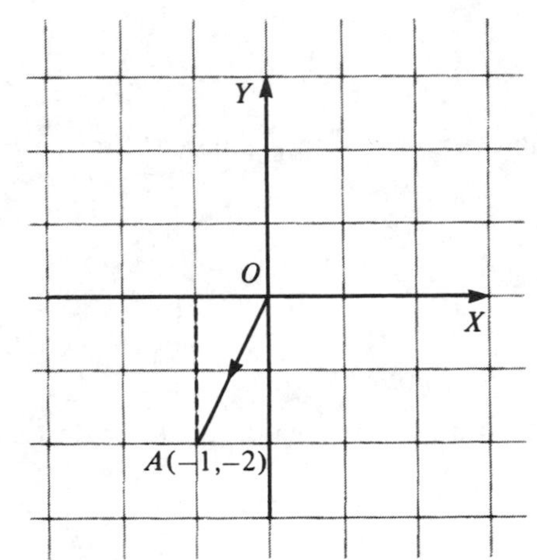

$$r_A = \begin{bmatrix} -1 \\ -2 \end{bmatrix}$$

(iii)

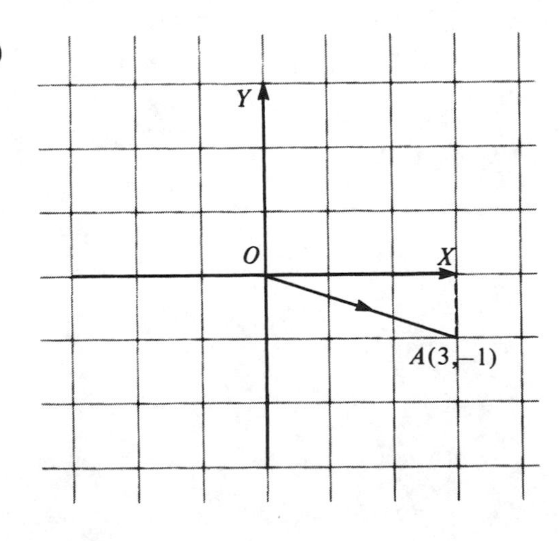

$$r_A = \begin{bmatrix} 3 \\ -1 \end{bmatrix}.$$

This relationship between algebraic vectors and directed lines is dependent on having some origin and coordinate axes, but it is important to realise that these laws which we have found are true irrespective of the choice of coordinate axes. We shall develop ideas and techniques which likewise are geometrical in character but which use the algebra of vectors in a convenient way without the necessity to refer to *particular* coordinate axes.

Before we proceed, we must take note of an important special case of all this, namely when the reference point (the first of the pair) is the origin.

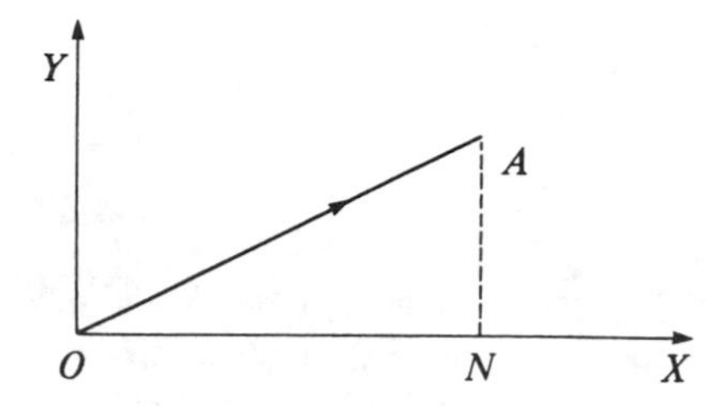

The construction which yields the components of the column vector $\overrightarrow{OA}$ involves the lines ON and NA. But now it is easy to see that the components of $\overrightarrow{OA}$ are just the coordinates of A. See Example 9.6 for particular cases, including cases with negative coordinates. The vector $\overrightarrow{OA}$ is called the *position vector* of A. It is customarily denoted by r_A. Every point has a uniquely determined position vector.

9.7

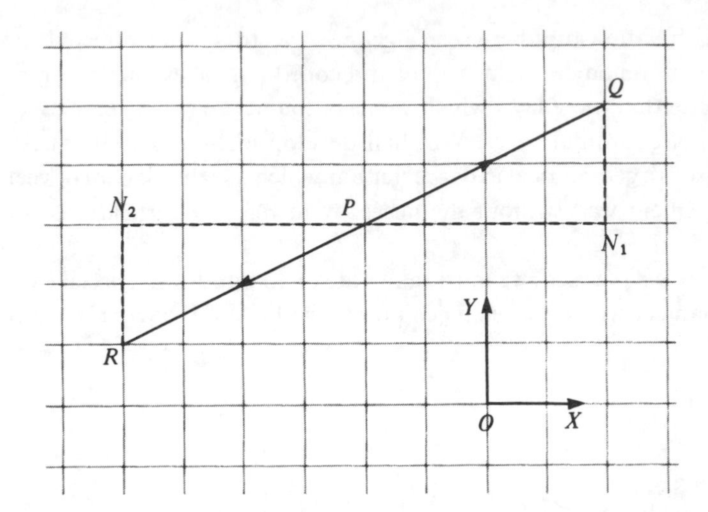

Produce QP back so that $|PR|=|PQ|$. Then the triangles PN_2R and PN_1Q are congruent, so PN_1 and PN_2 have equal lengths and opposite directions. The column vector associated with PR is thus the negative of the column vector associated with PQ. Notice that $\overrightarrow{PQ}+\overrightarrow{PR}=\mathbf{0}$, which can be thought of as an extension of the Parallelogram Law to an extreme case. (The parallelogram is flattened.)

9.8 Multiplication by a scalar.

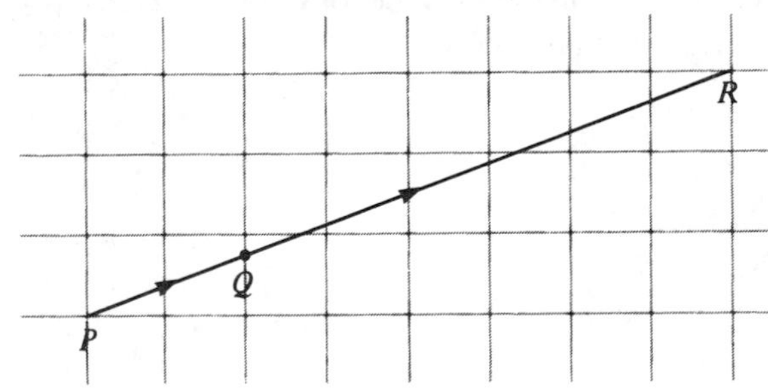

Here $\overrightarrow{PR}=4\,\overrightarrow{PQ}$.

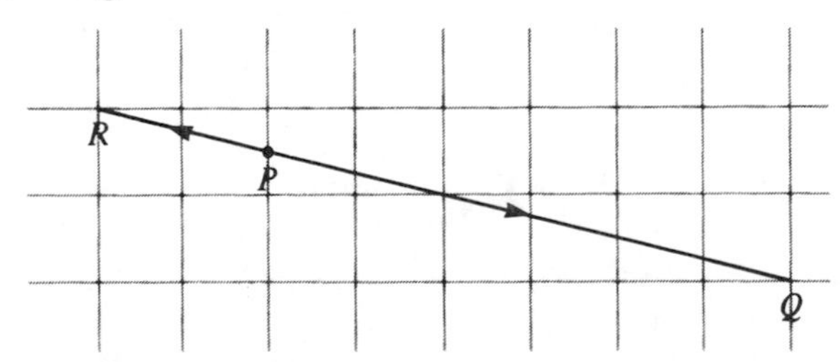

Here $\overrightarrow{PR}=-\frac{1}{3}\,\overrightarrow{PQ}$.

Rules

(i) The zero vector cannot be associated with any pair of distinct points. Nevertheless we can think of $\mathbf{0}$ as the position vector of the origin itself.

(ii) If $\mathbf{v}$ is a non-zero column vector, and $\overrightarrow{PQ} = \mathbf{v}$, then $-\mathbf{v} = \overrightarrow{PR}$, where R is the point obtained by producing the line QP so that PR and PQ have the same length. See Example 9.7.

(iii) If $\mathbf{v}$ is any non-zero column vector and k is any positive number, and if $\overrightarrow{PQ} = \mathbf{v}$, then $k\mathbf{v} = \overrightarrow{PR}$, where R is the point such that PQ and PR have the same direction and the length of PR is k times the length of PQ. Multiplying a vector by a negative number reverses the direction (as well as changing the length). See Example 9.8.

9.9 The difference of two vectors. Let PAB be a triangle.

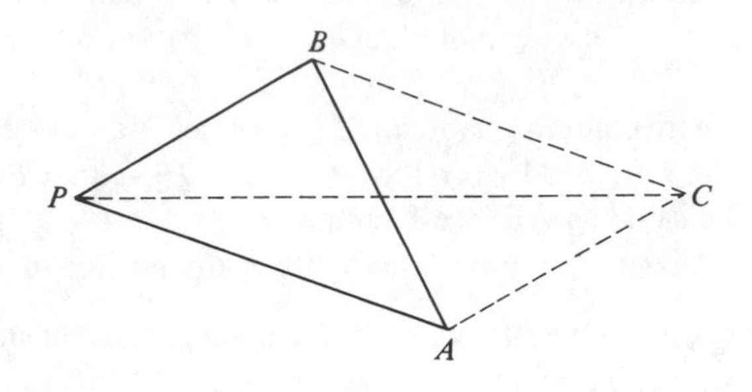

By the Triangle Law,

$$\overrightarrow{PA}+\overrightarrow{AB}=\overrightarrow{PB}.$$

Hence

$$\overrightarrow{AB}=\overrightarrow{PB}-\overrightarrow{PA}.$$

Notice that, in the parallelogram $PACB$, one diagonal (PC) represents the *sum* $\overrightarrow{PA}+\overrightarrow{PB}$, and the other diagonal (AB) represents the *difference* $\overrightarrow{PB}-\overrightarrow{PA}$. Of course the second diagonal may be taken in the opposite direction: $\overrightarrow{BA}=\overrightarrow{PA}-\overrightarrow{PB}$.

9.10 Illustrations of the ratio in which a point divides a line segment. In each case P divides AB in the given ratio.

(i) Ratio 1:1.

(ii) Ratio 3:1.

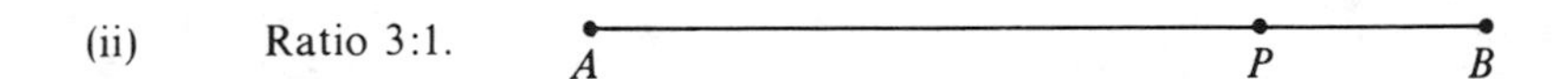

(iii) Ratio 2:3.

(iv) Ratio $-1:5$.

P A B

(v) Ratio $4:-1$.

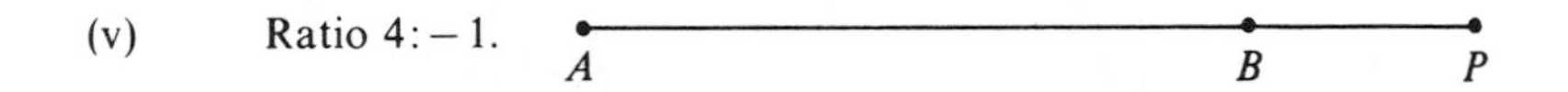

Subtraction of vectors also has an important geometrical interpretation. For a diagram, see Example 9.9. In a triangle PAB,

$$\overrightarrow{AB}=\overrightarrow{PB}-\overrightarrow{PA}.$$

This is a consequence of the Triangle Law. In particular, if A and B are any two distinct points then (taking the origin O as the point P) we have

$$\overrightarrow{AB}=\overrightarrow{OB}-\overrightarrow{OA}=\mathbf{r}_B-\mathbf{r}_A.$$

We shall use this repeatedly.

As mentioned earlier, all of these ideas apply equally well in three dimensions, where points have three coordinates and the vectors associated with directed lines have three components. One of the best features of this subject is the way in which the algebraic properties of 2-vectors and 3-vectors (which are substantially the same) can be used in the substantially different geometrical situations of two and three dimensions.

As an application of algebraic vector methods in geometry we shall derive the *Section Formula*. Consider a line segment AB. A point P on AB (possibly produced) divides AB in the ratio $m:n$ (with m and n both positive) if $|AP|/|PB|=m/n$. Here $|AP|$ and $|PB|$ denote the lengths of the lines AP and PB respectively. Extending this idea, we say that P divides AB in the ratio $-m:n$ (with m and n both positive) if AP and PB have opposite directions and $|AP|/|PB|=m/n$. See Examples 9.10 for illustrations of these.

9.11 Proof of the Section Formula (the case with m and n positive).

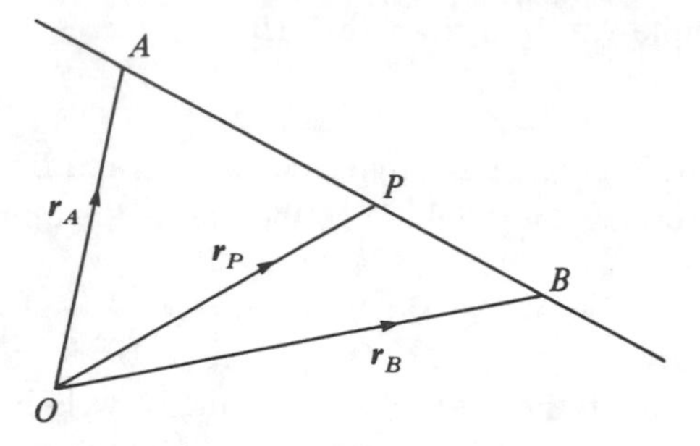

Let P divide AB in the ratio m:n, with $m>0$ and $n>0$. Then

$$\overrightarrow{AP}=(m/n)\,\overrightarrow{PB}.$$

Now

$$\overrightarrow{AP}=\boldsymbol{r}_P-\boldsymbol{r}_A \quad \text{and} \quad \overrightarrow{PB}=\boldsymbol{r}_B-\boldsymbol{r}_P,$$

so

$$n(\boldsymbol{r}_P-\boldsymbol{r}_A)=m(\boldsymbol{r}_B-\boldsymbol{r}_P).$$

From this we obtain

$$n\boldsymbol{r}_P-n\boldsymbol{r}_A-m\boldsymbol{r}_B+m\boldsymbol{r}_P=\boldsymbol{0},$$

$$(m+n)\boldsymbol{r}_P=n\boldsymbol{r}_A+m\boldsymbol{r}_B,$$

$$\boldsymbol{r}_P=\frac{1}{(m+n)}(n\boldsymbol{r}_A+m\boldsymbol{r}_B).$$

The cases where m and n have opposite signs require separate proofs, but the ideas are the same.

9.12 The medians of a triangle are concurrent at the centroid of the triangle.

Let ABC be a triangle, and let L, M and N be the midpoints of the sides BC, CA and AB, respectively. Let A, B and C have position vectors $\boldsymbol{a}$, $\boldsymbol{b}$ and $\boldsymbol{c}$. Then

L has position vector $\frac{1}{2}(\boldsymbol{b}+\boldsymbol{c})$,

M has position vector $\frac{1}{2}(\boldsymbol{c}+\boldsymbol{a})$, and

N has position vector $\frac{1}{2}(\boldsymbol{a}+\boldsymbol{b})$.

The point which divides AL in the ratio 2:1 has position vector

$$\frac{1}{(2+1)}(1\times\boldsymbol{a}+2\times\tfrac{1}{2}(\boldsymbol{b}+\boldsymbol{c})), \quad \text{i.e. } \tfrac{1}{3}(\boldsymbol{a}+\boldsymbol{b}+\boldsymbol{c}).$$

The point which divides BM in the ratio 2:1 has position vector

$$\frac{1}{(2+1)}(1\times\boldsymbol{b}+2\times\tfrac{1}{2}(\boldsymbol{c}+\boldsymbol{a})), \quad \text{i.e. } \tfrac{1}{3}(\boldsymbol{a}+\boldsymbol{b}+\boldsymbol{c}).$$

The point which divides CN in the ratio 2:1 has position vector

$$\frac{1}{(2+1)}(1\times\boldsymbol{c}+2\times\tfrac{1}{2}(\boldsymbol{a}+\boldsymbol{b})), \quad \text{i.e. } \tfrac{1}{3}(\boldsymbol{a}+\boldsymbol{b}+\boldsymbol{c}).$$

Hence the point with position vector $\frac{1}{3}(\boldsymbol{a}+\boldsymbol{b}+\boldsymbol{c})$ lies on all three medians. This is the centroid of the triangle.

Rule

Let P divide AB in the ratio $m:n$ (with $n \neq 0$ and $m+n \neq 0$). Then

$$\boldsymbol{r}_P = \frac{1}{m+n}(n\boldsymbol{r}_A + m\boldsymbol{r}_B).$$

See Example 9.11 for a proof.

An important special case of the Section Formula is when P is the midpoint of AB. In that case

$$\boldsymbol{r}_P = \tfrac{1}{2}(\boldsymbol{r}_A + \boldsymbol{r}_B).$$

The Section Formula may be used to give a convenient proof of a simple geometrical theorem: the medians of a triangle are concurrent at a point which trisects each median. A median of a triangle is a line which joins a vertex of the triangle to the midpoint of the opposite side. Let ABC be any triangle, and let $\boldsymbol{a}$, $\boldsymbol{b}$ and $\boldsymbol{c}$ be position vectors of A, B and C relative to some fixed origin. See Example 9.12. Let L, M and N be the midpoints of BC, CA and AB respectively. Then

$$\overrightarrow{OL} = \tfrac{1}{2}(\boldsymbol{b}+\boldsymbol{c}), \quad \overrightarrow{OM} = \tfrac{1}{2}(\boldsymbol{c}+\boldsymbol{a}), \quad \overrightarrow{ON} = \tfrac{1}{2}(\boldsymbol{a}+\boldsymbol{b}).$$

The Section Formula may now be used to find the position vectors of the points which divide AL, BM and CN respectively each in the ratio 2:1. It turns out to be the same point, which must therefore lie on all three medians. This point is called the *centroid* of the triangle ABC. It has position vector $\tfrac{1}{3}(\boldsymbol{a}+\boldsymbol{b}+\boldsymbol{c})$.

This result has been demonstrated using vectors in a geometrical style. The link with algebra has not been explicit, except in the algebraic operations on the vectors. Consequently we worked without reference to any particular coordinate system and obtained a purely geometrical result. These ideas can be taken considerably further, but we shall return now to the link with algebra, via coordinates.

9.13 Components of a vector in the directions of the coordinate axes.

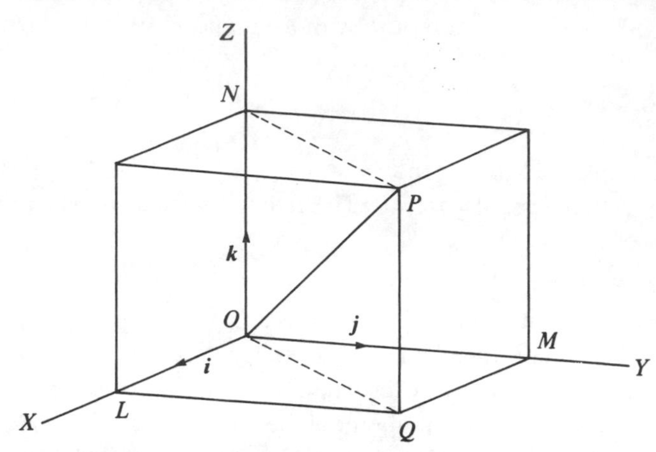

P has coordinates (x, y, z). Construct a rectangular solid figure by drawing perpendiculars from P to the three coordinate planes, and then to the coordinate axes from the feet of these perpendiculars. For example, PQ and then QL and QM, as shown in the diagram. The picture has been simplified by assuming that x, y and z are all positive. Our arguments work also if any or all of them are negative. You should try to visualise the various possible configurations.

$$|OL| = x, \quad |OM| = y, \quad |ON| = z.$$

So

$$\overrightarrow{OL} = x\boldsymbol{i}, \quad \overrightarrow{OM} = y\boldsymbol{j}, \quad \overrightarrow{ON} = z\boldsymbol{k}.$$

Now $\overrightarrow{OL} + \overrightarrow{OM} = \overrightarrow{OQ}$, by the Parallelogram Law. Also $OQPN$ is a parallelogram (it is a rectangle), so $\overrightarrow{OQ} + \overrightarrow{ON} = \overrightarrow{OP}$. Hence

$$\overrightarrow{OP} = \overrightarrow{OL} + \overrightarrow{OM} + \overrightarrow{ON} = x\boldsymbol{i} + y\boldsymbol{j} + z\boldsymbol{k}.$$

9.14 Examples of unit vectors.

(i) Let $\boldsymbol{a} = \begin{bmatrix} 1 \\ 2 \\ 1 \end{bmatrix}$. Then $\dfrac{1}{|\boldsymbol{a}|}\boldsymbol{a} = \begin{bmatrix} \dfrac{1}{\sqrt{6}} \\ \dfrac{2}{\sqrt{6}} \\ \dfrac{1}{\sqrt{6}} \end{bmatrix}$,

since $|\boldsymbol{a}| = \sqrt{1+4+1} = \sqrt{6}$.

(ii) Let $\boldsymbol{b} = \begin{bmatrix} 2 \\ 2 \\ -1 \end{bmatrix}$. Then $\dfrac{1}{|\boldsymbol{b}|}\boldsymbol{b} = \begin{bmatrix} \frac{2}{3} \\ \frac{2}{3} \\ -\frac{1}{3} \end{bmatrix}$,

since $|\boldsymbol{b}| = \sqrt{4+4+1} = 3$.

(iii) Let $\boldsymbol{c} = \begin{bmatrix} \frac{3}{5} \\ 0 \\ \frac{4}{5} \end{bmatrix}$. Then $|\boldsymbol{c}| = \sqrt{\frac{9}{25} + 0 + \frac{16}{25}} = 1$.

Thus $\boldsymbol{c}$ is already a unit vector.

Consider three dimensions. A point $P(x, y, z)$ has position vector

$$r_P = \begin{bmatrix} x \\ y \\ z \end{bmatrix}.$$

In particular:

the points $L(1,0,0)$, $M(0,1,0)$ and $N(0,0,1)$ have position vectors

$$\begin{bmatrix} 1 \\ 0 \\ 0 \end{bmatrix}, \quad \begin{bmatrix} 0 \\ 1 \\ 0 \end{bmatrix} \quad \text{and} \quad \begin{bmatrix} 0 \\ 0 \\ 1 \end{bmatrix}$$

respectively.

These three vectors are denoted respectively by $\boldsymbol{i}$, $\boldsymbol{j}$ and $\boldsymbol{k}$. Note that they are represented by directed lines OL, OM and ON in the directions of the coordinate axes. Notice also that

$$\begin{aligned} \begin{bmatrix} x \\ y \\ z \end{bmatrix} &= \begin{bmatrix} x \\ 0 \\ 0 \end{bmatrix} + \begin{bmatrix} 0 \\ y \\ 0 \end{bmatrix} + \begin{bmatrix} 0 \\ 0 \\ z \end{bmatrix} \\ &= x\begin{bmatrix} 1 \\ 0 \\ 0 \end{bmatrix} + y\begin{bmatrix} 0 \\ 1 \\ 0 \end{bmatrix} + z\begin{bmatrix} 0 \\ 0 \\ 1 \end{bmatrix} \\ &= x\boldsymbol{i} + y\boldsymbol{j} + z\boldsymbol{k}. \end{aligned}$$

The vectors $\boldsymbol{i}$, $\boldsymbol{j}$ and $\boldsymbol{k}$ are known as *standard basis vectors*. Every vector can be written (uniquely) as a sum of multiples of these, as above. The numbers x, y and z are called the *components* of the vector in the directions of the coordinate axes. See Example 9.13.

It is clear what should be meant by the *length* of a vector: the length of any line segment which represents it. Each of $\boldsymbol{i}$, $\boldsymbol{j}$ and $\boldsymbol{k}$ above has length equal to 1. A *unit* vector is a vector whose length is equal to 1. The length of an algebraic vector

$$\begin{bmatrix} x \\ y \\ z \end{bmatrix}$$

is (by definition) equal to $\sqrt{x^2+y^2+z^2}$, i.e. the length of the line OP, where P is the point with coordinates (x, y, z). Given any non-zero vector $\boldsymbol{a}$ we can always find a unit vector in the same direction as $\boldsymbol{a}$. If we denote the length of $\boldsymbol{a}$ by $|\boldsymbol{a}|$ then $(1/|\boldsymbol{a}|)\boldsymbol{a}$ is a unit vector in the direction of $\boldsymbol{a}$. See Example 9.14.

Summary

The way in which vectors are represented as directed line segments is explained, and algebraic operations on vectors are interpreted geometrically. The Section Formula is derived and used. The standard basis vectors $\boldsymbol{i}$, $\boldsymbol{j}$ and $\boldsymbol{k}$ are defined. The length of a vector (in three dimensions) is defined, and the notion of a unit vector is introduced.

Exercises

1. Let $ABCDEF$ be a regular hexagon whose centre is at the origin. Let A and B have position vectors $\boldsymbol{a}$ and $\boldsymbol{b}$. Find the position vectors of C, D, E and F in terms of $\boldsymbol{a}$ and $\boldsymbol{b}$.
2. Let $ABCD$ be a parallelogram, and let $\boldsymbol{a}, \boldsymbol{b}, \boldsymbol{c}$ and $\boldsymbol{d}$ be the position vectors of A, B, C and D respectively. Show that $\boldsymbol{a}+\boldsymbol{c}=\boldsymbol{b}+\boldsymbol{d}$.
3. Let L, M and N be the midpoints of BC, CA and AB respectively, and let O be any fixed origin. Show that

   (i) $\overrightarrow{OA}+\overrightarrow{OB}+\overrightarrow{OC}=\overrightarrow{OL}+\overrightarrow{OM}+\overrightarrow{ON}$, and
   (ii) $\overrightarrow{AL}+\overrightarrow{BM}+\overrightarrow{CN}=\mathbf{0}$.
4. Let $A_1, A_2, \ldots, A_k$ be any points in three dimensions. Show that

   $$\overrightarrow{A_1A_2}+\overrightarrow{A_2A_3}+\cdots+\overrightarrow{A_{k-1}A_k}+\overrightarrow{A_kA_1}=\mathbf{0}.$$
5. In each case below, write down the 3-vector $\overrightarrow{AB}$, where A and B are points with the given coordinates.

   (i) $A(0,0,0)$, $B(2,-1,3)$.
   (ii) $A(2,-1,3)$, $B(0,0,0)$.
   (iii) $A(3,4,1)$, $B(1,2,-1)$.
   (iv) $A(0,1,-1)$, $B(0,-1,0)$.
   (v) $A(2,2,2)$, $B(3,2,1)$.
6. In each case below, find the position vector of the point which divides the line segment AB in the given ratio.

   (i) $A(1,1,3)$, $B(-1,1,5)$, ratio 1:1.
   (ii) $A(-2,-1,1)$, $B(3,2,2)$, ratio $2:-1$.
   (iii) $A(0,0,0)$, $B(11,11,11)$, ratio 6:5.
   (iv) $A(3,-1,-2)$, $B(10,-8,12)$, ratio $9:-2$.
   (v) $A(2,1,-1)$, $B(-2,-1,1)$, ratio $-2:3$.

   Also in each case draw a rough diagram to indicate the relative positions of the three points.
7. Let $OABC$ be a parallelogram as shown, and let D divide OA in the ratio $m:n$ (where $n\neq 0$ and $m+n\neq 0$). Prove by vector methods that DC and OB intersect at a point P which divides each line in the ratio $m:(m+n)$.

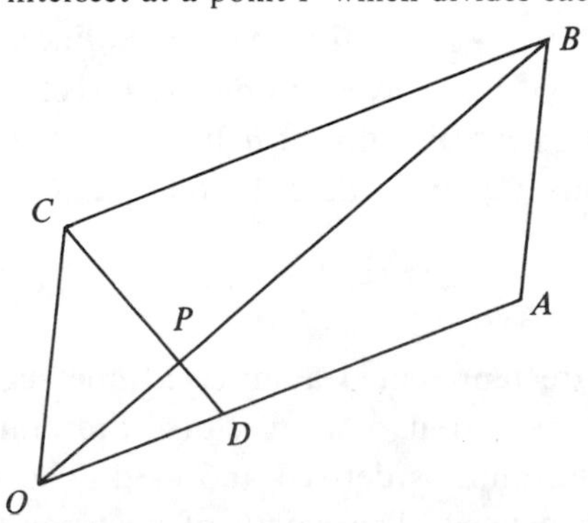

8. Let OAB be a triangle as shown. The midpoint of OA is M, P is the point which divides BM in the ratio 3:2, and S is the point where OP produced meets BA. Prove, using vector methods, that S divides BA in the ratio 3:4.

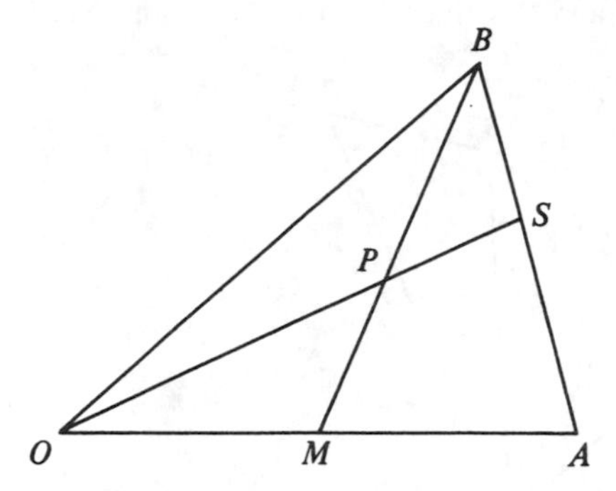

9. In each case below, find a unit vector in the same direction as the given vector.

   (i) $\begin{bmatrix} 1 \\ 0 \\ -1 \end{bmatrix}$. (ii) $\begin{bmatrix} 2 \\ 2 \\ 1 \end{bmatrix}$. (iii) $\begin{bmatrix} 1 \\ -2 \\ -1 \end{bmatrix}$. (iv) $\begin{bmatrix} 2 \\ 2 \\ 2 \end{bmatrix}$.

10. Let $\boldsymbol{a}$ and $\boldsymbol{b}$ be non-zero vectors. Prove that $|\boldsymbol{a}+\boldsymbol{b}| = |\boldsymbol{a}| + |\boldsymbol{b}|$ if and only if $\boldsymbol{a}$ and $\boldsymbol{b}$ have the same (or opposite) directions.

11. Let A and B be points with position vectors $\boldsymbol{a}$ and $\boldsymbol{b}$ respectively (with $\boldsymbol{a} \neq \boldsymbol{0}$ and $\boldsymbol{b} \neq \boldsymbol{0}$). Show that $|\boldsymbol{a}|\boldsymbol{b}$ and $|\boldsymbol{b}|\boldsymbol{a}$ are vectors with the same length, and deduce that the direction of the internal bisector of the angle AOB is given by the vector $|\boldsymbol{a}|\boldsymbol{b} + |\boldsymbol{b}|\boldsymbol{a}$.

Examples

10.1 Vector equation of a straight line.

(i) Given two points A and B with position vectors $\boldsymbol{a}$ and $\boldsymbol{b}$ respectively,

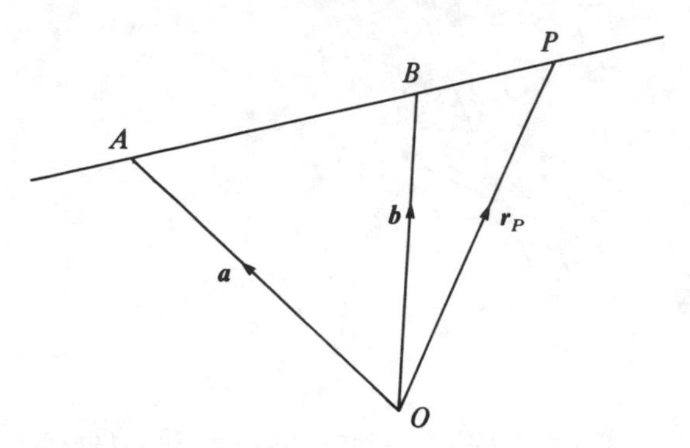

$\boldsymbol{r}_P = \boldsymbol{a} + t(\boldsymbol{b} - \boldsymbol{a}) \quad (t \in \mathbb{R})$.

(ii) Given one point A with position vector $\boldsymbol{a}$, and a vector $\boldsymbol{v}$ in the direction of the line.

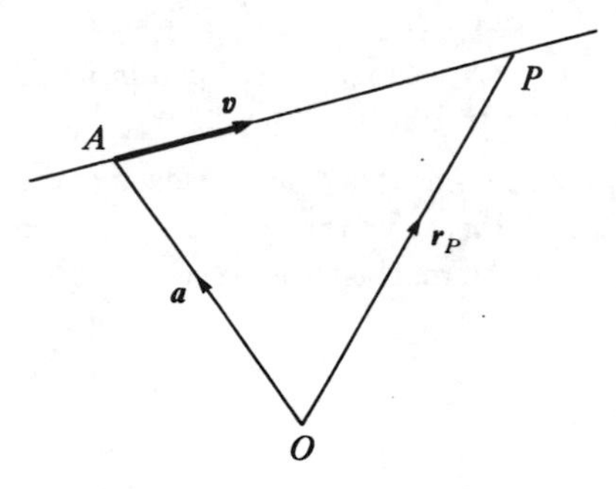

$\boldsymbol{r}_P = \boldsymbol{a} + t\boldsymbol{v} \quad (t \in \mathbb{R})$.

10

Straight lines and planes

In three dimensions a straight line may be specified by either

(i) two distinct points on it, or

(ii) a point on it and its direction.

These give rise to the vector form of equation for a line as follows. For diagrams, see Example 10.1. First, given points A and B, with position vectors $\boldsymbol{a}$ and $\boldsymbol{b}$, let P be any point on the straight line. The Triangle Law gives

$$\boldsymbol{r}_P = \overrightarrow{OA} + \overrightarrow{AP}.$$

Now $\overrightarrow{AP}$ is in the same direction as (or in the opposite direction to) $\overrightarrow{AB}$. Hence $\overrightarrow{AP}$ is a multiple of $\overrightarrow{AB}$, say $\overrightarrow{AP} = t\,\overrightarrow{AB}$. We know that $\overrightarrow{OA} = \boldsymbol{a}$ and $\overrightarrow{AB} = \boldsymbol{b} - \boldsymbol{a}$, so

$$\boldsymbol{r}_P = \boldsymbol{a} + t(\boldsymbol{b} - \boldsymbol{a}). \tag{1}$$

Second, given a point A with position vector $\boldsymbol{a}$, and a direction, say the direction specified by the vector $\boldsymbol{v}$, let P be any point on the straight line. Then

$$\boldsymbol{r}_P = \overrightarrow{OA} + \overrightarrow{AP}, \quad \text{as above,}$$

but here $\overrightarrow{AP}$ is a multiple of $\boldsymbol{v}$, say $\overrightarrow{AP} = t\boldsymbol{v}$. Hence

$$\boldsymbol{r}_P = \boldsymbol{a} + t\boldsymbol{v}. \tag{2}$$

10.2 To find parametric equations for given straight lines.

(i) Line through $A(2,3,-1)$ and $B(3,1,3)$.

$$\boldsymbol{a}=\begin{bmatrix}2\\3\\-1\end{bmatrix},\quad \boldsymbol{b}=\begin{bmatrix}3\\1\\3\end{bmatrix},\quad \text{so}\quad \boldsymbol{b}-\boldsymbol{a}=\begin{bmatrix}1\\-2\\4\end{bmatrix}.$$

An equation for the line is

$$\begin{bmatrix}x\\y\\z\end{bmatrix}=\begin{bmatrix}2\\3\\-1\end{bmatrix}+t\begin{bmatrix}1\\-2\\4\end{bmatrix}\quad (t\in\mathbb{R}).$$

In terms of coordinates, this becomes

$$x=2+t,\quad y=3-2t,\quad z=-1+4t\quad (t\in\mathbb{R}).$$

(ii) Line through $A(-2,5,1)$ in the direction of the vector

$$\boldsymbol{v}=\begin{bmatrix}1\\-1\\2\end{bmatrix}.$$

An equation for the line is

$$\begin{bmatrix}x\\y\\z\end{bmatrix}=\begin{bmatrix}-2\\5\\1\end{bmatrix}+t\begin{bmatrix}1\\-1\\2\end{bmatrix}\quad (t\in\mathbb{R}).$$

In terms of coordinates, this becomes

$$x=-2+t,\quad y=5-t,\quad z=1+2t\quad (t\in\mathbb{R}).$$

10.3 Find parametric equations for the straight line through $A(-2,5,1)$ in the direction of the unit vector

$$\boldsymbol{u}=\begin{bmatrix}\frac{1}{\sqrt{6}}\\-\frac{1}{\sqrt{6}}\\\frac{2}{\sqrt{6}}\end{bmatrix}.$$

An equation for the line is

$$\begin{bmatrix}x\\y\\z\end{bmatrix}=\begin{bmatrix}-2\\5\\1\end{bmatrix}+t\begin{bmatrix}\frac{1}{\sqrt{6}}\\-\frac{1}{\sqrt{6}}\\\frac{2}{\sqrt{6}}\end{bmatrix}\quad (t\in\mathbb{R}).$$

This of course is the *same line* as in Example 10.2 (ii), because the vector $\boldsymbol{u}$ is in the same direction as the vector $\boldsymbol{v}$ given there. The equations obtained look different, but as t varies in each case the sets of equations determine the same sets of values for x, y and z. For example, the point $(-1,4,3)$ arises from the equations in Example 10.2 (ii) with $t=1$, and from the equations above with $t=\sqrt{6}$.

Equations (1) and (2) are forms of vector equation for a straight line. The number t is a parameter: as t varies, the right-hand side gives the position vectors of the points on the line. The equation (1) is in fact a special case of (2). We may represent the vector $\boldsymbol{r}_P$ by

$$\begin{bmatrix} x \\ y \\ z \end{bmatrix},$$

where (x, y, z) are the coordinates of P, and then if

$$\boldsymbol{v}=\begin{bmatrix} v_1 \\ v_2 \\ v_3 \end{bmatrix},$$

(2) becomes

$$\begin{bmatrix} x \\ y \\ z \end{bmatrix}=\begin{bmatrix} a_1 \\ a_2 \\ a_3 \end{bmatrix}+t\begin{bmatrix} v_1 \\ v_2 \\ v_3 \end{bmatrix}$$

(say), or

$$\left.\begin{aligned} x&=a_1+tv_1 \\ y&=a_2+tv_2 \\ z&=a_3+tv_3 \end{aligned}\right\} \quad (t \in \mathbb{R}),$$

which is the coordinate form of parametric equations for a straight line. Examples 10.2 give specific cases.

The components of a vector such as $\boldsymbol{v}$ above, when used to specify a direction, are called *direction ratios*. In this situation *any* vector in the particular direction will serve the purpose. It is often convenient, however, to use a unit vector. See Example 10.3.

10.4 To find whether straight lines with given parametric equations intersect.

Line 1: $x=2-t, \quad y=3t, \quad z=2+2t \quad (t\in\mathbb{R})$.

Line 2: $x=-3-4u, \quad y=4+u, \quad z=1-3u \quad (u\in\mathbb{R})$.

At a point of intersection we would have a value of t and a value of u satisfying:

$$\left.\begin{aligned}2-\ t&=-3-4u\\ 3t&=\ \ 4+\ u\\ 2+2t&=\ \ 1-3u\end{aligned}\right\}.\quad \text{i.e.}\quad \left.\begin{aligned}t-4u&=\ \ 5\\ 3t-\ u&=\ \ 4\\ 2t+3u&=-1\end{aligned}\right\}.$$

Here we have three equations in two unknowns. We may expect them to be inconsistent, in which case the two lines have no point of intersection. Let us find out, using the GE process.

$$\begin{bmatrix}1 & -4 & 5\\ 3 & -1 & 4\\ 2 & 3 & -1\end{bmatrix}\rightarrow\begin{bmatrix}1 & -4 & 5\\ 0 & 11 & -11\\ 0 & 11 & -11\end{bmatrix}\rightarrow\begin{bmatrix}1 & -4 & 5\\ 0 & 1 & -1\\ 0 & 0 & 0\end{bmatrix}.$$

So the equations are *consistent*, and there is a unique solution $u=-1$, $t=1$. Substitute either of these into the equations for the appropriate line, to obtain

$$x=1,\quad y=3,\quad z=4,$$

the coordinates of the intersection point.

10.5 A proof that for any non-zero vectors $\boldsymbol{a}, \boldsymbol{b}$,

$$\boldsymbol{a}\cdot\boldsymbol{b}=|\boldsymbol{a}||\boldsymbol{b}|\cos\theta.$$

where θ is the angle between $\boldsymbol{a}$ and $\boldsymbol{b}$. Let

$$\boldsymbol{a}=\begin{bmatrix}a_1\\ a_2\\ a_3\end{bmatrix},\quad \text{and}\quad \boldsymbol{b}=\begin{bmatrix}b_1\\ b_2\\ b_3\end{bmatrix},$$

and let A and B be points with position vectors $\boldsymbol{a}$ and $\boldsymbol{b}$ respectively.

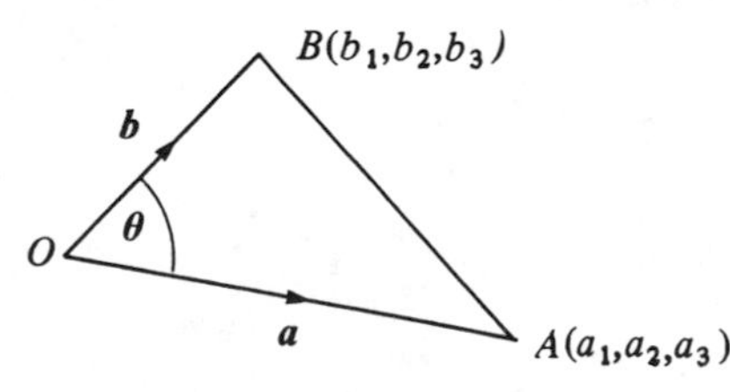

Then $A\hat{O}B=\theta$, and in triangle OAB the cosine formula gives

$$\begin{aligned}|AB|^2&=|OA|^2+|OB|^2-2|OA||OB|\cos\theta\\ &=|\boldsymbol{a}|^2+|\boldsymbol{b}|^2-2|\boldsymbol{a}||\boldsymbol{b}|\cos\theta.\end{aligned}$$

Hence

$$\begin{aligned}|\boldsymbol{a}||\boldsymbol{b}|\cos\theta&=\tfrac{1}{2}(|\boldsymbol{a}|^2+|\boldsymbol{b}|^2-|AB|^2)\\ &=\tfrac{1}{2}(a_1{}^2+a_2{}^2+a_3{}^2+b_1{}^2+b_2{}^2+b_3{}^2\\ &\qquad -(b_1-a_1)^2-(b_2-a_2)^2-(b_3-a_3)^2)\\ &=\tfrac{1}{2}(2a_1b_1+2a_2b_2+2a_3b_3)\\ &=a_1b_1+a_2b_2+a_3b_3\\ &=\boldsymbol{a}\cdot\boldsymbol{b}.\end{aligned}$$

Two straight lines in three dimensions need not intersect. Example 10.4 shows how, given vector equations for two lines, we can check whether they intersect and, if they do, how to find the point of intersection. Note that, to avoid confusion, we use different letters for the parameters in the equations for the two lines.

There is a way of dealing with angles in three dimensions, using vectors. This involves the idea of the dot product of two vectors. By the *angle between two vectors* $\boldsymbol{a}$ and $\boldsymbol{b}$ we mean the angle $A\hat{O}B$, where $\overrightarrow{OA}=\boldsymbol{a}$ and $\overrightarrow{OB}=\boldsymbol{b}$. (Note that this angle is the same as any angle $A\hat{P}B$ where $\overrightarrow{PA}=\boldsymbol{a}$ and $\overrightarrow{PB}=\boldsymbol{b}$.) This angle always lies between 0 and π radians (180°) inclusive. In algebraic terms, the *dot product* of two 3-vectors

$$\boldsymbol{a}=\begin{bmatrix} a_1 \\ a_2 \\ a_3 \end{bmatrix} \quad \text{and} \quad \boldsymbol{b}=\begin{bmatrix} b_1 \\ b_2 \\ b_3 \end{bmatrix}$$

is defined by

$$\boldsymbol{a}\cdot\boldsymbol{b}=a_1b_1+a_2b_2+a_3b_3.$$

Notice that the value of a dot product is a number, not a vector. In some books the dot product is referred to as the *scalar* product. What has this to do with angles?

Rule

Let $\boldsymbol{a}$ and $\boldsymbol{b}$ be non-zero vectors and let θ be the angle between them. Then

$$\boldsymbol{a}\cdot\boldsymbol{b}=|\boldsymbol{a}||\boldsymbol{b}|\cos\theta.$$

Consequently,

$$\cos\theta=\frac{\boldsymbol{a}\cdot\boldsymbol{b}}{|\boldsymbol{a}||\boldsymbol{b}|}.$$

A demonstration of this rule is given as Example 10.5.

10.6 Calculate the cosine of the angle θ between the given vectors in each case.

(i) $$\boldsymbol{a}=\begin{bmatrix}1\\1\\2\end{bmatrix},\quad \boldsymbol{b}=\begin{bmatrix}-1\\2\\1\end{bmatrix}.$$

$$|\boldsymbol{a}|=\sqrt{6},\qquad |\boldsymbol{b}|=\sqrt{6}.$$
$$\boldsymbol{a}\cdot\boldsymbol{b}=-1+2+2=3.$$

Hence

$$\cos\theta=\frac{3}{\sqrt{6}\times\sqrt{6}}=\frac{1}{2}.$$

(ii) $$\boldsymbol{a}=\begin{bmatrix}-2\\3\\1\end{bmatrix},\quad \boldsymbol{b}=\begin{bmatrix}1\\2\\2\end{bmatrix}.$$

$$|\boldsymbol{a}|=\sqrt{14},\qquad |\boldsymbol{b}|=3.$$
$$\boldsymbol{a}\cdot\boldsymbol{b}=-2+6+2=6.$$

Hence

$$\cos\theta=\frac{6}{\sqrt{14}\times 3}=\frac{2}{\sqrt{14}}.$$

(iii) $$\boldsymbol{a}=\begin{bmatrix}-1\\1\\-4\end{bmatrix},\quad \boldsymbol{b}=\begin{bmatrix}2\\3\\1\end{bmatrix}.$$

$$|\boldsymbol{a}|=\sqrt{18},\qquad |\boldsymbol{b}|=\sqrt{14}.$$
$$\boldsymbol{a}\cdot\boldsymbol{b}=-2+3-4=-3.$$

Hence

$$\cos\theta=\frac{-3}{\sqrt{18}\times\sqrt{14}}=-\frac{1}{2\times\sqrt{7}}.$$

(The negative sign indicates an *obtuse* angle.)

10.7 Proof that, for any three vectors $\boldsymbol{a}$, $\boldsymbol{b}$ and $\boldsymbol{c}$,

$$\boldsymbol{a}\cdot(\boldsymbol{b}+\boldsymbol{c})=\boldsymbol{a}\cdot\boldsymbol{b}+\boldsymbol{a}\cdot\boldsymbol{c}.$$

Let

$$\boldsymbol{a}=\begin{bmatrix}a_1\\a_2\\a_3\end{bmatrix},\quad \boldsymbol{b}=\begin{bmatrix}b_1\\b_2\\b_3\end{bmatrix},\quad \boldsymbol{c}=\begin{bmatrix}c_1\\c_2\\c_3\end{bmatrix}.$$

Then

$$\boldsymbol{b}+\boldsymbol{c}=\begin{bmatrix}b_1+c_1\\b_2+c_2\\b_3+c_3\end{bmatrix}.$$

So

$$\begin{aligned}\boldsymbol{a}\cdot(\boldsymbol{b}+\boldsymbol{c})&=a_1(b_1+c_1)+a_2(b_2+c_2)+a_3(b_3+c_3)\\&=a_1b_1+a_1c_1+a_2b_2+a_2c_2+a_3b_3+a_3c_3\\&=a_1b_1+a_2b_2+a_3b_3+a_1c_1+a_2c_2+a_3c_3\\&=\boldsymbol{a}\cdot\boldsymbol{b}+\boldsymbol{a}\cdot\boldsymbol{c}\end{aligned}$$

Using this rule, we are now able to calculate (the cosines of) angles between given vectors. See Examples 10.6. One of these examples illustrates a general rule. A dot product $\boldsymbol{a} \cdot \boldsymbol{b}$ may equal zero even though neither $\boldsymbol{a}$ nor $\boldsymbol{b}$ is itself 0.

Rule

(i) If the angle between two non-zero vectors $\boldsymbol{a}$ and $\boldsymbol{b}$ is a right angle, then $\boldsymbol{a} \cdot \boldsymbol{b} = 0$.

(ii) If $\boldsymbol{a}$ and $\boldsymbol{b}$ are non-zero vectors with $\boldsymbol{a} \cdot \boldsymbol{b} = 0$, the angle between $\boldsymbol{a}$ and $\boldsymbol{b}$ is a right angle.

Two non-zero vectors are said to be *perpendicular* (or *orthogonal*) if the angle between them is a right angle.

Rules

For dot products:

(i) $\boldsymbol{a} \cdot \boldsymbol{b} = \boldsymbol{b} \cdot \boldsymbol{a}$ for all vectors $\boldsymbol{a}$ and $\boldsymbol{b}$.

(ii) $\boldsymbol{a} \cdot \boldsymbol{a} = |\boldsymbol{a}|^2$ for all vectors $\boldsymbol{a}$.

(iii) $\boldsymbol{a} \cdot (\boldsymbol{b} + \boldsymbol{c}) = \boldsymbol{a} \cdot \boldsymbol{b} + \boldsymbol{a} \cdot \boldsymbol{c}$ for all vectors $\boldsymbol{a}$, $\boldsymbol{b}$, $\boldsymbol{c}$.

(iv) $(k\boldsymbol{a}) \cdot \boldsymbol{b} = \boldsymbol{a} \cdot (k\boldsymbol{b}) = k(\boldsymbol{a} \cdot \boldsymbol{b})$ for all vectors $\boldsymbol{a}$, $\boldsymbol{b}$, $\boldsymbol{c}$, and all $k \in \mathbb{R}$.

(v) $\boldsymbol{i} \cdot \boldsymbol{j} = \boldsymbol{j} \cdot \boldsymbol{k} = \boldsymbol{k} \cdot \boldsymbol{i} = 0$.

These are quite straightforward. Part (iii) is proved in Example 10.7. The others may be regarded as exercises.

10.8 A plane through a given point A, perpendicular to a given vector $\boldsymbol{n}$.

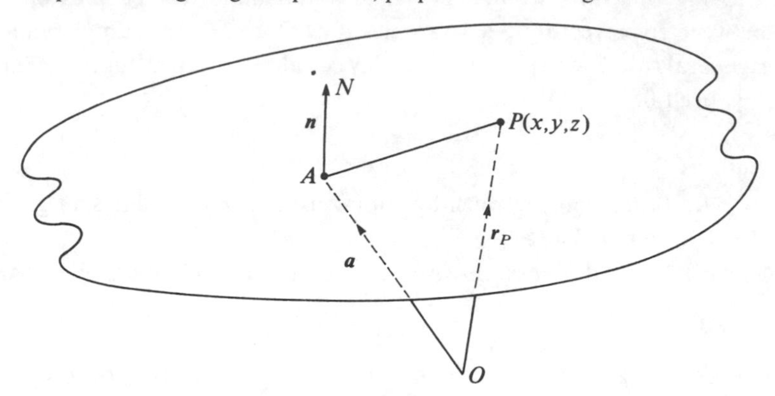

Let A be the point $(3, 1, -2)$, and let $\boldsymbol{n}$ be the vector

$$\begin{bmatrix} -1 \\ 2 \\ -1 \end{bmatrix}.$$

The point $P(x, y, z)$ lies on the plane if and only if AP is perpendicular to AN, i.e. if and only if $\overrightarrow{AP} \cdot \boldsymbol{n} = 0$. Now

$$\overrightarrow{AP} = \overrightarrow{OP} - \overrightarrow{OA} = \begin{bmatrix} x \\ y \\ z \end{bmatrix} - \begin{bmatrix} 3 \\ 1 \\ -2 \end{bmatrix} = \begin{bmatrix} x-3 \\ y-1 \\ z+2 \end{bmatrix}.$$

So $\overrightarrow{AP} \cdot \boldsymbol{n} = 0$ becomes

$$(x-3)(-1) + (y-1)2 + (z+2)(-1) = 0.$$

i.e. $$-x+3+2y-2-z-2=0.$$

i.e. $$-x+2y-z-1=0.$$

This is an equation for the plane in this case.

10.9 Show that the vector $\boldsymbol{n} = \begin{bmatrix} a \\ b \\ c \end{bmatrix}$ is perpendicular to the plane with equation $ax+by+cz+d=0$.

Let $P_1(x_1, y_1, z_1)$ and $P_2(x_2, y_2, z_2)$ be two points lying on the plane. Then

$$ax_1 + by_1 + cz_1 + d = 0$$

and $$ax_2 + by_2 + cz_2 + d = 0. \quad (*)$$

We show that P_1P_2 is perpendicular to $\boldsymbol{n}$.

$$\begin{aligned}
\boldsymbol{n} \cdot \overrightarrow{P_1P_2} &= a(x_2 - x_1) + b(y_2 - y_1) + c(z_2 - z_1) \\
&= ax_2 - ax_1 + by_2 - by_1 + cz_2 - cz_1 \\
&= ax_2 + by_2 + cz_2 - (ax_1 + by_1 + cz_1) \\
&= -d - (-d) \quad \text{by (*) above} \\
&= 0.
\end{aligned}$$

Hence $\boldsymbol{n}$ is perpendicular to P_1P_2 and, since P_1 and P_2 were chosen arbitrarily, $\boldsymbol{n}$ is perpendicular to the plane.

A *plane* in three dimensions may be specified by either
(i) three points on the plane, or
(ii) one point on the plane and a vector in the direction perpendicular to the plane.

We postpone consideration of (i) until later (see Example 11.9). The procedure for (ii) is as follows. A diagram is given in Example 10.8.

Let A be the point (a_1, a_2, a_3) and let $\boldsymbol{n}$ be a vector (which is to specify the direction perpendicular to the plane). Let $\overrightarrow{OA} = \boldsymbol{a}$, and let N be such that $\overrightarrow{AN} = \boldsymbol{n}$. The point $P(x, y, z)$ lies on the plane through A perpendicular to AN if and only if AP is perpendicular to AN, i.e. vector $\overrightarrow{AP}$ is perpendicular to vector $\overrightarrow{AN}$,

i.e. $$\overrightarrow{AP} \cdot \boldsymbol{n} = 0,$$

i.e. $$(\overrightarrow{OP} - \overrightarrow{OA}) \cdot \boldsymbol{n} = 0,$$

i.e. $$(\boldsymbol{r}_P - \boldsymbol{a}) \cdot \boldsymbol{n} = 0.$$

This last is a vector form of equation for the plane (not a parametric equation this time, though). The vector $\boldsymbol{n}$ is called a *normal* to the plane.

The equation which we have just derived can be written in coordinate form. Let

$$\boldsymbol{n} = \begin{bmatrix} n_1 \\ n_2 \\ n_3 \end{bmatrix}.$$

Then the equation becomes

$$(x - a_1)n_1 + (y - a_2)n_2 + (z - a_3)n_3 = 0.$$

Example 10.8 contains a specific case of this.

Rule

Equations of planes have the form

$$ax + by + cz + d = 0.$$

Example 10.9 shows that, given such an equation, we can read off a normal vector, namely

$$\begin{bmatrix} a \\ b \\ c \end{bmatrix}.$$

10.10 The angle between two planes.

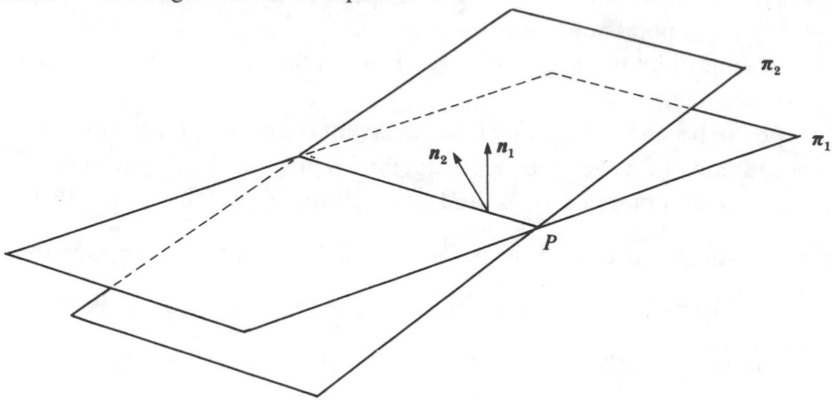

Plane π_1 has normal vector $\boldsymbol{n}_1$.
Plane π_2 has normal vector $\boldsymbol{n}_2$.
Here is an 'edge-on' view, looking along the line of intersection.

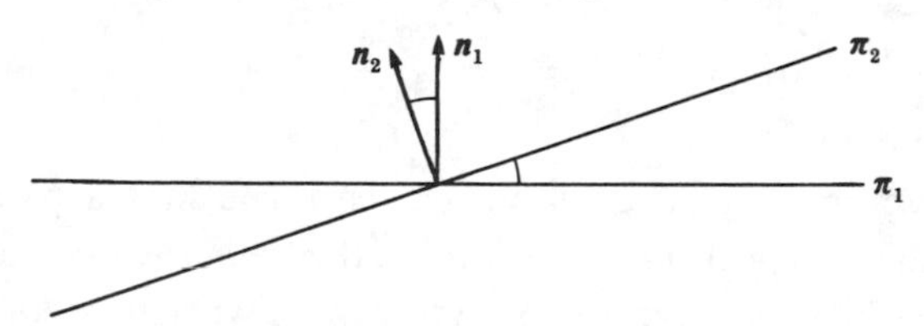

The angle observed here between the planes is the same (by a simple geometrical argument) as the angle between the normal vectors.

Let π_1 and π_2 have equations respectively

$$x - y + 3z + 2 = 0$$

and

$$-x + 2y + 2z - 3 = 0.$$

The two normal vectors can then be taken as

$$\boldsymbol{n}_1 = \begin{bmatrix} 1 \\ -1 \\ 3 \end{bmatrix} \quad \text{and} \quad \boldsymbol{n}_2 = \begin{bmatrix} -1 \\ 2 \\ 2 \end{bmatrix},$$

say. The angle between these vectors is θ, where

$$\cos\theta = \frac{\boldsymbol{n}_1 \cdot \boldsymbol{n}_2}{|\boldsymbol{n}_1||\boldsymbol{n}_2|} = \frac{-1-2+6}{\sqrt{11} \times \sqrt{3}} = \frac{3}{\sqrt{33}}.$$

10.11 Find parametric equations for the line of intersection of the two planes with equations:

$$x - y + 3z + 2 = 0,$$
$$-x + 2y + 2z - 3 = 0.$$

Try to solve the equations

Two planes will normally intersect in a straight line. The *angle* between two planes is defined to be the angle between their normal vectors. See Example 10.10.

Planes which do not intersect are parallel. We can tell immediately from their equations whether two given planes are parallel, since parallel planes have normal vectors in the same direction.

$$3x + y - 2z + 1 = 0$$

and

$$6x + 2y - 4z + 5 = 0$$

are equations of parallel planes. By inspection we can see that they have normal vectors

$$\begin{bmatrix} 3 \\ 1 \\ -2 \end{bmatrix} \quad \text{and} \quad \begin{bmatrix} 6 \\ 2 \\ -4 \end{bmatrix}$$

respectively. These vectors are clearly in the same direction, being multiples of one another. But we can also see quite easily that the two equations above are inconsistent. If we tried to solve them simultaneously we would find that there are no solutions. Of course this is to be expected: any solutions would yield coordinates for points common to the two planes.

Distinct planes which do intersect have a line in common. Parametric equations for the line of intersection are given by the standard process for solving sets of equations. We would expect a set of two equations with three unknowns (if it had solutions at all) to have infinitely many solutions, these being specified by expressions involving a single parameter. An example is given in Example 10.11.

Here is a clear situation where algebra and geometry impinge. And it becomes clearer when we consider the ways in which *three* planes might intersect. A point which is common to three planes has coordinates which satisfy three linear equations simultaneously. So finding the intersection of three planes amounts to solving a set of three linear equations in three unknowns. As we know, there can be three possible outcomes.

(i) The equations may be inconsistent. In this case the planes have no common point. Either they are all parallel or each is parallel to the line of intersection of the other two.

(ii) The equations may have a unique solution. This is the general case from the geometrical point of view. The line of intersection of two of the planes meets the third plane in a single point, common to all three planes.

(iii) The equations may have infinitely many solutions, which may be specified using one parameter or two parameters. In the first case the planes have a line in common, and in the second case the

$$x - y + 3z = -2,$$
$$-x + 2y + 2z = 3$$

simultaneously. The GE process leads to

$$\begin{bmatrix} 1 & -1 & 3 & -2 \\ 0 & 1 & 5 & 1 \end{bmatrix},$$

and hence to a solution

$$z = t, \quad y = 1 - 5t, \quad x = -1 - 8t \quad (t \in \mathbb{R}).$$

These are in fact just parametric equations for the line of intersection, as required.

10.12 Ways in which three planes can intersect.

(i) Three planes:

$$x + 2y + 3z = 0,$$
$$3x + y - z = 5,$$
$$x - y + z = 2.$$

The GE process leads to

$$\begin{bmatrix} 1 & 2 & 3 & 0 \\ 0 & 1 & 2 & -1 \\ 0 & 0 & 0 & 1 \end{bmatrix}.$$

Hence the set of equations is inconsistent. There is no point common to the three planes. Nevertheless each pair of these planes has a line of intersection. The three lines of intersection are parallel. What makes this so? If you are not sure, work out their parametric equations and confirm that they all have the same direction.

(ii) Three planes:

$$x - y - z = 4,$$
$$2x - 3y + 4z = -5,$$
$$-x + 2y - 2z = 3.$$

The GE process leads to

$$\begin{bmatrix} 1 & -1 & -1 & 4 \\ 0 & 1 & -6 & 13 \\ 0 & 0 & 1 & -2 \end{bmatrix}.$$

Thus there is a unique solution $x = 3$, $y = 1$, $z = -2$. These planes have a single point in common, namely $(3, 1, -2)$. Each pair of planes intersects in a line which meets the third plane at this point.

(iii) Three planes:

$$x + 2y + z = 6,$$
$$-x + y - 4z = 3,$$
$$x - 3y + 6z = -9.$$

The GE process leads to

$$\begin{bmatrix} 1 & 2 & 1 & 6 \\ 0 & 1 & -1 & 3 \\ 0 & 0 & 0 & 0 \end{bmatrix}.$$

Hence there are infinitely many solutions, which can be specified by parametric equations

$$x = -3t, \quad y = 3 + t, \quad z = t \quad (t \in \mathbb{R}).$$

The straight line with these parametric equations is common to all three planes.

planes are all the same plane. Illustrations are provided in Example 10.12.

There is a convenient formula for finding the perpendicular distance from a given point to a given plane. Let P be the point with coordinates (x_0, y_0, z_0), and let

$$ax+by+cz+d=0$$

be an equation of a plane. Not all of a, b and c can be zero, or else this equation would not be sensible.

Rule

The perpendicular distance from the point P to the plane given above is equal to

$$\frac{|ax_0+by_0+cz_0+d|}{\sqrt{a^2+b^2+c^2}}.$$

This formula can be derived using the methods of this chapter, and this is done in Example 10.13.

Summary

A vector form of equation for a straight line in three dimensions is derived and used in geometric deductions. The dot product of two vectors is defined, and properties of it are derived. A standard form of equation for a plane is established, and ideas of linear algebra are used in considering the nature of the intersection of two or three planes. Angles between lines and between planes are dealt with. A formula is given for the perpendicular distance from a point to a plane.

Exercises

1. Find parametric equations for the straight line passing through the points A and B in each case.
   (i) $A(0,1,3)$, $B(1,0,1)$.
   (ii) $A(1,1,-2)$, $B(1,2,0)$.
   (iii) $A(-1,2,4)$, $B(-1,2,-7)$.
   (iv) $A(1,1,1)$, $B(2,2,2)$.
   (v) $A(0,0,0)$, $B(3,-1,2)$.
2. In each case below, write down a vector in the direction of the straight line with the given parametric equations.
   (i) $x=3-t$, $y=-1+2t$, $z=4-5t$ $(t\in\mathbb{R})$.
   (ii) $x=2t$, $y=1-t$, $z=2+t$ $(t\in\mathbb{R})$.
   (iii) $x=1-3t$, $y=2$, $z=3-t$ $(t\in\mathbb{R})$.

10.13 Proof that the perpendicular distance from the point $P(x_0, y_0, z_0)$ to the plane with equation $ax+by+cz+d=0$ is

$$\frac{|ax_0+by_0+cz_0+d|}{\sqrt{a^2+b^2+c^2}}.$$

The vector

$$\begin{bmatrix} a \\ b \\ c \end{bmatrix}$$

is perpendicular to the plane, so the straight line through P perpendicular to the plane has parametric equations

$$x=x_0+at, \quad y=y_0+bt, \quad z=z_0+ct \quad (t\in\mathbb{R}).$$

This line meets the plane at the point M (say), whose coordinates are given by these parametric equations with the value of t given by

$$a(x_0+at)+b(y_0+bt)+c(z_0+ct)+d=0.$$

Solve for t, obtaining

$$(a^2+b^2+c^2)t=-ax_0-by_0-cz_0-d,$$

so

$$t=-\frac{ax_0+by_0+cz_0+d}{a^2+b^2+c^2}.$$

Now

$$\begin{aligned}|PM|^2&=(x_0+at-x_0)^2+(y_0+bt-y_0)^2+(z_0+ct-z_0)^2\\&=a^2t^2+b^2t^2+c^2t^2\\&=(a^2+b^2+c^2)t^2\\&=(a^2+b^2+c^2)\frac{(ax_0+by_0+cz_0+d)^2}{(a^2+b^2+c^2)^2}.\end{aligned}$$

Hence

$$|PM|=\frac{|ax_0+by_0+cz_0+d|}{\sqrt{a^2+b^2+c^2}},$$

as required.

3. In each case below, find whether the two lines with the given parametric equations intersect and, if they do, find the coordinates of the points of intersection.
   (i) $x=2+2t, \quad y=2+t, \quad z=-t \quad (t\in\mathbb{R}),$
   $x=-2+3u, \quad y=-3+6u, \quad z=7-9u \quad (u\in\mathbb{R}).$
   (ii) $x=1+t, \quad y=2-t, \quad z=-1-2t \quad (t\in\mathbb{R}),$
   $x=1+2u, \quad y=-6u, \quad z=1 \quad (u\in\mathbb{R}).$
   (iii) $x=2-t, \quad y=-1-3t, \quad z=2+2t \quad (t\in\mathbb{R}),$
   $x=u, \quad y=-2+u, \quad z=1-u \quad (u\in\mathbb{R}).$
   (iv) $x=2+t, \quad y=-t, \quad z=1+2t \quad (t\in\mathbb{R}),$
   $x=4-2u, \quad y=-2+2u, \quad z=5-4u \quad (u\in\mathbb{R}).$
4. Calculate the cosine of the angle $A\hat{P}B$ in each case, where A, P and B are the points given.
   (i) $A(1,1,1)$, $P(0,0,0)$, $B(1,1,0)$.
   (ii) $A(-2,1,-1)$, $P(0,0,0)$, $B(1,2,1)$.
   (iii) $A(3,-1,2)$, $P(1,2,3)$, $B(0,1,-2)$.
   (iv) $A(6,7,8)$, $P(0,1,2)$, $B(0,0,1)$.
5. Find the cosines of the internal angles of the triangle whose vertices are $A(1,3,-1)$, $B(0,2,-1)$, and $C(2,5,1)$, and find the radian measure of the largest of these angles.
6. Find a vector perpendicular to both $\boldsymbol{a}$ and $\boldsymbol{b}$, where
   $$\boldsymbol{a}=\begin{bmatrix}2\\-1\\0\end{bmatrix} \quad \text{and} \quad \boldsymbol{b}=\begin{bmatrix}1\\1\\-1\end{bmatrix}.$$
7. Find the length of the vector $2\boldsymbol{a}+\boldsymbol{b}$, where $\boldsymbol{a}$ and $\boldsymbol{b}$ are unit vectors and the angle between them is $\pi/3$ (i.e. 60°).
8. In each case below, find the length of the perpendicular from the given point to the straight line with the given parametric equations.
   (i) $(2,1,-1)$, $x=3-t, \quad y=1+2t, \quad z=t \quad (t\in\mathbb{R}).$
   (ii) $(0,1,4)$, $x=-5+t, \quad y=3, \quad z=4-2t \quad (t\in\mathbb{R}).$
   (iii) $(2,3,1)$, $x=2t, \quad y=3t, \quad z=2-t \quad (t\in\mathbb{R}).$
9. In each case below, find an equation for the plane through the given point A, with normal in the direction of the given vector $\boldsymbol{n}$.
   (i) $A(3,2,1)$, $\boldsymbol{n}=\begin{bmatrix}-1\\1\\-2\end{bmatrix}$. (ii) $A(0,1,-1)$, $\boldsymbol{n}=\begin{bmatrix}4\\5\\6\end{bmatrix}$.
   (iii) $A(-2,3,5)$, $\boldsymbol{n}=\begin{bmatrix}0\\1\\3\end{bmatrix}$. (iv) $A(1,1,1)$, $\boldsymbol{n}=\begin{bmatrix}1\\1\\1\end{bmatrix}$.

10. Find the cosine of the acute angle between the planes with given equations in each case.

(i) $x + y + z - 3 = 0$,
$2x - y - z + 1 = 0$.

(ii) $3x + y + 2 = 0$,
$x + y - 4z - 3 = 0$.

(iii) $x + y = 0$,
$y + z = 0$.

(iv) $3x - y + 2z - 7 = 0$,
$z = 0$.

11. In each case below, determine whether the intersection of the three planes with given equations is empty, is a single point, or is a straight line.

(i) $x - 2y - 3z = -1$,
$y + 2z = 1$,
$2x + y + 4z = 3$.

(ii) $x + 3z = -1$,
$2x - y + z = 2$
$x + 2y - z = 5$.

(iii) $x + 3y + 5z = 0$,
$x + 2y + 3z = 1$,
$x - z = -1$.

(iv) $x - 2y + z = -6$,
$-2x + z = 8$,
$x + 2y + 2z = 1$.

12. In each case below, find the perpendicular distance from the given point P to the plane with the given equation.

(i) $P(2, 1, 2)$, $x - 2y - z + 2 = 0$.

(ii) $P(-1, 0, 1)$, $3x - y + 2z - 5 = 0$.

(iii) $P(1, 1, 1)$, $x + y + z = 0$.

(iv) $P(0, 0, 0)$, $5x - y - 4z + 6 = 0$.

(v) $P(0, 0, 0)$, $5x - y - 4z + 1 = 0$.

Examples

11.1 Evaluation of cross products.

(i) $$\boldsymbol{a}=\begin{bmatrix}1\\2\\3\end{bmatrix},\quad \boldsymbol{b}=\begin{bmatrix}2\\3\\5\end{bmatrix},\quad \boldsymbol{a}\times\boldsymbol{b}=\begin{bmatrix}10-9\\6-5\\3-4\end{bmatrix}=\begin{bmatrix}1\\1\\-1\end{bmatrix}.$$

(ii) $$\boldsymbol{a}=\begin{bmatrix}2\\1\\0\end{bmatrix},\quad \boldsymbol{b}=\begin{bmatrix}-1\\3\\2\end{bmatrix},\quad \boldsymbol{a}\times\boldsymbol{b}=\begin{bmatrix}2-0\\0-4\\6+1\end{bmatrix}=\begin{bmatrix}2\\-4\\7\end{bmatrix}.$$

(iii) $$\boldsymbol{a}=\begin{bmatrix}2\\-1\\3\end{bmatrix},\quad \boldsymbol{b}=\begin{bmatrix}-4\\2\\-6\end{bmatrix},\quad \boldsymbol{a}\times\boldsymbol{b}=\begin{bmatrix}6-6\\-12+12\\4-4\end{bmatrix}=\begin{bmatrix}0\\0\\0\end{bmatrix}.$$

(iv) $$\boldsymbol{a}=\begin{bmatrix}-1\\3\\2\end{bmatrix},\quad \boldsymbol{b}=\begin{bmatrix}2\\1\\0\end{bmatrix},\quad \boldsymbol{a}\times\boldsymbol{b}=\begin{bmatrix}0-2\\4-0\\-1-6\end{bmatrix}=\begin{bmatrix}-2\\4\\-7\end{bmatrix}.$$

Compare (ii) with (iv).

11.2 Use of the determinant mnemonic in evaluating cross products.

(i) $\boldsymbol{a}=-\boldsymbol{i}+2\boldsymbol{j}-\boldsymbol{k},\quad \boldsymbol{b}=3\boldsymbol{i}-\boldsymbol{j}.$

$$\boldsymbol{a}\times\boldsymbol{b}=\begin{vmatrix}\boldsymbol{i} & -1 & 3\\ \boldsymbol{j} & 2 & 1\\ \boldsymbol{k} & -1 & 0\end{vmatrix}=\boldsymbol{i}\begin{vmatrix}2 & 1\\ -1 & 0\end{vmatrix}-\boldsymbol{j}\begin{vmatrix}-1 & 3\\ -1 & 0\end{vmatrix}+\boldsymbol{k}\begin{vmatrix}-1 & 3\\ 2 & 1\end{vmatrix}$$
$$=\boldsymbol{i}-3\boldsymbol{j}-7\boldsymbol{k}.$$

(ii) $\boldsymbol{a}=3\boldsymbol{i}-4\boldsymbol{j}+2\boldsymbol{k},\quad \boldsymbol{b}=\boldsymbol{j}+2\boldsymbol{k}.$

$$\boldsymbol{a}\times\boldsymbol{b}=\begin{vmatrix}\boldsymbol{i} & 3 & 0\\ \boldsymbol{j} & -4 & 1\\ \boldsymbol{k} & 2 & 2\end{vmatrix}=\boldsymbol{i}\begin{vmatrix}-4 & 1\\ 2 & 2\end{vmatrix}-\boldsymbol{j}\begin{vmatrix}3 & 0\\ 2 & 2\end{vmatrix}+\boldsymbol{k}\begin{vmatrix}3 & 0\\ -4 & 1\end{vmatrix}$$
$$=-10\boldsymbol{i}-6\boldsymbol{j}+3\boldsymbol{k}.$$

11.3 Proof that for any 3-vectors $\boldsymbol{a}$ and $\boldsymbol{b}$, $\boldsymbol{a}\times\boldsymbol{b}$ is perpendicular to $\boldsymbol{a}$ and to $\boldsymbol{b}$.

Let

$$\boldsymbol{a}=\begin{bmatrix}a_1\\a_2\\a_3\end{bmatrix}\quad\text{and}\quad \boldsymbol{b}=\begin{bmatrix}b_1\\b_2\\b_3\end{bmatrix}.$$

Then

$$\boldsymbol{a}\times\boldsymbol{b}=\begin{bmatrix}a_2b_3-a_3b_2\\a_3b_1-a_1b_3\\a_1b_2-a_2b_1\end{bmatrix},$$

so

$$\begin{aligned}\boldsymbol{a}\cdot(\boldsymbol{a}\times\boldsymbol{b})&=a_1(a_2b_3-a_3b_2)+a_2(a_3b_1-a_1b_3)+a_3(a_1b_2-a_2b_1)\\&=a_1a_2b_3-a_1a_3b_2+a_2a_3b_1-a_2a_1b_3+a_3a_1b_2-a_3a_2b_1\\&=0.\end{aligned}$$

Similarly $\boldsymbol{b}\cdot(\boldsymbol{a}\times\boldsymbol{b})=0.$

11

Cross product

We have dealt with the dot product of two vectors. There is another way of combining vectors which is useful in geometric applications.

Let

$$\boldsymbol{a}=\begin{bmatrix} a_1 \\ a_2 \\ a_3 \end{bmatrix} \quad \text{and} \quad \boldsymbol{b}=\begin{bmatrix} b_1 \\ b_2 \\ b_3 \end{bmatrix}$$

be two 3-vectors. The *cross product* of $\boldsymbol{a}$ and $\boldsymbol{b}$ is defined by

$$\boldsymbol{a}\times\boldsymbol{b}=\begin{bmatrix} a_2b_3-a_3b_2 \\ a_3b_1-a_1b_3 \\ a_1b_2-a_2b_1 \end{bmatrix}.$$

Note that the result is a vector this time. Example 11.1 gives some calculations. Example 11.2 shows how to apply the following mnemonic which is useful in calculating cross products. Write

$$\boldsymbol{a}=a_1\boldsymbol{i}+a_2\boldsymbol{j}+a_3\boldsymbol{k} \quad \text{and} \quad \boldsymbol{b}=b_1\boldsymbol{i}+b_2\boldsymbol{j}+b_3\boldsymbol{k},$$

where $\boldsymbol{i}$, $\boldsymbol{j}$ and $\boldsymbol{k}$ are the standard basis vectors. Then

$$\boldsymbol{a}\times\boldsymbol{b}=(a_2b_3-a_3b_2)\boldsymbol{i}+(a_3b_1-a_1b_3)\boldsymbol{j}+(a_1b_2-a_2b_1)\boldsymbol{k}.$$

This is reminiscent of the expansion of a determinant, and we can stretch the idea of a determinant to write

$$\boldsymbol{a}\times\boldsymbol{b}=\begin{vmatrix} \boldsymbol{i} & a_1 & b_1 \\ \boldsymbol{j} & a_2 & b_2 \\ \boldsymbol{k} & a_3 & b_3 \end{vmatrix}.$$

Expanding this 'determinant' by the first column gives the correct expression for $\boldsymbol{a}\times\boldsymbol{b}$.

The definition above of the cross product involved two 3-vectors explicitly. It is important to realise that, unlike the dot product, the cross product applies only to 3-vectors, and its application is of use in three-dimensional geometry.

11.4 Proof that $|\boldsymbol{a}\times\boldsymbol{b}|=|\boldsymbol{a}||\boldsymbol{b}|\sin\theta$, where $\boldsymbol{a}$ and $\boldsymbol{b}$ are non-zero 3-vectors and θ is the angle between them, for the case when $\theta\neq 0$ and $\theta\neq\pi$.

Suppose that $\boldsymbol{a}$ and $\boldsymbol{b}$ are 3-vectors, as in Example 11.3.

$$\begin{aligned}|\boldsymbol{a}\times\boldsymbol{b}|^2&=(a_2b_3-a_3b_2)^2+(a_3b_1-a_1b_3)^2+(a_1b_2-a_2b_1)^2\\&=a_2{}^2b_3{}^2+a_3{}^2b_2{}^2+a_3{}^2b_1{}^2+a_1{}^2b_3{}^2+a_1{}^2b_2{}^2+a_2{}^2b_1{}^2\\&\quad-2a_2b_3a_3b_2-2a_3b_1a_1b_3-2a_1b_2a_2b_1.\end{aligned}$$

Also

$$\begin{aligned}(|\boldsymbol{a}||\boldsymbol{b}|\sin\theta)^2&=|\boldsymbol{a}|^2|\boldsymbol{b}|^2(1-\cos^2\theta)\\&=|\boldsymbol{a}|^2|\boldsymbol{b}|^2(1-(\boldsymbol{a}\cdot\boldsymbol{b})^2/|\boldsymbol{a}|^2|\boldsymbol{b}|^2)\\&=|\boldsymbol{a}|^2|\boldsymbol{b}|^2-(\boldsymbol{a}\cdot\boldsymbol{b})^2\\&=(a_1{}^2+a_2{}^2+a_3{}^2)(b_1{}^2+b_2{}^2+b_3{}^2)\\&\quad-(a_1b_1+a_2b_2+a_3b_3)^2\\&=a_1{}^2b_1{}^2+a_1{}^2b_2{}^2+a_1{}^2b_3{}^2+a_2{}^2b_1{}^2+a_2{}^2b_2{}^2+a_2{}^2b_3{}^2\\&\quad+a_3{}^2b_1{}^2+a_3{}^2b_2{}^2+a_3{}^2b_3{}^2-a_1{}^2b_1{}^2-a_2{}^2b_2{}^2\\&\quad-a_3{}^2b_3{}^2-2a_1b_1a_2b_2-2a_1b_1a_3b_3-2a_2b_2a_3b_3\\&=|\boldsymbol{a}\times\boldsymbol{b}|^2.\end{aligned}$$

Now $\sin\theta$ is positive, as are $|\boldsymbol{a}|$, $|\boldsymbol{b}|$ and $|\boldsymbol{a}\times\boldsymbol{b}|$, so it follows that

$$|\boldsymbol{a}\times\boldsymbol{b}|=|\boldsymbol{a}||\boldsymbol{b}|\sin\theta.$$

11.5 Proof that $\boldsymbol{a}\times(\boldsymbol{b}+\boldsymbol{c})=(\boldsymbol{a}\times\boldsymbol{b})+(\boldsymbol{a}\times\boldsymbol{c})$, for any 3-vectors $\boldsymbol{a}$, $\boldsymbol{b}$ and $\boldsymbol{c}$. Let

$$\boldsymbol{a}=\begin{bmatrix}a_1\\a_2\\a_3\end{bmatrix},\quad \boldsymbol{b}=\begin{bmatrix}b_1\\b_2\\b_3\end{bmatrix},\quad \boldsymbol{c}=\begin{bmatrix}c_1\\c_2\\c_3\end{bmatrix}.$$

Then

$$\begin{aligned}\boldsymbol{a}\times(\boldsymbol{b}+\boldsymbol{c})&=\begin{bmatrix}a_2(b_3+c_3)-a_3(b_2+c_2)\\a_3(b_1+c_1)-a_1(b_3+c_3)\\a_1(b_2+c_2)-a_2(b_1+c_1)\end{bmatrix}\\&=\begin{bmatrix}a_2b_3-a_3b_2+a_2c_3-a_3c_2\\a_3b_1-a_1b_3+a_3c_1-a_1c_3\\a_1b_2-a_2b_1+a_1c_2-a_2c_1\end{bmatrix}\\&=(\boldsymbol{a}\times\boldsymbol{b})+(\boldsymbol{a}\times\boldsymbol{c}).\end{aligned}$$

11.6 Remember that $\boldsymbol{a}\times\boldsymbol{b}=-(\boldsymbol{b}\times\boldsymbol{a})$, for any 3-vectors $\boldsymbol{a}$ and $\boldsymbol{b}$. See Example 11.1, parts (ii) and (iv).

Here is another illustration. Let

$$\boldsymbol{a}=\begin{bmatrix}2\\-3\\1\end{bmatrix}\quad\text{and}\quad\boldsymbol{b}=\begin{bmatrix}5\\-4\\1\end{bmatrix}.$$

Then

$$\boldsymbol{a}\times\boldsymbol{b}=\begin{bmatrix}1\\3\\7\end{bmatrix}\quad\text{and}\quad\boldsymbol{b}\times\boldsymbol{a}=\begin{bmatrix}-1\\-3\\-7\end{bmatrix}=-(\boldsymbol{a}\times\boldsymbol{b}).$$

Let us now explore these geometrical uses. Some properties of the cross product will emerge as we proceed.

Rule
The product vector $\boldsymbol{a}\times\boldsymbol{b}$ is perpendicular to both $\boldsymbol{a}$ and $\boldsymbol{b}$.

To see this we just write down expanded expressions for $\boldsymbol{a}\cdot(\boldsymbol{a}\times\boldsymbol{b})$ and $\boldsymbol{b}\cdot(\boldsymbol{a}\times\boldsymbol{b})$. Details are in Example 11.3. Both expressions are identically zero, and the rule is therefore justified, using a result from Chapter 10.

The above rule is in fact the most useful property of the cross product, and we shall see applications of it shortly. But consider now the *length* of $\boldsymbol{a}\times\boldsymbol{b}$. It has a convenient and useful geometrical interpretation.

Rule
If $\boldsymbol{a}$ and $\boldsymbol{b}$ are non-zero vectors then

$$|\boldsymbol{a}\times\boldsymbol{b}|=|\boldsymbol{a}||\boldsymbol{b}|\sin\theta,$$

where θ is the angle between $\boldsymbol{a}$ and $\boldsymbol{b}$.

Justification of the general case (when θ is not 0 or π) is given in Example 11.4. The special case is also significant, so let us formulate it into a separate rule.

Rule
(i) $\boldsymbol{a}\times\boldsymbol{a}=\boldsymbol{0}$ for every 3-vector $\boldsymbol{a}$.
(ii) $\boldsymbol{a}\times(k\boldsymbol{a})=\boldsymbol{0}$ for every 3-vector $\boldsymbol{a}$ and any number k.

Justification of these is straightforward verification, which is left as an exercise. This rule is perhaps surprising. It suggests that the cross product behaves in ways which we might not expect. This is indeed so, and we must be careful when using it.

Rules (Properties of the cross product)
(i) $\boldsymbol{a}\times(k\boldsymbol{b})=(k\boldsymbol{a})\times\boldsymbol{b}=k(\boldsymbol{a}\times\boldsymbol{b})$, for all 3-vectors $\boldsymbol{a}$ and $\boldsymbol{b}$, and any number k.
(ii) $\boldsymbol{a}\times\boldsymbol{b}=-(\boldsymbol{b}\times\boldsymbol{a})$, for all 3-vectors $\boldsymbol{a}$ and $\boldsymbol{b}$.
(iii) $\boldsymbol{a}\times(\boldsymbol{b}+\boldsymbol{c})=(\boldsymbol{a}\times\boldsymbol{b})+(\boldsymbol{a}\times\boldsymbol{c})$, for all 3-vectors $\boldsymbol{a}$, $\boldsymbol{b}$ and $\boldsymbol{c}$.
(iv) $\boldsymbol{i}\times\boldsymbol{j}=\boldsymbol{k}$, $\boldsymbol{j}\times\boldsymbol{k}=\boldsymbol{i}$, $\boldsymbol{k}\times\boldsymbol{i}=\boldsymbol{j}$.

Demonstration of these is not difficult, using the definition. See Example 11.5 for (iii). Take note of (ii)! See Example 11.6.

Now let us see how these geometrical interpretations can be used. First we consider areas.

11.7 Areas of triangles and parallelograms.

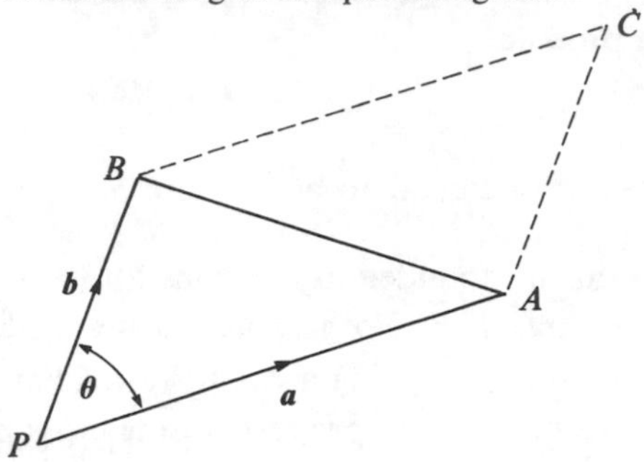

Let PAB be a triangle, with $\overrightarrow{PA}=\boldsymbol{a}$ and $\overrightarrow{PB}=\boldsymbol{b}$.
Then the area of triangle PAB is $\frac{1}{2}|\boldsymbol{a}\times\boldsymbol{b}|$.
Let C be such that $PACB$ is a parallelogram. Then the area of $PACB$ is $|\boldsymbol{a}\times\boldsymbol{b}|$.

11.8 Calculation of areas.
(i) Let

$$\boldsymbol{a}=\begin{bmatrix}2\\0\\-3\end{bmatrix},\quad \boldsymbol{b}=\begin{bmatrix}4\\1\\4\end{bmatrix}.\quad \text{Then } \boldsymbol{a}\times\boldsymbol{b}=\begin{bmatrix}3\\4\\2\end{bmatrix},$$

and $|\boldsymbol{a}\times\boldsymbol{b}|=\sqrt{9+16+4}=\sqrt{29}$.

Hence the area of the triangle PAB in the above diagram would be $\frac{1}{2}\sqrt{29}$ units^2.
(ii) Let the three points be $P(3,-1,1)$, $A(1,1,0)$ and $B(0,3,1)$. Then

$$\overrightarrow{PA}=\begin{bmatrix}-2\\2\\-1\end{bmatrix},\quad \overrightarrow{PB}=\begin{bmatrix}-3\\4\\0\end{bmatrix}\quad\text{and}\quad \overrightarrow{PA}\times\overrightarrow{PB}=\begin{bmatrix}4\\4\\-2\end{bmatrix}.$$

Hence the area of triangle PAB is equal to $\frac{1}{2}\sqrt{16+16+4}$, i.e. 3 units^2.

11.9 Find an equation for the plane through the three points $A(5,3,-1)$, $B(2,-2,0)$ and $C(3,1,1)$.

$$\overrightarrow{AB}=\begin{bmatrix}-3\\-5\\1\end{bmatrix},\quad \overrightarrow{AC}=\begin{bmatrix}-2\\-2\\2\end{bmatrix}.$$

The vector $\overrightarrow{AB}\times\overrightarrow{AC}$ is perpendicular to both $\overrightarrow{AB}$ and $\overrightarrow{AC}$, so is a normal vector to the plane of A, B and C.

$$\overrightarrow{AB}\times\overrightarrow{AC}=\begin{bmatrix}-8\\4\\-4\end{bmatrix}.$$

Any vector in the direction of this vector is a normal vector for the plane. We can choose this one or any multiple of it. Taking $\begin{bmatrix}2\\-1\\1\end{bmatrix}$, as the normal vector, we obtain an equation for the plane:

$$2(x-5)-1(y-3)+1(z+1)=0,$$

i.e. $2x-y+z-6=0$. (See Chapter 10 for the method used.)

Rule

Let $\overrightarrow{PA}$ and $\overrightarrow{PB}$ represent 3-vectors $\boldsymbol{a}$ and $\boldsymbol{b}$. Then the area of triangle PAB is equal to $\frac{1}{2}|\boldsymbol{a}\times\boldsymbol{b}|$.

This follows from our knowledge that

$$|\boldsymbol{a}\times\boldsymbol{b}|=|\boldsymbol{a}||\boldsymbol{b}|\sin\theta,$$

and the familiar rule that

$$\text{area of } \triangle PAB=\tfrac{1}{2}|PA||PB|\sin A\hat{P}B.$$

For a diagram see Example 11.7.

Further, if C is such that $PACB$ is a parallelogram, the area of the parallelogram is equal to $|\boldsymbol{a}\times\boldsymbol{b}|$. Some calculations of areas are given in Examples 11.8.

Example 11.9 gives an application of another use of cross products. A plane may be specified by giving the position vectors (or coordinates) of three points on it, say $A(\boldsymbol{a})$, $B(\boldsymbol{b})$ and $C(\boldsymbol{c})$. We know from Chapter 10 how to derive an equation for a plane given a point on it and the direction of a normal vector to it. We obtain a normal vector in the present case by using the cross product. AB represents the vector $\boldsymbol{b}-\boldsymbol{a}$ and AC represents the vector $\boldsymbol{c}-\boldsymbol{a}$. Consequently $(\boldsymbol{b}-\boldsymbol{a})\times(\boldsymbol{c}-\boldsymbol{a})$ is perpendicular to both AB and AC, and so must be perpendicular to the plane containing A, B and C. It will serve as a normal vector, and the method of Chapter 10 can now be applied.

11.10 Calculation of volumes of parallelepipeds.

(i) Find the volume of the parallelepiped with one vertex at $P(1,2,-1)$ and adjacent vertices at $A(3,-1,0)$, $B(2,1,1)$ and $C(4,0,-2)$.

The volume is $|\boldsymbol{a}.(\boldsymbol{b}\times\boldsymbol{c})|$, where $\boldsymbol{a}=\overrightarrow{PA}$, $\boldsymbol{b}=\overrightarrow{PB}$ and $\boldsymbol{c}=\overrightarrow{PC}$.

$$\overrightarrow{PA}=\begin{bmatrix}2\\-3\\1\end{bmatrix},\quad \overrightarrow{PB}=\begin{bmatrix}1\\-1\\2\end{bmatrix},\quad \overrightarrow{PC}=\begin{bmatrix}3\\-2\\-1\end{bmatrix}.$$

$$\overrightarrow{PB}\times\overrightarrow{PC}=\begin{bmatrix}5\\7\\1\end{bmatrix}.$$

so

$$\overrightarrow{PA}\cdot(\overrightarrow{PB}\times\overrightarrow{PC})=10-21+1=-10.$$

Hence the volume required is 10 units^3.

(ii) Repeat (i) with the points $P(0,0,0)$, $A(2,1,0)$, $B(1,2,0)$ and $C(3,3,2)$.

Here

$$\overrightarrow{PA}=\begin{bmatrix}2\\1\\0\end{bmatrix},\quad \overrightarrow{PB}=\begin{bmatrix}1\\2\\0\end{bmatrix},\quad \overrightarrow{PC}=\begin{bmatrix}3\\3\\2\end{bmatrix}.$$

$$\overrightarrow{PB}\times\overrightarrow{PC}=\begin{bmatrix}4\\-2\\-3\end{bmatrix},$$

so

$$\overrightarrow{PA}\cdot(\overrightarrow{PB}\times\overrightarrow{PC})=8-2+0=6.$$

Hence the volume of the parallelepiped is 6 units^3.

Besides areas, volumes can be calculated using the cross product: in particular, volumes of parallelepipeds. A parallelepiped is a solid figure with six faces such that opposite faces are congruent parallelograms.

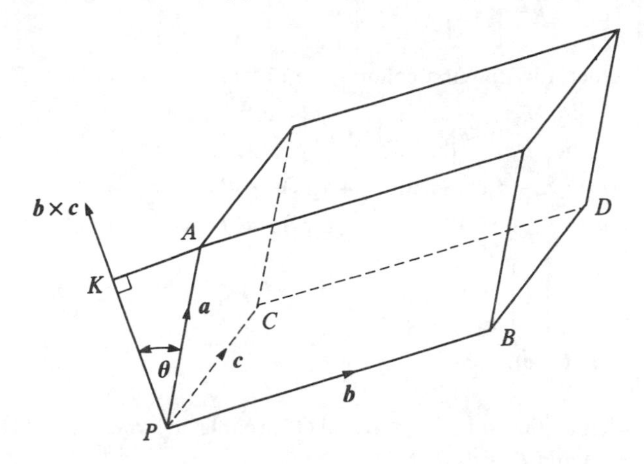

The volume of such a solid is equal to the area of the base multiplied by the height. Take $PBDC$ as the base, draw a line through P perpendicular to the base and let K be the foot of the perpendicular from A to this line. Then $|PK|$ is the height of the parallelepiped. The area of $PBDC$ is equal to $|\boldsymbol{b} \times \boldsymbol{c}|$, and the height $|PK|$ is equal to $|PA| \cos A\hat{P}K$, i.e. $|\boldsymbol{a}| \cos A\hat{P}K$. In our diagram, $A\hat{P}K$ is the angle between $\boldsymbol{a}$ and $(\boldsymbol{b} \times \boldsymbol{c})$, so

$$\boldsymbol{a} \cdot (\boldsymbol{b} \times \boldsymbol{c}) = |\boldsymbol{a}||\boldsymbol{b} \times \boldsymbol{c}| \cos A\hat{P}K = \text{volume of parallelepiped}.$$

It may happen that the direction of $\boldsymbol{b} \times \boldsymbol{c}$ is opposite to the direction of PK, in which case the angle between $\boldsymbol{a}$ and $(\boldsymbol{b} \times \boldsymbol{c})$ is $\pi - A\hat{P}K$, and

$$\begin{aligned} \boldsymbol{a} \cdot (\boldsymbol{b} \times \boldsymbol{c}) &= |\boldsymbol{a}||\boldsymbol{b} \times \boldsymbol{c}| \cos(\pi - A\hat{P}K) \\ &= |\boldsymbol{a}||\boldsymbol{b} \times \mathbf{c}|(-\cos A\hat{P}K), \end{aligned}$$

which is the negative of what we obtained above for the other case. A volume is normally taken as a positive number, so we may combine both cases in the result

$$\text{Volume of the parallelepiped} = |\boldsymbol{a} \cdot (\boldsymbol{b} \times \boldsymbol{c})|.$$

Notice that $\boldsymbol{a} \cdot (\boldsymbol{b} \times \boldsymbol{c})$ is a *number*, since it is the dot product of two vectors. This form of product of three vectors is quite important, and it has a name. It is called the *scalar triple product* of $\boldsymbol{a}$, $\boldsymbol{b}$ and $\boldsymbol{c}$. The appearance of $\boldsymbol{a} \cdot (\boldsymbol{b} \times \boldsymbol{c})$ in the formula for the volume of a parallelepiped enables us to see an unexpected property. Because the parallelepiped is the same no matter what order the three vectors are taken, we have

$$\begin{aligned} |\boldsymbol{a} \cdot (\boldsymbol{b} \times \boldsymbol{c})| &= |\boldsymbol{b} \cdot (\boldsymbol{c} \times \boldsymbol{a})| = |\boldsymbol{c} \cdot (\boldsymbol{a} \times \boldsymbol{b})| = |\boldsymbol{a} \cdot (\boldsymbol{c} \times \boldsymbol{b})| \\ &= |\boldsymbol{b} \cdot (\boldsymbol{a} \times \boldsymbol{c})| = |\boldsymbol{c} \cdot (\boldsymbol{b} \times \boldsymbol{a})|. \end{aligned}$$

11.11 Proof that $\det A = \boldsymbol{a}\cdot(\boldsymbol{b}\times\boldsymbol{c})$, where A is the 3×3 matrix having the vectors $\boldsymbol{a}$, $\boldsymbol{b}$ and $\boldsymbol{c}$ as its columns. Let

$$\boldsymbol{a}=\begin{bmatrix}a_1\\a_2\\a_3\end{bmatrix},\quad \boldsymbol{b}=\begin{bmatrix}b_1\\b_2\\b_3\end{bmatrix},\quad \boldsymbol{c}=\begin{bmatrix}c_1\\c_2\\c_3\end{bmatrix}.$$

Then, expanding $\det A$ by the first column, we obtain

$$\begin{aligned}\det A &= a_1\begin{vmatrix}b_2 & c_2\\ b_3 & c_3\end{vmatrix} - a_2\begin{vmatrix}b_1 & c_1\\ b_3 & c_3\end{vmatrix} + a_3\begin{vmatrix}b_1 & c_1\\ b_2 & c_2\end{vmatrix}\\ &= a_1(b_2c_3-b_3c_2)-a_2(b_1c_3-b_3c_1)+a_3(b_1c_2-b_2c_1)\\ &= a_1(b_2c_3-b_3c_2)+a_2(b_3c_1-b_1c_3)+a_3(b_1c_2-b_2c_1)\\ &= \begin{bmatrix}a_1\\a_2\\a_3\end{bmatrix}\cdot\begin{bmatrix}b_2c_3-b_3c_2\\ b_3c_1-b_1c_3\\ b_1c_2-b_2c_1\end{bmatrix}\\ &= \boldsymbol{a}\cdot(\boldsymbol{b}\times\boldsymbol{c}).\end{aligned}$$

11.12 Find whether the points O, A, B and C are coplanar, where A is $(1,3,0)$, B is $(0, 1, 1)$ and C is $(-2, 1, 7)$.

They are coplanar if and only if the three vectors $\overrightarrow{OA}$, $\overrightarrow{OB}$ and $\overrightarrow{OC}$ are coplanar. We therefore have to test whether these three vectors are linearly dependent. Use the standard GE process.

$$\overrightarrow{OA}=\begin{bmatrix}1\\3\\0\end{bmatrix},\quad \overrightarrow{OB}=\begin{bmatrix}0\\1\\1\end{bmatrix},\quad \overrightarrow{OC}=\begin{bmatrix}-2\\1\\7\end{bmatrix}.$$

$$\begin{bmatrix}1 & 0 & -2\\ 3 & 1 & 1\\ 0 & 1 & 7\end{bmatrix}\rightarrow\begin{bmatrix}1 & 0 & -2\\ 0 & 1 & 7\\ 0 & 1 & 7\end{bmatrix}\rightarrow\begin{bmatrix}1 & 0 & -2\\ 0 & 1 & 7\\ 0 & 0 & 0\end{bmatrix}.$$

Hence the three vectors form a linearly dependent list (see Chapter 6), and so the four points are coplanar.

11.13 Find whether the points $P(-1,1,-1)$, $A(2,3,0)$, $B(0,1,1)$ and $C(-2,2,2)$ are coplanar.

They are coplanar if and only if the vectors

$$\overrightarrow{PA}=\begin{bmatrix}3\\2\\1\end{bmatrix},\quad \overrightarrow{PB}=\begin{bmatrix}1\\0\\2\end{bmatrix},\quad \overrightarrow{PC}=\begin{bmatrix}-1\\1\\3\end{bmatrix}$$

form a linearly dependent list. The GE process yields

$$\begin{bmatrix}3 & 1 & -1\\ 2 & 0 & 1\\ 1 & 2 & 3\end{bmatrix}\rightarrow\begin{bmatrix}1 & \frac{1}{3} & -\frac{1}{3}\\ 2 & 0 & 1\\ 1 & 2 & 3\end{bmatrix}\rightarrow\begin{bmatrix}1 & \frac{1}{3} & -\frac{1}{3}\\ 0 & -\frac{2}{3} & \frac{5}{3}\\ 0 & \frac{5}{3} & \frac{10}{3}\end{bmatrix}$$

$$\rightarrow\begin{bmatrix}1 & \frac{1}{3} & -\frac{1}{3}\\ 0 & 1 & -\frac{5}{2}\\ 0 & \frac{5}{3} & \frac{10}{3}\end{bmatrix}\rightarrow\begin{bmatrix}1 & \frac{1}{3} & -\frac{1}{3}\\ 0 & 1 & -\frac{5}{2}\\ 0 & 0 & 1\end{bmatrix}.$$

Hence the three vectors form a linearly *in*dependent list, and so the given points are *not* coplanar.

These six products, however, do not all have the same value. Three of them have one value and the other three have the negative of that value. As an exercise, find out how they are grouped in this way.

We end with a further link-up between geometry and algebra. Given three 3-vectors $\boldsymbol{a}$, $\boldsymbol{b}$ and $\boldsymbol{c}$ we can construct line segments OA, OB and OC respectively representing them. We say that the vectors are *coplanar* if the three lines OA, OB and OC lie in a single plane.

Rule

Three vectors $\boldsymbol{a}$, $\boldsymbol{b}$ and $\boldsymbol{c}$ in three dimensions are coplanar if and only if $\boldsymbol{a}\cdot(\boldsymbol{b}\times\boldsymbol{c})=0$.

To see this we need consider only the parallelepiped with one vertex at O and with A, B and C as the vertices adjacent to O. The vectors are coplanar if and only if the volume of this parallelepiped is zero (i.e. the parallelepiped is squashed flat).

Recall that three vectors $\boldsymbol{a}$, $\boldsymbol{b}$ and $\boldsymbol{c}$ form a linearly dependent list if there exist numbers l, m and n, not all zero, such that

$$l\boldsymbol{a}+m\boldsymbol{b}+n\boldsymbol{c}=\boldsymbol{0}.$$

Recall also the Equivalence Theorem, part of which stated (in the case of a 3×3 matrix A): the columns of A form a linearly independent list if and only if $\det A\neq 0$. This is logically equivalent to: the columns of A form a linearly dependent list if and only if $\det A=0$. To make the connection between this and the ideas of coplanarity and scalar triple product, consider three 3-vectors $\boldsymbol{a}$, $\boldsymbol{b}$ and $\boldsymbol{c}$. Let A be the matrix with these vectors as its columns. Then

$$\det A=\boldsymbol{a}\cdot(\boldsymbol{b}\times\boldsymbol{c}).$$

To see this, it is necessary only to evaluate both sides. This is done in Example 11.11.

We can now see that the conditions

$$\boldsymbol{a}\cdot(\boldsymbol{b}\times\boldsymbol{c})=0 \quad \text{and} \quad \det A=0$$

are *the same*. We therefore have:

Rule

Three vectors $\boldsymbol{a}$, $\boldsymbol{b}$ and $\boldsymbol{c}$ in three dimensions are coplanar if and only if they form a linearly dependent list.

See Examples 11.12 and 11.13 for applications of this.

As a final remark, let us note that the rule $\boldsymbol{a}\cdot(\boldsymbol{b}\times\boldsymbol{c})=\det A$ can be a convenient way of evaluating scalar triple products and volumes of parallelepipeds.

Summary
The cross product of two 3-vectors is defined, and algebraic and geometrical properties are derived. The use of cross products in finding areas and volumes is discussed, leading to the idea of scalar triple product. The equivalence between coplanarity and linear dependence is established using the link between the scalar triple product and determinants.

Exercises

1. Evaluate the cross product of each of the following pairs of vectors (in the order given).

   (i) $\begin{bmatrix} 3 \\ 1 \\ 2 \end{bmatrix}, \begin{bmatrix} -1 \\ 1 \\ -1 \end{bmatrix}$.
   (ii) $\begin{bmatrix} 2 \\ 0 \\ -1 \end{bmatrix}, \begin{bmatrix} 1 \\ 3 \\ 3 \end{bmatrix}$.

   (iii) $\begin{bmatrix} 0 \\ 1 \\ -2 \end{bmatrix}, \begin{bmatrix} 3 \\ -1 \\ -2 \end{bmatrix}$.
   (iv) $\begin{bmatrix} -1 \\ 1 \\ -1 \end{bmatrix}, \begin{bmatrix} 3 \\ 1 \\ 2 \end{bmatrix}$.

   (v) $\begin{bmatrix} -4 \\ 4 \\ -4 \end{bmatrix}, \begin{bmatrix} 3 \\ 1 \\ 2 \end{bmatrix}$.
   (vi) $\begin{bmatrix} -1 \\ 2 \\ -1 \end{bmatrix}, \begin{bmatrix} 1 \\ 1 \\ 1 \end{bmatrix}$.

2. Write down the areas of the triangles OAB, where A and B are points with the pairs of vectors given in Exercise 1 as their position vectors.

3. Find the area of the triangle ABC in each case below, where A, B and C have the coordinates given.

   (i) $A(2, 1, 3)$, $B(1, 1, 0)$, $C(0, 2, 2)$.
   (ii) $A(-1, 2, -2)$, $B(-3, 0, 1)$, $C(0, 1, 0)$.

4. For each of the following pairs of planes, find a vector in the direction of the line of intersection.

   (i) $x - y + 3z - 4 = 0.$
   $2x + y - z + 5 = 0.$

   (ii) $3x + y - 2 = 0.$
   $x - 3y + z + 1 = 0.$

5. Find an equation for the plane containing the three points A, B and C, where A, B and C have the given coordinates.

   (i) $A(2, 1, 1)$, $B(4, 0, -2)$, $C(1, 1, 1)$.
   (ii) $A(0, 1, 1)$, $B(1, 0, 1)$, $C(1, 1, 0)$.
   (iii) $A(-1, 1, 2)$, $B(0, 0, 0)$, $C(3, 2, -1)$.

6. Find the volume of the parallelepiped which has $P(-1, 1, 2)$ as one vertex and $A(1, 1, -1)$, $B(0, 0, 1)$ and $C(1, 2, 0)$ as the vertices adjacent to P.

7. Repeat the calculation of Exercise 6, where P is the point $(-1,0,0)$, A is the point $(1,0,-1)$, B is the point $(1,0,1)$ and C is the point $(0,1,0)$.
8. Using the scalar triple product, find whether the given four points are coplanar.

   (i) $O(0,0,0)$, $A(1,3,5)$, $B(1,2,3)$, $C(1,0,-1)$.

   (ii) $O(0,0,0)$, $A(2,1,1)$, $B(1,1,2)$, $C(-1,2,7)$.

   (iii) $O(0,0,0)$, $A(1,-1,1)$, $B(2,2,-1)$, $C(-1,1,3)$.

   (iv) $P(2,-1,0)$, $A(3,2,2)$, $B(2,-2,-1)$, $C(4,0,-1)$.

   (v) $P(1,2,3)$, $A(-1,3,4)$, $B(3,4,7)$, $C(4,0,4)$.

Part 2

Examples

12.1 Find all solutions to the simultaneous equations:

$$\begin{aligned} x_1 + x_2 - x_3 &= 3 \\ 2x_1 - x_2 + 4x_3 &= 3 \\ 3x_1 + 4x_2 - 5x_3 &= 10. \end{aligned}$$

The Gaussian elimination process yields:

$$\begin{array}{rrrrl} 1 & 1 & -1 & 3 & \\ 2 & -1 & 4 & 3 & \\ 3 & 4 & -5 & 10 & \\ \hline 1 & 1 & -1 & 3 & \\ 0 & -3 & 6 & -3 & (2)-2\times(1) \\ 0 & 1 & -2 & 1 & (3)-3\times(1) \\ \hline 1 & 1 & -1 & 3 & \\ 0 & 1 & -2 & 1 & (2)\div(-3) \\ 0 & 1 & -2 & 1 & \\ \hline 1 & 1 & -1 & 3 & \\ 0 & 1 & -2 & 1 & \\ 0 & 0 & 0 & 0 & (3)-(2) \\ \hline \end{array}$$

Here set $x_3 = t$ (say), so that substituting gives

$$x_2 = 1 + 2t$$

$$x_1 = 3 - (1 + 2t) + t = 2 - t.$$

There are infinitely many solutions, one for each value of $t \in \mathbb{R}$.

12.2 Product of a matrix with a column vector.

$$\begin{bmatrix} 1 & 2 & 3 \\ 4 & 5 & 6 \\ 7 & 8 & 9 \end{bmatrix}\begin{bmatrix} x_1 \\ x_2 \\ x_3 \end{bmatrix} = \begin{bmatrix} x_1 + 2x_2 + 3x_3 \\ 4x_1 + 5x_2 + 6x_3 \\ 7x_1 + 8x_2 + 9x_3 \end{bmatrix}$$

$$= \begin{bmatrix} x_1 \\ 4x_1 \\ 7x_1 \end{bmatrix} + \begin{bmatrix} 2x_2 \\ 5x_2 \\ 8x_2 \end{bmatrix} + \begin{bmatrix} 3x_3 \\ 6x_3 \\ 9x_3 \end{bmatrix}$$

$$= \begin{bmatrix} 1 \\ 4 \\ 7 \end{bmatrix} x_1 + \begin{bmatrix} 2 \\ 5 \\ 8 \end{bmatrix} x_2 + \begin{bmatrix} 3 \\ 6 \\ 9 \end{bmatrix} x_3.$$

12

Basic ideas

Part 2 is about the algebra of real n-vectors. Throughout, $\mathbb{R}^n$ will denote the set of real n-vectors, i.e. the set of all $n \times 1$ matrices with entries from the set $\mathbb{R}$ of real numbers. We shall develop ideas and techniques which are applicable for all values of n, however large. Most of our examples, however, will be restricted (for practical reasons) to the two-, three- and four-dimensional cases.

The purpose of this chapter is to re-establish some of the basic ideas about solutions of sets of equations, about matrices, and about linear dependence, which will be essential for what follows. In this process we shall also introduce two new notions which will be significant later, namely the column space of a matrix and the solution space of a set of homogeneous linear equations.

Let us review some facts about the solution of sets of simultaneous linear equations. Let A be a $p \times q$ matrix and let $\boldsymbol{h}$ be a p-vector. Then attempting to solve the equation $A\boldsymbol{x} = \boldsymbol{h}$ will have one of the following outcomes.

(i) There may be no solution, i.e. the equation is inconsistent.
(ii) There may be a unique solution.
(iii) There may be infinitely many solutions. In this case the set of solutions is generally specified by taking one or more of the unknowns as parameters and expressing the other unknowns in terms of these parameters. Example 12.1 provides illustration of this.

Now let us think of sets of simultaneous equations in a different way. See Example 12.2. The left-hand side $A\boldsymbol{x}$ (when multiplied out) is a p-vector, and its ith entry is

$$a_{i1}x_1 + a_{i2}x_2 + \cdots + a_{iq}x_q,$$

where a_{ij} denotes the entry in the ith row and jth column in the matrix A, and $x_1, \ldots, x_q$ are the entries in the q-vector $\boldsymbol{x}$.

12.3 Solutions to simultaneous equations.

(i)
$$\begin{aligned} x_1 + x_2 + x_3 &= 2 \\ 2x_1 - 2x_2 + 2x_3 &= 4 \\ x_1 + 2x_2 + 4x_3 &= 3 \end{aligned}$$

or, alternatively:

$$\begin{bmatrix}1\\2\\1\end{bmatrix}x_1 + \begin{bmatrix}1\\-2\\2\end{bmatrix}x_2 + \begin{bmatrix}1\\2\\4\end{bmatrix}x_3 = \begin{bmatrix}2\\4\\3\end{bmatrix}.$$

The GE process leads to

$$\begin{bmatrix}1 & 1 & 3 & 2\\0 & 1 & 1 & 0\\0 & 0 & 0 & 1\end{bmatrix}.$$

In this case the given equations are inconsistent. This may be regarded as a consequence of the fact that the vector $\begin{bmatrix}2\\0\\1\end{bmatrix}$ is not a linear combination of the vectors $\begin{bmatrix}1\\0\\0\end{bmatrix}$, $\begin{bmatrix}1\\1\\0\end{bmatrix}$ and $\begin{bmatrix}3\\1\\0\end{bmatrix}$.

(ii)
$$\begin{aligned} x_1 + 2x_2 &= 1 \\ 2x_1 + 3x_2 + x_3 &= 2 \\ -x_1 + x_2 + x_3 &= -6 \end{aligned}$$

or, alternatively:

$$\begin{bmatrix}1\\2\\-1\end{bmatrix}x_1 + \begin{bmatrix}2\\3\\1\end{bmatrix}x_2 + \begin{bmatrix}0\\1\\1\end{bmatrix}x_3 = \begin{bmatrix}1\\2\\-6\end{bmatrix}.$$

The GE process leads to

$$\begin{bmatrix}1 & 2 & 0 & 1\\0 & 1 & -1 & 1\\0 & 0 & 1 & -2\end{bmatrix},$$

and there is a unique solution for x_1, x_2 and x_3. So $\begin{bmatrix}1\\2\\-6\end{bmatrix}$ may be written uniquely as a linear combination of $\begin{bmatrix}1\\2\\-1\end{bmatrix}$, $\begin{bmatrix}2\\3\\1\end{bmatrix}$ and $\begin{bmatrix}0\\1\\1\end{bmatrix}$.

(iii)
$$\begin{aligned} x_1 + 4x_2 - x_3 &= 0 \\ x_1 + 2x_2 + 3x_3 &= 2 \\ -x_1 + 2x_2 - 11x_3 &= -6 \end{aligned}$$

or, alternatively:

$$\begin{bmatrix}1\\1\\-1\end{bmatrix}x_1 + \begin{bmatrix}4\\2\\2\end{bmatrix}x_2 + \begin{bmatrix}-1\\3\\-11\end{bmatrix}x_3 = \begin{bmatrix}0\\2\\-6\end{bmatrix}.$$

Thus each entry of $A\boldsymbol{x}$ is a sum of q terms, so $A\boldsymbol{x}$ itself can be written as a sum of vectors:

$$A\boldsymbol{x} = \boldsymbol{a}_1 x_1 + \boldsymbol{a}_2 x_2 + \cdots + \boldsymbol{a}_q x_q,$$

where, for each j ($1 \leqslant j \leqslant q$), $\boldsymbol{a}_j$ is the jth column of the matrix A.

One way to interpret this is to say: for different values of $\boldsymbol{x}$ (i.e. of $x_1, \ldots, x_q$), the p-vector $A\boldsymbol{x}$ will always be a linear combination of the columns of A. What does this mean in terms of the three possible outcomes listed above for the solution of the equation $A\boldsymbol{x} = \boldsymbol{h}$? See Example 12.3.

(i) No solution. This arises when $\boldsymbol{h}$ cannot be written as a linear combination of the columns of A.
(ii) Unique solution. This happens when $\boldsymbol{h}$ can be written as a linear combination of the columns of A in only one way.
(iii) Infinitely many solutions. This happens when there are infinitely many ways of writing $\boldsymbol{h}$ as a linear combination of the columns of A.

Definition
Given any $p \times q$ matrix A, the set of all p-vectors which can be written as linear combinations of the columns of A is called the *column space* of A.

Obviously a sufficient condition for the equation $A\boldsymbol{x} = \boldsymbol{h}$ to be consistent (i.e. to have at least one solution) is that $\boldsymbol{h}$ belongs to the column space of A. A moment's thought should convince the reader that it is also a necessary condition.

Let us now recall the notion of rank, defined as follows. The rank of a matrix A is the number of non-zero rows in the matrix obtained (in row-echelon form) after applying the Gaussian elimination process to A. The following are then equivalent.

(i) $A\boldsymbol{x} = \boldsymbol{h}$ is consistent.
(ii) The rank of $[A|\boldsymbol{h}]$ (the augmented matrix) is equal to the rank of A.
(iii) $\boldsymbol{h}$ belongs to the column space of A.

The second of these is dealt with in Chapter 8. See Example 12.4.

The GE process leads to

$$\begin{bmatrix} 1 & 4 & -1 & 0 \\ 0 & 1 & -2 & 1 \\ 0 & 0 & 0 & 0 \end{bmatrix},$$

and there are infinitely many solutions. So $\begin{bmatrix} 0 \\ 2 \\ -6 \end{bmatrix}$ may be written as a linear combination of $\begin{bmatrix} 1 \\ 1 \\ -1 \end{bmatrix}$, $\begin{bmatrix} 4 \\ 2 \\ 2 \end{bmatrix}$ and $\begin{bmatrix} -1 \\ 3 \\ -11 \end{bmatrix}$ in infinitely many ways.

12.4 The rank of the augmented matrix.

In Example 12.3(i),

$$A = \begin{bmatrix} 1 & 1 & 3 \\ 2 & -2 & 2 \\ 1 & 2 & 4 \end{bmatrix},$$

and $$[A|\boldsymbol{h}] = \begin{bmatrix} 1 & 1 & 3 & 2 \\ 2 & -2 & 2 & 4 \\ 1 & 2 & 4 & 3 \end{bmatrix}.$$

The GE process applied to A yields

$$\begin{bmatrix} 1 & 1 & 3 \\ 0 & 1 & 1 \\ 0 & 0 & 0 \end{bmatrix},$$

so the rank of A is equal to 2.

The GE process applied to $[A|\boldsymbol{h}]$ yields

$$\begin{bmatrix} 1 & 1 & 3 & 2 \\ 0 & 1 & 1 & 0 \\ 0 & 0 & 0 & 1 \end{bmatrix},$$

so the rank of $[A|\boldsymbol{h}]$ is equal to 3.

But notice that both ranks may be deduced from the result of the GE process applied to $[A|\boldsymbol{h}]$.

In Example 12.3(ii), the rank of A is 3 and the rank of $[A|\boldsymbol{h}]$ is 3 also.

In Example 12.3(iii), the rank of A is 2 and the rank of $[A|\boldsymbol{h}]$ is 2 also.

12.5 Tests for linear dependence or independence.

(i) $$\left(\begin{bmatrix} 1 \\ 2 \\ 5 \end{bmatrix}, \begin{bmatrix} -2 \\ 4 \\ -2 \end{bmatrix}, \begin{bmatrix} 3 \\ -2 \\ 7 \end{bmatrix}\right).$$

Apply the GE process to the matrix $\begin{bmatrix} 1 & -2 & 3 & 0 \\ 2 & 4 & -2 & 0 \\ 5 & -2 & 7 & 0 \end{bmatrix}$. This yields $\begin{bmatrix} 1 & -2 & 3 & 0 \\ 0 & 1 & -1 & 0 \\ 0 & 0 & 0 & 0 \end{bmatrix}$, so there are solutions other than $x_1 = x_2 = x_3 = 0$ to the

Next consider homogeneous linear equations, i.e. equations of the form

$$A\boldsymbol{x} = \boldsymbol{0},$$

where A is a $p \times q$ matrix, $\boldsymbol{x}$ is a q-vector of unknowns, and $\boldsymbol{0}$ is the zero p-vector. Such equations are obviously consistent, because $\boldsymbol{x} = \boldsymbol{0}$ is certainly a solution. Does $\boldsymbol{0}$ belong to the column space of A? Yes, in a trivial way. We may write

$$\boldsymbol{0} = \boldsymbol{a}_1 0 + \boldsymbol{a}_2 0 + \cdots + \boldsymbol{a}_q 0,$$

a linear combination of the columns of A. In what circumstances is $\boldsymbol{x} = \boldsymbol{0}$ the only solution? Only if we cannot write

$$\boldsymbol{0} = \boldsymbol{a}_1 x_1 + \boldsymbol{a}_2 x_2 + \cdots + \boldsymbol{a}_q x_q,$$

with any of the numbers $x_1, \ldots, x_q$ different from 0. What this means is that the columns of A (i.e. $\boldsymbol{a}_1, \ldots, \boldsymbol{a}_q$) form a linearly independent list of p-vectors. The definition of linear independence was given in Chapter 6, but we include it here also.

Definition

A list $(\boldsymbol{v}_1, \ldots, \boldsymbol{v}_q)$ of p-vectors (for any p) is *linearly dependent* if there is some non-trivial linear combination

$$x_1 \boldsymbol{v}_1 + x_2 \boldsymbol{v}_2 + \cdots + x_q \boldsymbol{v}_q$$

which is equal to the zero p-vector. A list of p-vectors is *linearly independent* if it is not linearly dependent.

We shall customarily use the abbreviations LD and LI for these terms.

A routine procedure for testing whether a given list of vectors is LD or LI is given in Chapter 6. This procedure is illustrated in Examples 12.5.

We have established above that in the case when the columns of A form a LI list, the equation $A\boldsymbol{x} = \boldsymbol{0}$ has no solution other than $\boldsymbol{x} = \boldsymbol{0}$. If the columns of A form a LD list, then $A\boldsymbol{x} = \boldsymbol{0}$ does have another solution, and so must have infinitely many solutions.

In the special case where A is a square matrix, there is an important theorem connecting several different aspects of linear algebra. This was discussed in Chapters 6 and 7, and there called the Equivalence Theorem. We shall refer to this again later, so it is reproduced below.

Theorem

Let A be a $p \times p$ matrix. The following are equivalent.

(i) A is invertible.
(ii) The rank of A is equal to p.
(iii) The columns of A form a LI list.

equation

$$\begin{bmatrix}1\\2\\5\end{bmatrix}x_1+\begin{bmatrix}-2\\4\\-2\end{bmatrix}x_2+\begin{bmatrix}3\\-2\\7\end{bmatrix}x_3=\begin{bmatrix}0\\0\\0\end{bmatrix}.$$

Thus the given list of vectors is LD.

(ii) $\left(\begin{bmatrix}1\\-2\\3\end{bmatrix},\begin{bmatrix}1\\2\\1\end{bmatrix},\begin{bmatrix}-2\\0\\2\end{bmatrix}\right).$

The GE process gives

$$\begin{bmatrix}1&1&-2&0\\-2&2&0&0\\3&1&2&0\end{bmatrix}\rightarrow\begin{bmatrix}1&1&-2&0\\0&1&-1&0\\0&0&1&0\end{bmatrix},$$

so $x_1=0$, $x_2=0$, $x_3=0$ is the only solution to the equation

$$\begin{bmatrix}1\\-2\\3\end{bmatrix}x_1+\begin{bmatrix}1\\2\\1\end{bmatrix}x_2+\begin{bmatrix}-2\\0\\2\end{bmatrix}x_3=\begin{bmatrix}0\\0\\0\end{bmatrix}.$$

Thus the given list of vectors is LI.

12.6 Find all solutions to the equation $A\boldsymbol{x}=\mathbf{0}$, where

$$A=\begin{bmatrix}1&0&2&-1\\3&1&-2&4\\-1&2&-2&3\\2&1&0&2\end{bmatrix}\quad\text{and}\quad\boldsymbol{x}=\begin{bmatrix}x_1\\x_2\\x_3\\x_4\end{bmatrix}.$$

The GE process gives

$$\begin{bmatrix}1&0&2&-1&0\\3&1&-2&4&0\\-1&2&-2&3&0\\2&1&0&2&0\end{bmatrix}\rightarrow\begin{bmatrix}1&0&2&1&0\\0&1&-8&7&0\\0&0&1&-\frac{3}{4}&0\\0&0&0&0&0\end{bmatrix}.$$

Hence there are infinitely many solutions. These may be described by setting $x_4=t$ (say), whence $x_3=\frac{3}{4}t$, $x_2=-t$ and $x_1=-\frac{1}{2}t$ $(t\in\mathbb{R})$.

(iv) The equation $A\mathbf{x}=\mathbf{0}$ has no solution other than $\mathbf{x}=\mathbf{0}$.
(v) The determinant of A is not equal to zero.

The justification for this theorem was that in each of these circumstances, the Gaussian elimination procedure applied to the matrix A will yield an (upper triangular) matrix with p non-zero rows, with 1s down the main diagonal. At this stage the reader should be familiar enough with the ideas involved here so that this theorem is quite clear. If not, the later chapters will cause difficulties.

Summary

The possible forms of solution to a set of simultaneous linear equations are discussed and related to the Gaussian elimination process of transforming a matrix to row-echelon form. The column space of a matrix is introduced and the ideas of rank and linear dependence are recalled. Finally the Equivalence Theorem (from Chapter 7) is re-stated. This relates the various ideas of the chapter.

Exercises

1. Solve the following sets of simultaneous equations.

(i)
$$\begin{aligned} x-2y+2z&=2\\ 3x+y+z&=8\\ 2x-2y-z&=1. \end{aligned}$$

(ii)
$$\begin{aligned} 2x+y+4z&=0\\ x+2y-z&=6\\ 2x+2y+2z&=6. \end{aligned}$$

(iii)
$$\begin{aligned} x+2y+3z&=2\\ -3x+3z&=0\\ -3x+y+5z&=1. \end{aligned}$$

(iv)
$$\begin{aligned} x-3z&=-4\\ y+z+2w&=1\\ 2x+y+2w&=-2\\ -x+y+z-w&=5. \end{aligned}$$

2. Re-write the sets of equations in Exercise 1 above as vector equations (by regarding the left-hand sides as linear combinations of column vectors).

3. In each case below, find the rank of the given matrix and find whether the given vector belongs to the column space of the matrix.

(i) $\begin{bmatrix} 1 & 0 & 1 \\ 0 & 1 & 1 \\ 1 & 1 & 0 \end{bmatrix}, \quad \begin{bmatrix} 1 \\ 2 \\ 3 \end{bmatrix}.$

(ii) $\begin{bmatrix} 0 & -1 & 3 \\ 4 & 1 & 1 \\ 2 & 3 & -1 \end{bmatrix}, \quad \begin{bmatrix} 1 \\ 1 \\ 1 \end{bmatrix}.$

(iii) $\begin{bmatrix} 3 & 2 & -2 \\ 2 & -2 & 12 \\ -3 & 4 & -22 \end{bmatrix}, \quad \begin{bmatrix} 1 \\ 0 \\ 1 \end{bmatrix}.$

(iv) $\begin{bmatrix} 1 & 0 & 1 & 1 \\ -1 & 1 & 0 & -1 \\ 1 & 0 & 1 & -1 \\ -1 & 1 & 0 & 1 \end{bmatrix}, \quad \begin{bmatrix} -1 \\ 1 \\ -3 \\ 1 \end{bmatrix}.$

4. In each case below find whether the columns of the given matrix form a linearly dependent or independent list. Where possible, find the inverse of the given matrix.

(i) $\begin{bmatrix} 1 & 1 & 3 \\ 0 & 2 & 1 \\ -1 & 1 & -2 \end{bmatrix},$ (ii) $\begin{bmatrix} 2 & 1 & 4 \\ 2 & 0 & 2 \\ -3 & 1 & -1 \end{bmatrix},$

(iii) $\begin{bmatrix} 1 & -1 & 2 \\ -1 & 2 & -2 \\ 1 & 2 & 3 \end{bmatrix},$ (iv) $\begin{bmatrix} 1 & 1 & 0 & 1 \\ -1 & 0 & 2 & -2 \\ 1 & 0 & -1 & 3 \\ -1 & -3 & -2 & 4 \end{bmatrix},$

(v) $\begin{bmatrix} 2 & 0 & 1 & 5 \\ 1 & 1 & 2 & 0 \\ 0 & 1 & 3 & -4 \\ 1 & 3 & 0 & 0 \end{bmatrix},$

(vi) $\begin{bmatrix} 1 & 0 & 3 & 2 \\ -1 & 1 & 1 & 2 \\ 2 & 3 & 1 & 0 \end{bmatrix}.$

Examples

13.1 Column spaces of matrices.

(i) Let $A=\begin{bmatrix}3 & 2\\ 1 & -1\end{bmatrix}$. The column space of A is the set of vectors $\begin{bmatrix}3\\1\end{bmatrix}x_1+\begin{bmatrix}2\\-1\end{bmatrix}x_2$ for $x_1, x_2 \in \mathbb{R}$. For example, taking some values for x_1 and x_2:

$$x_1=1,\quad x_2=0:\qquad \begin{bmatrix}3\\1\end{bmatrix}\quad \text{itself.}$$

$$x_1=0,\quad x_2=1:\qquad \begin{bmatrix}2\\-1\end{bmatrix}\quad \text{itself.}$$

$$x_1=1,\quad x_2=\tfrac{1}{2}:\qquad \begin{bmatrix}5\\0\end{bmatrix}.$$

$$x_1=\tfrac{1}{3},\quad x_2=\tfrac{1}{2}:\qquad \begin{bmatrix}2\\\tfrac{1}{6}\end{bmatrix}.$$

(ii) Let $A=\begin{bmatrix}2 & 1\\ 1 & 2\\ -1 & 3\end{bmatrix}$. The column space of A is the set of all vectors $\begin{bmatrix}2\\1\\-1\end{bmatrix}x_1+\begin{bmatrix}1\\2\\3\end{bmatrix}x_2$ for $x_1, x_2 \in \mathbb{R}$.

(iii) Let $A=\begin{bmatrix}1 & 1 & 1 & 0\\ 1 & 1 & 0 & 1\\ 1 & 0 & 1 & 1\\ 0 & 1 & 1 & 1\end{bmatrix}$. The column space of A is the set of all vectors

$$\begin{bmatrix}1\\1\\1\\0\end{bmatrix}x_1+\begin{bmatrix}1\\1\\0\\1\end{bmatrix}x_2+\begin{bmatrix}1\\0\\1\\1\end{bmatrix}x_3+\begin{bmatrix}0\\1\\1\\1\end{bmatrix}x_4,$$

for $x_1, x_2, x_3, x_4 \in \mathbb{R}$.

13

Subspaces of $\mathbb{R}^n$

We shall latch onto, and generalise from, two ideas which arose in Chapter 12. These are the column space of a matrix and the set of all solutions to an equation $A\mathbf{x}=\mathbf{0}$. They represent different aspects of the idea of a subspace of $\mathbb{R}^n$.

First, the column space of a matrix is a special case of a space spanned by a list of vectors. This is a very important notion, and here is the definition.

Definition

Let X be a finite list of n-vectors. The *space spanned by* X is the set of all vectors which can be written as linear combinations of the vectors in X.

If we take the vectors in the list X as the columns of a matrix A, then the space spanned by X is the same thing as the column space of A. Some simple cases are given in Examples 13.1. An example which is of great significance is the space spanned by the list

$$\left(\begin{bmatrix}1\\0\\0\end{bmatrix},\begin{bmatrix}0\\1\\0\end{bmatrix},\begin{bmatrix}0\\0\\1\end{bmatrix}\right)$$

of vectors in $\mathbb{R}^3$. These vectors are called the standard basis vectors in $\mathbb{R}^3$, and it is easy to see that every 3-vector may be written as a linear combination of them:

$$\begin{bmatrix}a\\b\\c\end{bmatrix}=a\begin{bmatrix}1\\0\\0\end{bmatrix}+b\begin{bmatrix}0\\1\\0\end{bmatrix}+c\begin{bmatrix}0\\0\\1\end{bmatrix}.$$

The space spanned by this list of 3-vectors is $\mathbb{R}^3$ itself. For each n there is a list of standard basis vectors in $\mathbb{R}^n$, namely the list consisting of all n-vectors having a 1 in one position and 0s in the other positions. The space spanned by this list of vectors in $\mathbb{R}^n$ is $\mathbb{R}^n$ itself.

13.2 The space spanned by a list of vectors.

Does $\begin{bmatrix}0\\0\\1\end{bmatrix}$ belong to the space spanned by the list $\left(\begin{bmatrix}1\\2\\3\end{bmatrix}, \begin{bmatrix}-1\\3\\4\end{bmatrix}, \begin{bmatrix}3\\1\\2\end{bmatrix}\right)$?

We seek solutions x_1, x_2, x_3 to the equation

$$\begin{bmatrix}1\\2\\3\end{bmatrix}x_1 + \begin{bmatrix}-1\\3\\4\end{bmatrix}x_2 + \begin{bmatrix}3\\1\\2\end{bmatrix}x_3 = \begin{bmatrix}0\\0\\1\end{bmatrix}.$$

The GE process gives:

$$\begin{bmatrix}1 & -1 & 3 & 0\\2 & 3 & 1 & 0\\3 & 4 & 2 & 1\end{bmatrix} \to \begin{bmatrix}1 & -1 & 3 & 0\\0 & 1 & 1 & 0\\0 & 0 & 0 & 1\end{bmatrix}.$$

Hence there are no solutions, and the given vector is not in the space spanned by the given list.

13.3 Lists which span $\mathbb{R}^3$.

(i) Does $\left(\begin{bmatrix}1\\2\\0\end{bmatrix}, \begin{bmatrix}-1\\1\\1\end{bmatrix}, \begin{bmatrix}0\\3\\-1\end{bmatrix}\right)$ span $\mathbb{R}^3$?

Let $\boldsymbol{h} = \begin{bmatrix}a\\b\\c\end{bmatrix}$. We seek solutions to the equation

$$\begin{bmatrix}1\\2\\0\end{bmatrix}x_1 + \begin{bmatrix}-1\\1\\1\end{bmatrix}x_2 + \begin{bmatrix}0\\3\\-1\end{bmatrix}x_3 = \begin{bmatrix}a\\b\\c\end{bmatrix}.$$

The GE process gives:

$$\begin{bmatrix}1 & -1 & 0 & a\\2 & 1 & 3 & b\\0 & 1 & -1 & c\end{bmatrix} \to \begin{bmatrix}1 & -1 & 0 & a\\0 & 1 & 1 & \frac{1}{3}(b-2a)\\0 & 0 & 1 & \frac{1}{6}(b-2a)-\frac{1}{2}c\end{bmatrix}.$$

Hence there do exist solutions to the equation, whatever the values of a, b and c. Hence the given list spans $\mathbb{R}^3$.

(ii) Does $\left(\begin{bmatrix}1\\1\\-2\end{bmatrix}, \begin{bmatrix}3\\0\\1\end{bmatrix}, \begin{bmatrix}-2\\4\\1\end{bmatrix}\right)$ span $\mathbb{R}^3$?

The following skeleton of the argument given in (i) above is sufficient.

The GE process gives: $\begin{bmatrix}1 & 3 & -2\\1 & 0 & 4\\-2 & 1 & 1\end{bmatrix} \to \begin{bmatrix}1 & 3 & -2\\0 & 1 & -2\\0 & 0 & 1\end{bmatrix}.$

Even without including the fourth column, it is apparent from this that there would be solutions in this case, irrespective of the values in the fourth column. (If the last row had been all zeros then we could not say this.) Hence the given list spans $\mathbb{R}^3$ in this case also.

How can we tell whether a given vector is in the space spanned by a given list of vectors? Answer: solve equations. See Example 13.2. In fact, we need not go as far as solving equations: the knowledge that solutions exist (or do not exist) is sufficient.

We can even tell by means of an elementary process whether the space spanned by a given list of n-vectors is $\mathbb{R}^n$ itself. This is done for $n=3$ in Example 13.3. The vectors in the given list are taken as the columns of a matrix A. Then we try to solve the equation $A\boldsymbol{x}=\boldsymbol{h}$, where $\boldsymbol{h}$ is an unspecified 3-vector. Our standard process will tell us whether solutions exist. Of course this will depend on what value $\boldsymbol{h}$ has. If there are some vectors $\boldsymbol{h}$ for which no solution $\boldsymbol{x}$ can be found, then the column space of A is not the whole of $\mathbb{R}^3$, so the space spanned by the given list of vectors is not the whole of $\mathbb{R}^3$. If there are solutions $\boldsymbol{x}$ irrespective of the choice of $\boldsymbol{h}$, then the space spanned by the given list of vectors is the whole of $\mathbb{R}^3$. Think about this: the space spanned by the given list of 3-vectors is the whole of $\mathbb{R}^3$ if and only if the rank of A is equal to 3.

Next, let us return to the idea of the set of solutions to an equation $A\boldsymbol{x}=\boldsymbol{0}$. There may be one solution (namely $\boldsymbol{x}=\boldsymbol{0}$) or infinitely many solutions. For what follows it is best to bear in mind the second of these possibilities, although what we do also makes sense for the first.

If $\boldsymbol{x}=\boldsymbol{u}$ is a solution then so is $\boldsymbol{x}=k\boldsymbol{u}$, for any real number k. This is not difficult:

$$\text{if} \quad A\boldsymbol{u}=\boldsymbol{0} \quad \text{then} \quad A(k\boldsymbol{u})=k(A\boldsymbol{u})=k\boldsymbol{0}=\boldsymbol{0}.$$

Further, if $\boldsymbol{x}=\boldsymbol{u}$ and $\boldsymbol{x}=\boldsymbol{v}$ are solutions, then so is $\boldsymbol{x}=\boldsymbol{u}+\boldsymbol{v}$. To see this, suppose that $A\boldsymbol{u}=\boldsymbol{0}$ and $A\boldsymbol{v}=\boldsymbol{0}$. Then

$$A(\boldsymbol{u}+\boldsymbol{v})=A\boldsymbol{u}+A\boldsymbol{v}=\boldsymbol{0}+\boldsymbol{0}=\boldsymbol{0}.$$

The set of solutions to the equation $A\boldsymbol{x}=\boldsymbol{0}$ contains $\boldsymbol{0}$ and is closed under addition and under multiplication by a scalar.

13.4 Let S be the space spanned by a list of vectors $(\boldsymbol{v}_1, \ldots, \boldsymbol{v}_r)$ in $\mathbb{R}^n$. Show that

(i) $\mathbf{0} \in S$,

(ii) for each $\boldsymbol{u}, \boldsymbol{v} \in S$, $\boldsymbol{u} + \boldsymbol{v} \in S$ also, and

(iii) for each $\boldsymbol{u} \in S$ and $k \in \mathbb{R}$, $k\boldsymbol{u} \in S$.

For (i): $\mathbf{0} \in S$ because $\mathbf{0} = 0\boldsymbol{v}_1 + \cdots + 0\boldsymbol{v}_r$.

For (ii): Let $\boldsymbol{u} = a_1\boldsymbol{v}_1 + \cdots + a_r\boldsymbol{v}_r$, and
$\boldsymbol{v} = b_1\boldsymbol{v}_1 + \cdots + b_r\boldsymbol{v}_r$.

Then $\boldsymbol{u} + \boldsymbol{v} = (a_1 + b_1)\boldsymbol{v}_1 + \cdots + (a_r + b_r)\boldsymbol{v}_r$,
which belongs to S.

For (iii): Let $\boldsymbol{u} = c_1\boldsymbol{v}_1 + \cdots + c_r\boldsymbol{v}_r$, and let $k \in \mathbb{R}$.

Then $k\boldsymbol{u} = kc_1\boldsymbol{v}_1 + \cdots + kc_r\boldsymbol{v}_r$,
which belongs to S.

13.5 To find whether a given set is a subspace of $\mathbb{R}^n$, for an appropriate value of n.

(i) Let $S = \left\{ \begin{bmatrix} x \\ y \\ z \end{bmatrix} \in \mathbb{R}^3 : x + y = z \right\}$.

First, $\mathbf{0} \in S$ because $0 + 0 = 0$.

Let $\boldsymbol{u} = \begin{bmatrix} x_1 \\ y_1 \\ z_1 \end{bmatrix}$ and $\boldsymbol{v} = \begin{bmatrix} x_2 \\ y_2 \\ z_2 \end{bmatrix}$ belong to S,

so that $x_1 + y_1 = z_1$ and $x_2 + y_2 = z_2$. Then $\boldsymbol{u} + \boldsymbol{v} = \begin{bmatrix} x_1 + x_2 \\ y_1 + y_2 \\ z_1 + z_2 \end{bmatrix}$. Does $\boldsymbol{u} + \boldsymbol{v}$ belong to S?

Test the condition for membership of S:

$$\begin{aligned} (x_1 + x_2) + (y_1 + y_2) &= (x_1 + y_1) + (x_2 + y_2) \\ &= z_1 + z_2, \end{aligned}$$

because $x_1 + y_1 = z_1$ and $x_2 + y_2 = z_2$. So $\boldsymbol{u} + \boldsymbol{v}$ does belong to S.

Last, let $\boldsymbol{u} = \begin{bmatrix} x \\ y \\ z \end{bmatrix}$ belong to S, so that $x + y = z$, and let $k \in \mathbb{R}$. Then $k\boldsymbol{u} = \begin{bmatrix} kx \\ ky \\ kz \end{bmatrix}$.

Does $k\boldsymbol{u}$ belong to S? Test the condition for membership of S:

$$kx + ky = k(x + y) = kz,$$

because $x + y = z$. So $k\boldsymbol{u}$ does belong to S. Hence the set S is a subspace of $\mathbb{R}^3$.

(ii) Let $S = \left\{ \begin{bmatrix} x \\ y \\ z \end{bmatrix} \in \mathbb{R}^3 : x^2 + y^2 = z^2 \right\}$.

Certainly $\mathbf{0} \in S$.

Let $\boldsymbol{u} = \begin{bmatrix} x_1 \\ y_1 \\ z_1 \end{bmatrix}$ and $\boldsymbol{v} = \begin{bmatrix} x_2 \\ y_2 \\ z_2 \end{bmatrix}$ belong to S,

Now we make the connection. A space spanned by a list of vectors has these same properties: contains $\mathbf{0}$ and is closed under addition and multiplication by a scalar. See Examples 13.4. These properties characterise an important notion.

Definition

A *subspace* of $\mathbb{R}^n$ is a subset S of $\mathbb{R}^n$ with the following properties:

(i) $\mathbf{0} \in S$.
(ii) If $\boldsymbol{u} \in S$ then $k\boldsymbol{u} \in S$, for all real numbers k.
(iii) If $\boldsymbol{u} \in S$ and $\boldsymbol{v} \in S$ then $\boldsymbol{u} + \boldsymbol{v} \in S$.

We can now say that the following sorts of set are subspaces of $\mathbb{R}^n$:

(a) any space spanned by a list of vectors in $\mathbb{R}^n$,
(b) the set of all solutions to an equation $A\boldsymbol{x} = \mathbf{0}$, where A is a $p \times n$ matrix, for any number p.

Perhaps it should be made clear that $\mathbb{R}^n$ itself is classed as a subspace of $\mathbb{R}^n$. It satisfies the requirements of the definition, and it falls into the first category above. Also, in a logically trivial way, the set $\{\mathbf{0}\}$ satisfies the definition and is a subspace of $\mathbb{R}^n$.

Subspaces of $\mathbb{R}^n$ can be specified in other ways. In Examples 13.5 some subsets of $\mathbb{R}^n$ (for various values of n) are given. Some of these are shown to be subspaces of $\mathbb{R}^n$, and others are shown not to be.

In the geometry of three dimensions, the notion of subspace has an interpretation which it may be useful to bear in mind. Recall that a 3-vector

$$\begin{bmatrix} x \\ y \\ z \end{bmatrix}$$

can be thought of as the position vector associated with a point P (relative to some given coordinate axes). P has coordinates (x, y, z). Addition of vectors and multiplication of a vector by a scalar have geometrical interpretations which should be familiar (see Chapter 9). The three requirements above for a subspace translate into geometrical requirements for a locus in three dimensions:

(i) The locus contains the origin O.
(ii) If P is on the locus (and P is not O), then the locus contains all points on the line OP produced indefinitely in both directions.
(iii) If P and Q (neither of which is O) lie on the locus, then R lies on the locus, where R is the point such that $OPRQ$ is a parallelogram.

so that $x_1^2+y_1^2=z_1^2$ and $x_2^2+y_2^2=z_2^2$. Does $\boldsymbol{u}+\boldsymbol{v}$ belong to S?

$$\begin{aligned}(x_1+x_2)^2+(y_1+y_2)^2 &= x_1^2+2x_1x_2+x_2^2+y_1^2+2y_1y_2+y_2^2\\ &= x_1^2+y_1^2+x_2^2+y_2^2+2x_1x_2+2y_1y_2\\ &= z_1^2+z_2^2+2x_1x_2+2y_1y_2,\end{aligned}$$

which appears to be different from $z_1^2+z_2^2+2z_1z_2$, which is what we seek. To confirm that it is different, we must find particular values for x_1, x_2, y_1, y_2, z_1 and z_2 such that

$$x_1^2+y_1^2=z_1^2 \quad \text{and} \quad x_2^2+y_2^2=z_2^2,$$

but $\quad (x_1+x_2)^2+(y_1+y_2)^2 \neq (z_1+z_2)^2.$

Take $x=1$, $y=0$, $z=1$, $x=0$, $y=1$, $z=1$. So $\begin{bmatrix}1\\0\\1\end{bmatrix}\in S$, $\begin{bmatrix}0\\1\\1\end{bmatrix}\in S$ but the sum $\begin{bmatrix}1\\1\\2\end{bmatrix}\notin S$.

Hence S is not a subspace of $\mathbb{R}^3$. Of course finding this counterexample is sufficient, irrespective of the route by which it is arrived at. The above working shows what happens if we just proceed normally, but it is strictly not necessary.

(iii) Let S be the set

$$\left\{\begin{bmatrix}x\\y\\z\end{bmatrix}\in\mathbb{R}^3 : x,\ y \text{ and } z \text{ are even integers}\right\}.$$

Certainly $\mathbf{0}\in S$, since 0 is an even integer.

Let $\quad \boldsymbol{u}=\begin{bmatrix}x_1\\y_1\\z_1\end{bmatrix}$ and $\boldsymbol{v}=\begin{bmatrix}x_2\\y_2\\z_2\end{bmatrix}$ belong to S.

Then $\boldsymbol{u}+\boldsymbol{v}$ must belong to S because x_1+x_2, y_1+y_2 and z_1+z_2 must all be even integers.

Let $\quad \boldsymbol{u}=\begin{bmatrix}x\\y\\z\end{bmatrix}$ belong to S, and let $k\in\mathbb{R}$.

Then $k\boldsymbol{u}$ need not belong to S. In cases where k is an irrational number, for example, kx, ky and kz will not be even integers.
Hence this set S is *not* a subspace of $\mathbb{R}^3$.

(iv) Let S be the set

$$\left\{\begin{bmatrix}x\\y\end{bmatrix}\in\mathbb{R}^2 : x=0 \text{ or } y=0 \text{ or both}\right\}.$$

Certainly $\mathbf{0}\in S$.

Without following the routine procedure, we can see that if $\boldsymbol{u}=\begin{bmatrix}1\\0\end{bmatrix}$ and $\boldsymbol{v}=\begin{bmatrix}0\\1\end{bmatrix}$, then $\boldsymbol{u}\in S$ and $\boldsymbol{v}\in S$ but $\boldsymbol{u}+\boldsymbol{v}\notin S$. So this set S is *not* a subspace of $\mathbb{R}^2$.

(v) Let S be the set

$$\left\{\begin{bmatrix}x\\y\\z\\w\end{bmatrix}\in\mathbb{R}^4 : x=z \text{ and } y=w\right\}.$$

A locus satisfying these requirements must be one of the following.

(a) The set consisting of the origin only. Here (ii) and (iii) are trivially satisfied, since such points P and Q do not exist.

(b) A straight line passing through the origin, extending indefinitely in both directions. Here (iii) is trivially satisfied.

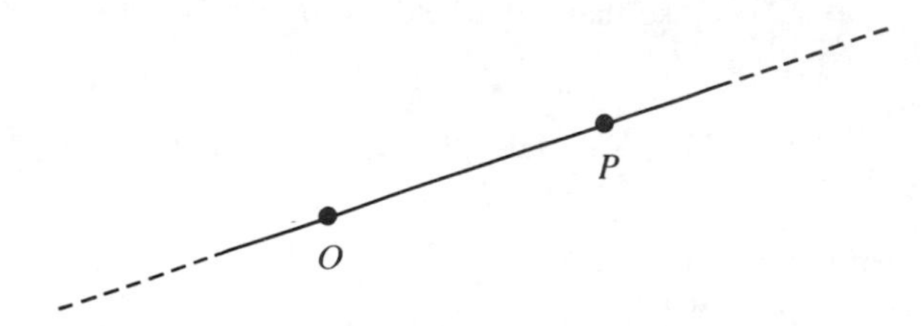

(c) A plane containing the origin.

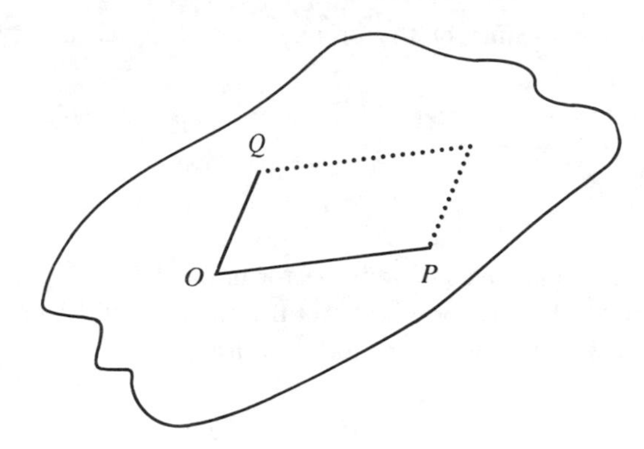

(d) The whole of three-dimensional space.

Certainly $\mathbf{0} \in S$.

Let $\boldsymbol{u} = \begin{bmatrix} x_1 \\ y_1 \\ z_1 \\ w_1 \end{bmatrix}$ and $\boldsymbol{v} = \begin{bmatrix} x_2 \\ y_2 \\ z_2 \\ w_2 \end{bmatrix}$ belong to S.

Then $x_1 = z_1$ and $y_1 = w_1$, and $x_2 = z_2$ and $y_2 = w_2$. Consequently $x_1 + x_2 = z_1 + z_2$, and $y_1 + y_2 = w_1 + w_2$, so that $\boldsymbol{u} + \boldsymbol{v}$ belongs to S.

Last, let $\boldsymbol{u} = \begin{bmatrix} x \\ y \\ z \\ w \end{bmatrix}$ belong to S, and let $k \in \mathbb{R}$. Then $x = z$ and $y = w$, so $kx = kz$ and $ky = kw$. Consequently $k\boldsymbol{u}$ belongs to S.

Hence this set is a subspace of $\mathbb{R}^4$.

13.6 Examples of loci in three dimensions.

(i) $\{\mathbf{0}\}$. The locus consisting of the origin only. This has already been noted as a subspace.

(ii) A straight line through the origin. This may be represented by equations

$$\left.\begin{aligned} x &= lt \\ y &= mt \\ z &= nt \end{aligned}\right\} \quad (t \in \mathbb{R}),$$

where l, m and n are the components of some vector in the direction of the line. Varying the parameter t yields the coordinates of the points on the line.

The vector $\mathbf{0}$ corresponds to a point on the locus (with $t = 0$).

Let $\boldsymbol{u} = \begin{bmatrix} lt_1 \\ mt_1 \\ nt_1 \end{bmatrix}$ and $\boldsymbol{v} = \begin{bmatrix} lt_2 \\ mt_2 \\ nt_2 \end{bmatrix}$ correspond to points on the line.

Then $\boldsymbol{u} + \boldsymbol{v} = \begin{bmatrix} l(t_1 + t_2) \\ m(t_1 + t_2) \\ n(t_1 + t_2) \end{bmatrix}$, which certainly corresponds to a point on the line.

Last, let $\boldsymbol{u} = \begin{bmatrix} lt \\ mt \\ nt \end{bmatrix}$ correspond to a point on the line, and let $k \in \mathbb{R}$. Then $k\boldsymbol{u} = \begin{bmatrix} lkt \\ mkt \\ nkt \end{bmatrix}$, which also corresponds to a point on the line.

Hence the position vectors of all points on the straight line constitute a subspace of $\mathbb{R}^3$.

(iii) A plane through the origin. Such a plane can be shown to have an equation of the form $ax + by + cz = 0$.

Let $S = \left\{ \begin{bmatrix} x \\ y \\ z \end{bmatrix} \in \mathbb{R}^3 : ax + by + cz = 0 \right\}$.

This may be shown to be a subspace of $\mathbb{R}^3$ by our standard methods from Example 13.5.

Examples 13.6 show that all such loci are in fact subspaces of $\mathbb{R}^3$. It is important to remember that lines and planes which do not contain the origin are not subspaces of $\mathbb{R}^3$, however.

13.7 Subspaces of $\mathbb{R}^3$ spanned by LI lists of vectors.

(i) The space spanned by the list $\left(\begin{bmatrix}1\\2\\3\end{bmatrix}\right)$ is the set of all multiples of this vector. It corresponds with the straight line whose equations are $x=t$, $y=2t$, $z=3t$ $(t\in\mathbb{R})$.

(ii) The space spanned by the list $\left(\begin{bmatrix}1\\2\\3\end{bmatrix}, \begin{bmatrix}2\\0\\1\end{bmatrix}\right)$ is the set of all vectors

$$\begin{bmatrix}1\\2\\3\end{bmatrix}x_1 + \begin{bmatrix}2\\0\\1\end{bmatrix}x_2 \qquad (x_1, x_2 \in \mathbb{R}).$$

Let P_1 and P_2 have position vectors $\begin{bmatrix}1\\2\\3\end{bmatrix}$ and $\begin{bmatrix}2\\0\\1\end{bmatrix}$ respectively. Because of the parallelogram rule, any sum of multiples of these two vectors must be the position vector of some point in the plane of O, P_1 and P_2. Conversely, the position vector of any point P in the plane of O, P_1 and P_2 is a linear combination of $\overrightarrow{OP_1}$ and $\overrightarrow{OP_2}$. To see this, draw straight lines through P parallel to $\overrightarrow{OP_1}$ and $\overrightarrow{OP_2}$, thus constructing a parallelogram.

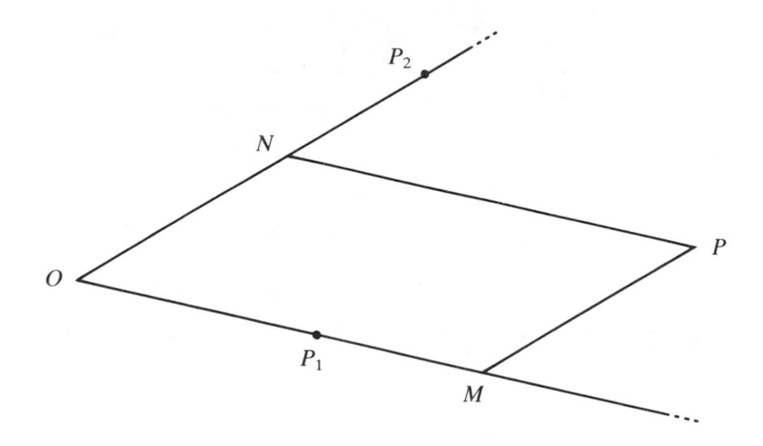

Now $\overrightarrow{OP}=\overrightarrow{OM}+\overrightarrow{ON}$, $\overrightarrow{OM}$ is a multiple of $\overrightarrow{OP_1}$, and $\overrightarrow{ON}$ is a multiple of $\overrightarrow{OP_2}$. So $\overrightarrow{OP}=\begin{bmatrix}1\\2\\3\end{bmatrix}x_1+\begin{bmatrix}2\\0\\1\end{bmatrix}x_2$ for some $x_1, x_2 \in \mathbb{R}$. We have therefore shown that the space spanned by the given list of two vectors corresponds to a plane through the origin.

Now we can see that there are geometrical interpretations of the idea of the space spanned by a list of vectors in $\mathbb{R}^3$. The space spanned by a list containing a single (non-zero) vector in $\mathbb{R}^3$ corresponds to a straight line through the origin. The space spanned by a list containing two non-zero vectors (neither a multiple of the other) in $\mathbb{R}^3$ corresponds to a plane through the origin. The space spanned by a LI list of three vectors in $\mathbb{R}^3$ corresponds to the whole of three-dimensional space. See Examples 13.7.

The three requirements of the definition of a subspace are somewhat unwieldy in practice. The next rule provides a means of easing this burden somewhat.

(iii) The space spanned by a LI list of three vectors is the whole of $\mathbb{R}^3$. Form a matrix A with the three vectors as columns. Then A is invertible, and an equation $A\boldsymbol{x}=\boldsymbol{h}$ has a (unique) solution, irrespective of the choice of vector $\boldsymbol{h}$. Hence every vector in $\mathbb{R}^3$ is in the column space of A, i.e. the columns of A form a spanning list for $\mathbb{R}^3$.

13.8 Let S_1 and S_2 be subspaces of $\mathbb{R}^n$. Show that $S_1 \cap S_2$ is also a subspace of $\mathbb{R}^n$.

We use the rule given opposite, rather than the definition of subspace. Certainly $S_1 \cap S_2$ is non-empty, because $\mathbf{0} \in S_1$ and $\mathbf{0} \in S_2$, so $\mathbf{0} \in S_1 \cap S_2$. Now let $\boldsymbol{u}, \boldsymbol{v} \in S_1 \cap S_2$ and let $k, l \in \mathbb{R}$. We show that $k\boldsymbol{u} + l\boldsymbol{v} \in S_1 \cap S_2$.

$$\boldsymbol{u}, \boldsymbol{v} \in S_1, \quad \text{so} \quad k\boldsymbol{u} + l\boldsymbol{v} \in S_1,$$

and
$$\boldsymbol{u}, \boldsymbol{v} \in S_2, \quad \text{so} \quad k\boldsymbol{u} + l\boldsymbol{v} \in S_2,$$

since both S_1 and S_2 are subspaces.
Consequently, $k\boldsymbol{u} + l\boldsymbol{v} \in S_1 \cap S_2$.
By the rule, then, $S_1 \cap S_2$ is a subspace of $\mathbb{R}^n$.

Rule

A subset S of $\mathbb{R}^n$ is a subspace of $\mathbb{R}^n$ if and only if

(a) S is non-empty, and
(b) for every $\boldsymbol{u}, \boldsymbol{v} \in S$ and $k, l \in \mathbb{R}$, $k\boldsymbol{u} + l\boldsymbol{v} \in S$.

It is not hard to justify this rule, from the definition above of what is meant by a subspace. This justification is left as an exercise.

As another example of the application of this rule, we show in Example 13.8 that the intersection of two subspaces of $\mathbb{R}^n$ is again a subspace of $\mathbb{R}^n$.

Summary

The space spanned by a finite list of vectors in $\mathbb{R}^n$ is introduced as a generalisation of the idea of the column space of a matrix. The definition is given of a subspace of $\mathbb{R}^n$. This is discussed fully in the context of $\mathbb{R}^3$ and appropriate geometrical ideas of lines and planes are brought in. Methods are given for determining whether given subsets of $\mathbb{R}^n$ are subspaces.

Exercises

1. In each case find whether the given single vector is in the space spanned by the given list of vectors.

(i) $\begin{bmatrix}1\\1\\1\end{bmatrix}$, $\left(\begin{bmatrix}1\\2\\-1\end{bmatrix}, \begin{bmatrix}3\\0\\1\end{bmatrix}, \begin{bmatrix}0\\1\\2\end{bmatrix}\right)$.

(ii) $\begin{bmatrix}0\\0\\0\end{bmatrix}$, $\left(\begin{bmatrix}0\\1\\1\end{bmatrix}, \begin{bmatrix}3\\1\\-1\end{bmatrix}, \begin{bmatrix}3\\2\\0\end{bmatrix}\right)$.

(iii) $\begin{bmatrix}2\\-1\\2\end{bmatrix}$, $\left(\begin{bmatrix}1\\2\\-1\end{bmatrix}, \begin{bmatrix}0\\1\\1\end{bmatrix}\right)$.

(iv) $\begin{bmatrix}1\\0\\1\\0\end{bmatrix}$, $\left(\begin{bmatrix}1\\1\\1\\1\end{bmatrix}, \begin{bmatrix}0\\1\\1\\1\end{bmatrix}, \begin{bmatrix}1\\1\\1\\0\end{bmatrix}\right)$.

(v) $\begin{bmatrix}1\\1\\1\\1\end{bmatrix}$, $\left(\begin{bmatrix}0\\1\\1\\2\end{bmatrix}, \begin{bmatrix}2\\1\\1\\0\end{bmatrix}\right)$.

(vi) $\begin{bmatrix}1\\2\\3\\4\end{bmatrix}$, $\left(\begin{bmatrix}0\\1\\1\\0\end{bmatrix}, \begin{bmatrix}-1\\2\\-1\\0\end{bmatrix}, \begin{bmatrix}1\\0\\0\\1\end{bmatrix}, \begin{bmatrix}1\\1\\1\\1\end{bmatrix}\right)$.

(vii) $\begin{bmatrix}1\\2\\3\\4\end{bmatrix}$, $\left(\begin{bmatrix}0\\1\\1\\0\end{bmatrix}, \begin{bmatrix}-1\\2\\-1\\0\end{bmatrix}, \begin{bmatrix}1\\0\\0\\1\end{bmatrix}, \begin{bmatrix}2\\0\\3\\1\end{bmatrix}\right)$.

2. Which of the following lists of vectors span $\mathbb{R}^2$?

(i) $\left(\begin{bmatrix}1\\1\end{bmatrix}, \begin{bmatrix}1\\2\end{bmatrix}\right)$. (ii) $\left(\begin{bmatrix}1\\0\end{bmatrix}, \begin{bmatrix}0\\0\end{bmatrix}\right)$.

(iii) $\left(\begin{bmatrix}3\\-1\end{bmatrix}, \begin{bmatrix}-1\\3\end{bmatrix}\right)$. (iv) $\left(\begin{bmatrix}1\\2\end{bmatrix}, \begin{bmatrix}3\\6\end{bmatrix}\right)$.

3. Which of the following lists of vectors span $\mathbb{R}^3$?

(i) $\left(\begin{bmatrix}2\\1\\1\end{bmatrix}, \begin{bmatrix}1\\2\\1\end{bmatrix}, \begin{bmatrix}1\\1\\2\end{bmatrix}\right)$.

(ii) $\left(\begin{bmatrix}1\\-1\\0\end{bmatrix}, \begin{bmatrix}2\\1\\1\end{bmatrix}, \begin{bmatrix}0\\-3\\-1\end{bmatrix}, \begin{bmatrix}5\\-1\\1\end{bmatrix}\right)$.

(iii) $\left(\begin{bmatrix}1\\1\\1\end{bmatrix}, \begin{bmatrix}2\\2\\2\end{bmatrix}, \begin{bmatrix}0\\1\\1\end{bmatrix}, \begin{bmatrix}1\\1\\0\end{bmatrix}\right)$.

(iv) $\left(\begin{bmatrix}1\\-1\\3\end{bmatrix}, \begin{bmatrix}1\\1\\1\end{bmatrix}, \begin{bmatrix}2\\1\\3\end{bmatrix}\right)$.

4. Show that each of the following sets is a subspace of $\mathbb{R}^3$.

(i) $\left\{\begin{bmatrix}x\\y\\z\end{bmatrix} \in \mathbb{R}^3 : x - y = z\right\}$.

(ii) $\left\{\begin{bmatrix}x\\y\\z\end{bmatrix} \in \mathbb{R}^3 : x = 0\right\}$.

(iii) $\left\{\begin{bmatrix}x\\y\\z\end{bmatrix} \in \mathbb{R}^3 : 2x - y + 3z = 0\right\}$.

(iv) $\left\{\begin{bmatrix}x\\y\\z\end{bmatrix} \in \mathbb{R}^3 : y = 2x - z \text{ and } x = y + 2z\right\}$.

5. Show that each of the following subsets of $\mathbb{R}^3$ is not a subspace of $\mathbb{R}^3$.

   (i) $\left\{\begin{bmatrix} x \\ y \\ z \end{bmatrix} \in \mathbb{R}^3 : x > 0\right\}$.

   (ii) $\left\{\begin{bmatrix} x \\ y \\ z \end{bmatrix} \in \mathbb{R}^3 : x, y \text{ and } z \text{ are integers}\right\}$.

   (iii) $\left\{\begin{bmatrix} x \\ y \\ z \end{bmatrix} \in \mathbb{R}^3 : x^2 = y^2\right\}$.

   (iv) $\left\{\begin{bmatrix} x \\ y \\ z \end{bmatrix} \in \mathbb{R}^3 : y = 1\right\}$.

   (v) $\left\{\begin{bmatrix} x \\ y \\ z \end{bmatrix} \in \mathbb{R}^3 : x + y + z = 3\right\}$.

6. Let A be a $n \times n$ matrix and let k be a fixed real number. Show that the set $\{\boldsymbol{x} \in \mathbb{R}^n : A\boldsymbol{x} = k\boldsymbol{x}\}$ is a subspace of $\mathbb{R}^n$.
7. Which of the following sets are subspaces (of $\mathbb{R}^2$, $\mathbb{R}^3$ or $\mathbb{R}^4$, as appropriate)? Justify your answers.

   (i) $\left\{\begin{bmatrix} x \\ y \\ z \end{bmatrix} \in \mathbb{R}^3 : x \leqslant y\right\}$.

   (ii) $\left\{\begin{bmatrix} x \\ y \\ z \\ w \end{bmatrix} \in \mathbb{R}^4 : x + y = z + w\right\}$.

   (iii) $\left\{\begin{bmatrix} x \\ y \end{bmatrix} \in \mathbb{R}^2 : x^2 + y^2 = 0\right\}$.

   (iv) $\left\{\begin{bmatrix} x \\ y \\ z \end{bmatrix} \in \mathbb{R}^3 : x = y \text{ and } y = z\right\}$.

   (v) $\left\{\begin{bmatrix} x \\ y \\ z \\ w \end{bmatrix} \in \mathbb{R}^4 : x + y = z \text{ or } y + z = w\right\}$.

8. For all of the sets in Exercise 4 above, and for all of the sets in Exercise 7 which are subspaces of $\mathbb{R}^3$, give a geometrical description, i.e. say whether the subspace is the origin itself, a straight line through the origin, or a plane through the origin. In each case find also a matrix A such that the subspace is the set of all vectors $\boldsymbol{x}$ such that $A\boldsymbol{x} = \boldsymbol{0}$.

Examples

14.1 Spanning lists.

(i) $\left(\begin{bmatrix}1\\0\\0\end{bmatrix},\begin{bmatrix}0\\1\\0\end{bmatrix},\begin{bmatrix}0\\0\\1\end{bmatrix}\right)$ spans the whole of $\mathbb{R}^3$. So does $\left(\begin{bmatrix}1\\1\\-2\end{bmatrix},\begin{bmatrix}3\\0\\1\end{bmatrix},\begin{bmatrix}-2\\4\\1\end{bmatrix}\right)$. (See Example 13.3.)

(ii) $\left(\begin{bmatrix}2\\1\\1\end{bmatrix},\begin{bmatrix}1\\0\\3\end{bmatrix},\begin{bmatrix}3\\1\\4\end{bmatrix}\right)$ spans the same subspace of $\mathbb{R}^3$ as does $\left(\begin{bmatrix}2\\1\\1\end{bmatrix},\begin{bmatrix}1\\0\\3\end{bmatrix}\right)$. This is because

$$\begin{bmatrix}3\\1\\4\end{bmatrix}=\begin{bmatrix}2\\1\\1\end{bmatrix}+\begin{bmatrix}1\\0\\3\end{bmatrix}.$$

(iii) For any two vectors $\boldsymbol{v}_1, \boldsymbol{v}_2 \in \mathbb{R}^n$, $(\boldsymbol{v}_1, \boldsymbol{v}_2)$ spans the same subspace as $(\boldsymbol{v}_1 + \boldsymbol{v}_2, \boldsymbol{v}_1 - \boldsymbol{v}_2)$. To see this:

$$a_1\boldsymbol{v}_1 + a_2\boldsymbol{v}_2 = \tfrac{1}{2}(a_1 + a_2)(\boldsymbol{v}_1 + \boldsymbol{v}_2) + \tfrac{1}{2}(a_1 - a_2)(\boldsymbol{v}_1 - \boldsymbol{v}_2),$$

so every linear combination of $\boldsymbol{v}_1$ and $\boldsymbol{v}_2$ is a linear combination of $\boldsymbol{v}_1 + \boldsymbol{v}_2$ and $\boldsymbol{v}_1 - \boldsymbol{v}_2$. And, conversely,

$$b_1(\boldsymbol{v}_1 + \boldsymbol{v}_2) + b_2(\boldsymbol{v}_1 - \boldsymbol{v}_2) = (b_1 + b_2)\boldsymbol{v}_1 + (b_1 - b_2)\boldsymbol{v}_2.$$

14.2 Extending a spanning list.

The list $\left(\begin{bmatrix}1\\1\\-2\end{bmatrix},\begin{bmatrix}3\\0\\1\end{bmatrix},\begin{bmatrix}-2\\4\\1\end{bmatrix}\right)$ is a spanning list for the whole of $\mathbb{R}^3$. This means that every vector in $\mathbb{R}^3$ may be written as a linear combination of these vectors. For example,

$$\begin{bmatrix}1\\1\\9\end{bmatrix}=-3\begin{bmatrix}1\\1\\-2\end{bmatrix}+2\begin{bmatrix}3\\0\\1\end{bmatrix}+\begin{bmatrix}-2\\4\\1\end{bmatrix}.$$

Let us append another vector to the spanning list, say $\begin{bmatrix}2\\1\\-5\end{bmatrix}$.

Now we have a list of four vectors, which is certainly still a spanning list, because every vector still may be written as a linear combination of the original list in the same way as before, and all we need do is add another term with a zero coefficient:

$$\begin{bmatrix}1\\1\\9\end{bmatrix}=-3\begin{bmatrix}1\\1\\-2\end{bmatrix}+2\begin{bmatrix}3\\0\\1\end{bmatrix}+\begin{bmatrix}-2\\4\\1\end{bmatrix}+0\begin{bmatrix}2\\1\\-5\end{bmatrix}.$$

14

Spanning lists, bases, dimension

In the previous chapter we gave various ways in which subspaces of $\mathbb{R}^n$ may be described. One of these was as the space spanned by a given (finite) list of vectors. This idea of a spanning list is indeed a general one. In this chapter we consider the nature and properties of spanning lists, and how to find them. These ideas will lead to the notions of basis and dimension, which are among the most important in the book.

Rule

Let S be a subspace of $\mathbb{R}^n$. Then there is a finite list X of vectors from S such that S is the space spanned by X.

We shall not give a justification for this, even though it is a fundamental result. This is because the rest of the chapter contains ideas and methods which will assist us in particular cases to construct spanning lists, and these will be of more significance than the general rule above. This will be clearer towards the end of the chapter, so we shall wait until then before giving practical illustration. (See Example 14.12.)

Before going any further, we should note that a subspace S may have many different spanning lists. This is illustrated by Example 14.1. Indeed there are some general rules about this, for example the following. Let B be a finite list of vectors in $\mathbb{R}^n$, and let S be the subspace spanned by B. Now extend B by appending some other vector from S. Then the extended list is still a spanning list for S. This is justified in Example 14.2. Notice that the extended list is LD, and that this must be so. Extending a spanning list will not be a useful procedure, however. What will be useful is reducing a given spanning list. This is harder, and we must be rather careful about it, but shortly we shall describe an algorithm which may be used to do this.

14.3 Let $U = \left(\begin{bmatrix}2\\1\\3\end{bmatrix}, \begin{bmatrix}-2\\2\\0\end{bmatrix}\right)$, and let $\boldsymbol{x} = \begin{bmatrix}3\\2\\5\end{bmatrix}$. Then U is LI and (as can be easily verified) the list obtained by appending $\boldsymbol{x}$ on to U is LD. In fact we have

$$10\begin{bmatrix}2\\1\\3\end{bmatrix} + \begin{bmatrix}-2\\2\\0\end{bmatrix} - 6\begin{bmatrix}3\\2\\5\end{bmatrix} = \begin{bmatrix}0\\0\\0\end{bmatrix}.$$

This equation may be rearranged, to give

$$6\begin{bmatrix}3\\2\\5\end{bmatrix} = 10\begin{bmatrix}2\\1\\3\end{bmatrix} + \begin{bmatrix}-2\\2\\0\end{bmatrix},$$

so that $$\begin{bmatrix}3\\2\\5\end{bmatrix} = (10/6)\begin{bmatrix}2\\1\\3\end{bmatrix} + (1/6)\begin{bmatrix}-2\\2\\0\end{bmatrix}.$$

This shows that $\boldsymbol{x}$ may be written as a linear combination of the vectors in U, or, in other words, that $\boldsymbol{x}$ belongs to the subspace spanned by U. This argument works in the same way whatever $\boldsymbol{x}$ is chosen, so long as the original list is LI and the extended list is LD.

14.4 Find a LI list which spans the same subspace of $\mathbb{R}^3$ as the list

$$\left(\begin{bmatrix}1\\-1\\2\end{bmatrix}, \begin{bmatrix}2\\2\\1\end{bmatrix}, \begin{bmatrix}3\\1\\3\end{bmatrix}, \begin{bmatrix}-1\\-3\\1\end{bmatrix}\right).$$

Take $\boldsymbol{v}_1$ to be $\begin{bmatrix}1\\-1\\2\end{bmatrix}$. Next consider $\begin{bmatrix}2\\2\\1\end{bmatrix}$. We choose $\boldsymbol{v}_2$ to be $\begin{bmatrix}2\\2\\1\end{bmatrix}$ because $\left(\begin{bmatrix}1\\-1\\2\end{bmatrix}, \begin{bmatrix}2\\2\\1\end{bmatrix}\right)$ is LI. Next, $\left(\begin{bmatrix}1\\-1\\2\end{bmatrix}, \begin{bmatrix}2\\2\\1\end{bmatrix}, \begin{bmatrix}3\\1\\3\end{bmatrix}\right)$ is LD, so ignore $\begin{bmatrix}3\\1\\3\end{bmatrix}$. Last, $\left(\begin{bmatrix}1\\-1\\2\end{bmatrix}, \begin{bmatrix}2\\2\\1\end{bmatrix}, \begin{bmatrix}-1\\-3\\1\end{bmatrix}\right)$ is LD, so ignore $\begin{bmatrix}-1\\-3\\1\end{bmatrix}$. This leaves $\left(\begin{bmatrix}1\\-1\\2\end{bmatrix}, \begin{bmatrix}2\\2\\1\end{bmatrix}\right)$ as a LI spanning list.

Last, let us follow the argument opposite which shows that this list actually is a spanning list. Here we have two rejected elements of the original list.

$$\left(\begin{bmatrix}1\\-1\\2\end{bmatrix}, \begin{bmatrix}2\\2\\1\end{bmatrix}, \begin{bmatrix}3\\1\\3\end{bmatrix}\right) \text{ is LD,}$$

and we can write

$$\begin{bmatrix}3\\1\\3\end{bmatrix} = \begin{bmatrix}1\\-1\\2\end{bmatrix} + \begin{bmatrix}2\\2\\1\end{bmatrix}.$$

Before that, we require to make an observation about LI lists. We do this in general terms here, and Example 14.3 provides the details of a particular case. Let U $(=(\boldsymbol{u}_1, \ldots, \boldsymbol{u}_r))$ be a LI list of vectors in $\mathbb{R}^n$, and let $\boldsymbol{x}$ be some vector in $\mathbb{R}^n$ such that the list $(\boldsymbol{u}_1, \ldots, \boldsymbol{u}_r, \boldsymbol{x})$ is LD. Then $\boldsymbol{x}$ may be written as a linear combination of the elements of U. Here is why. We must have

$$a_1\boldsymbol{u}_1 + \cdots + a_r\boldsymbol{u}_r + b\boldsymbol{x} = \boldsymbol{0}, \qquad (*)$$

say, where not all of the coefficients are zero. Now b cannot be zero because, if it were, the equation (*) would be

$$a_1\boldsymbol{u}_1 + \cdots + a_r\boldsymbol{u}_r = \boldsymbol{0},$$

which would contradict the fact that U is LI. Hence the equation (*) may be rearranged, to give

$$\boldsymbol{x} = (-a_1/b)\boldsymbol{u}_1 + \cdots + (-a_r/b)\boldsymbol{u}_r,$$

as required.

We can now describe a procedure for reducing a spanning list. Of course there will certainly be a limit to the number of elements we can remove from a list and still leave a list which spans the same subspace. As the following rule suggests, this limit is achieved when the list is LI. The justification of the rule is given below, and is illustrated by Example 14.4. The process described is a useful one in practice.

Rule

Let S be the subspace of $\mathbb{R}^n$ spanned by the list $(\boldsymbol{u}_1, \ldots, \boldsymbol{u}_r)$. Then there is a LI sublist of $(\boldsymbol{u}_1, \ldots, \boldsymbol{u}_r)$ which also spans S.

Denote $(\boldsymbol{u}_1, \ldots, \boldsymbol{u}_r)$ by U. Instead of removing elements from U, we build up a sublist. Consider each element of U in turn. We choose or reject each one according to the following procedure. First choose $\boldsymbol{u}_1$ to belong to the sublist. Say let $\boldsymbol{v}_1 = \boldsymbol{u}_1$. Next, if $(\boldsymbol{v}_1, \boldsymbol{u}_2)$ is LI then we choose $\boldsymbol{v}_2$ to be $\boldsymbol{u}_2$. Otherwise we reject $\boldsymbol{u}_2$ and move on to consider $\boldsymbol{u}_3$ in the same way. Continue thus, at each stage asking whether the next element of U, when appended to the list of vectors already chosen, would yield a LI list. If so, we include this element of U in the list of chosen vectors, and if not we reject it. The process is complete when the last element of U has been considered. Denote the list of chosen vectors by V. Then V is certainly a LI sublist of U, because of the way that it has been constructed. It remains, therefore, to show that V spans S, i.e. that every vector in S may be written as a linear combination of the vectors in V. The first step in doing this is to show that all of the rejected elements of U may be so written. Suppose that $\boldsymbol{u}_i$ is one of the rejected elements of U. It was rejected

Similarly we can write

$$\begin{bmatrix}-1\\-3\\1\end{bmatrix}=\begin{bmatrix}1\\-1\\2\end{bmatrix}-\begin{bmatrix}2\\2\\1\end{bmatrix}.$$

Now take any vector in the space spanned by the original list, say

$$2\begin{bmatrix}1\\-1\\2\end{bmatrix}+\begin{bmatrix}2\\2\\1\end{bmatrix}-\begin{bmatrix}3\\1\\3\end{bmatrix}+2\begin{bmatrix}-1\\-3\\1\end{bmatrix}.$$

This is equal to

$$2\begin{bmatrix}1\\-1\\2\end{bmatrix}+\begin{bmatrix}2\\2\\1\end{bmatrix}-\left(\begin{bmatrix}1\\-1\\2\end{bmatrix}+\begin{bmatrix}2\\2\\1\end{bmatrix}\right)+2\left(\begin{bmatrix}1\\-1\\2\end{bmatrix}-\begin{bmatrix}2\\2\\1\end{bmatrix}\right)$$

and so (collecting terms together) equal to

$$3\begin{bmatrix}1\\-1\\2\end{bmatrix}-2\begin{bmatrix}2\\2\\1\end{bmatrix}.$$

14.5 Find a LI list which spans the column space of the matrix

$$A=\begin{bmatrix}3&6&1&2&1\\1&2&0&1&1\\-1&-2&1&-2&1\end{bmatrix}.$$

Let $$X=\left(\begin{bmatrix}3\\1\\-1\end{bmatrix},\begin{bmatrix}6\\2\\-2\end{bmatrix},\begin{bmatrix}1\\0\\1\end{bmatrix},\begin{bmatrix}2\\1\\-2\end{bmatrix},\begin{bmatrix}1\\1\\1\end{bmatrix}\right),$$

the list of the columns of A, and let S be the column space of A, i.e. the subspace of $\mathbb{R}^3$ spanned by X.

Pick $\boldsymbol{v}_1=\begin{bmatrix}3\\1\\-1\end{bmatrix}$. Pick $\boldsymbol{v}_2$ to be the next element of X such that $(\boldsymbol{v}_1,\boldsymbol{v}_2)$ is LI.

This is not $\begin{bmatrix}6\\2\\-2\end{bmatrix}$, but $\begin{bmatrix}1\\0\\1\end{bmatrix}$.

Pick $\boldsymbol{v}_3$ to be the next element of X such that $(\boldsymbol{v}_1,\boldsymbol{v}_2,\boldsymbol{v}_3)$ is LI. The standard process shows that the list $\left(\begin{bmatrix}3\\1\\-1\end{bmatrix},\begin{bmatrix}1\\0\\1\end{bmatrix},\begin{bmatrix}2\\1\\-2\end{bmatrix}\right)$ is LD, so $\begin{bmatrix}2\\1\\-2\end{bmatrix}$ is ignored. However,

$\left(\begin{bmatrix}3\\1\\-1\end{bmatrix},\begin{bmatrix}1\\0\\1\end{bmatrix},\begin{bmatrix}1\\1\\1\end{bmatrix}\right)$ is LI, so take $\boldsymbol{v}_3=\begin{bmatrix}1\\1\\1\end{bmatrix}$.

The list X is now exhausted, and the list

$$\left(\begin{bmatrix}3\\1\\-1\end{bmatrix},\begin{bmatrix}1\\0\\1\end{bmatrix},\begin{bmatrix}1\\1\\1\end{bmatrix}\right)$$

is a LI spanning list for S.

because (say) $(\boldsymbol{v}_1, \ldots, \boldsymbol{v}_j, \boldsymbol{u}_i)$ was LD. But $(\boldsymbol{v}_1, \ldots, \boldsymbol{v}_j)$ is LI so, by the discussion earlier in this chapter, $\boldsymbol{u}_i$ may be written as a linear combination of $(\boldsymbol{v}_1, \ldots, \boldsymbol{v}_j)$, i.e. of vectors in V. Finally we can now show that V spans S. Let $\boldsymbol{x} \in S$. Because U spans S, we may write

$$\boldsymbol{x} = c_1\boldsymbol{u}_1 + \cdots + c_r\boldsymbol{u}_r.$$

Now replace each of the vectors $\boldsymbol{u}_i$ which does not belong to V by its expression as a linear combination of the elements of V. Collecting terms together on the right-hand side then gives $\boldsymbol{x}$ as a linear combination of the vectors in V.

The column space of a matrix is an example of a space spanned by a list of vectors. Example 14.5 shows the process above applied to the columns of a matrix.

We have gone to considerable trouble over all this. LI spanning lists are so important, however, that it is essential to know not only that they must exist, but also how to construct them. The remainder of this chapter is devoted to discussion of those properties of LI spanning lists which are the reason for their significance.

14.6 Bases for $\mathbb{R}^3$.

(i) The standard basis: $\left(\begin{bmatrix}1\\0\\0\end{bmatrix}, \begin{bmatrix}0\\1\\0\end{bmatrix}, \begin{bmatrix}0\\0\\1\end{bmatrix}\right)$.

(ii) (See Example 14.1) $\left(\begin{bmatrix}1\\1\\-2\end{bmatrix}, \begin{bmatrix}3\\0\\1\end{bmatrix}, \begin{bmatrix}-2\\4\\1\end{bmatrix}\right)$.

(This is a LI list which spans the whole of $\mathbb{R}^3$.)

(iii) Show that $\left(\begin{bmatrix}0\\1\\1\end{bmatrix}, \begin{bmatrix}1\\0\\1\end{bmatrix}, \begin{bmatrix}1\\1\\0\end{bmatrix}\right)$ is a basis for $\mathbb{R}^3$. First show that this list is LI, by the usual process (the GE process).

Next show that this list spans the whole of $\mathbb{R}^3$. Try to solve

$$\begin{bmatrix}0\\1\\1\end{bmatrix}x_1 + \begin{bmatrix}1\\0\\1\end{bmatrix}x_2 + \begin{bmatrix}1\\1\\0\end{bmatrix}x_3 = \begin{bmatrix}a\\b\\c\end{bmatrix},$$

where a, b and c are arbitrary real numbers. As we saw in Example 13.3 we may suppress the right-hand side and apply the GE process to the array

$$\begin{bmatrix}0 & 1 & 1\\1 & 0 & 1\\1 & 1 & 0\end{bmatrix},$$

leading to

$$\begin{bmatrix}1 & 0 & 1\\0 & 1 & 1\\0 & 0 & 1\end{bmatrix},$$

and we deduce that there is a solution for x_1, x_2 and x_3, irrespective of the values of a, b and c. (Alternatively, we may argue that a solution exists because the matrix

$$\begin{bmatrix}0 & 1 & 1\\1 & 0 & 1\\1 & 1 & 0\end{bmatrix}$$

is invertible.)

14.7 Find a basis for the subspace of $\mathbb{R}^4$ spanned by the list

$$\left(\begin{bmatrix}1\\0\\1\\0\end{bmatrix}, \begin{bmatrix}2\\1\\0\\1\end{bmatrix}, \begin{bmatrix}1\\1\\-1\\1\end{bmatrix}, \begin{bmatrix}0\\-1\\2\\-1\end{bmatrix}\right).$$

Let $\mathbf{v}_1 = \begin{bmatrix}1\\0\\1\\0\end{bmatrix}$. Let $\mathbf{v}_2 = \begin{bmatrix}2\\1\\0\\1\end{bmatrix}$ (so that $(\mathbf{v}_1, \mathbf{v}_2)$ is LI).

Definition
Let S be a subspace of $\mathbb{R}^n$ (for some n). A list X is a *basis* for S if X is LI and S is the subspace spanned by X.

The two rules of this chapter so far, taken together, amount to the assertion that every subspace of $\mathbb{R}^n$ has a basis. As a first example, consider $\mathbb{R}^n$ itself. The standard basis for $\mathbb{R}^n$ is the list of n-vectors which have one entry equal to 1 and all the other entries equal to 0, normally taken in order so that the ith member of the list has the 1 in the ith position. It is not difficult to see that this is a basis for $\mathbb{R}^n$ as defined above (verification is left as an exercise). Example 14.6 gives other bases for $\mathbb{R}^3$.

The method described above and used in Examples 14.4 and 14.5 will find a basis for a given subspace, provided that we have a spanning list to start with. Example 14.7 is an application of this process.

The most important property of a basis is the following. Given a subspace S of $\mathbb{R}^n$ and a basis X for S, every vector in S may be written as a linear combination of the vectors in X, because X is a spanning list, but we can say more.

Find whether $\left(\begin{bmatrix}1\\0\\1\\0\end{bmatrix}, \begin{bmatrix}2\\1\\0\\1\end{bmatrix}, \begin{bmatrix}1\\1\\-1\\1\end{bmatrix}\right)$ is LI. It is not, so ignore $\begin{bmatrix}1\\1\\-1\\1\end{bmatrix}$, and go on to find whether $\left(\begin{bmatrix}1\\0\\1\\0\end{bmatrix}, \begin{bmatrix}2\\1\\0\\1\end{bmatrix}, \begin{bmatrix}0\\-1\\2\\-1\end{bmatrix}\right)$ is LI. It is not, so ignore $\begin{bmatrix}0\\-1\\2\\-1\end{bmatrix}$. Conclusion: $\left(\begin{bmatrix}1\\0\\1\\0\end{bmatrix}, \begin{bmatrix}2\\1\\0\\1\end{bmatrix}\right)$ is a basis as required.

14.8 Proof that if S is a subspace of $\mathbb{R}^n$, $(\boldsymbol{v}_1, \ldots, \boldsymbol{v}_r)$ $(=X)$ is a basis for S, and $\boldsymbol{u} \in S$, then $\boldsymbol{u}$ may be expressed uniquely as

$$\boldsymbol{u} = a_1\boldsymbol{v}_1 + \cdots + a_r\boldsymbol{v}_r.$$

Let $\boldsymbol{u} \in S$. Then $\boldsymbol{u}$ may be written in the given form, since the basis X spans S. Suppose that

$$\boldsymbol{u} = b_1\boldsymbol{v}_1 + \cdots + b_r\boldsymbol{v}_r$$

also. Subtracting then gives

$$\boldsymbol{0} = (a_1 - b_1)\boldsymbol{v}_1 + \cdots + (a_r - b_r)\boldsymbol{v}_r.$$

But $(\boldsymbol{v}_1, \ldots, \boldsymbol{v}_r)$ is LI, so

$$a_1 - b_1 = 0, \ldots, a_r - b_r = 0,$$

i.e. $a_1 = b_1, \ldots, a_r = b_r$.

Thus $\boldsymbol{u} = a_1\boldsymbol{v}_1 + \cdots + a_r\boldsymbol{v}_r$ uniquely.

14.9 Proof that if S is a subspace of $\mathbb{R}^n$ and $(\boldsymbol{v}_1, \ldots, \boldsymbol{v}_r)$ $(=X)$ is a basis for S then every list of vectors from S which contains more than r vectors is LD.

Let $Y = (\boldsymbol{u}_1, \ldots, \boldsymbol{u}_s)$ be a list of vectors from S, where $s > r$. Since X is a basis for S, each element of Y may be expressed as a linear combination thus:

$$\boldsymbol{u}_i = a_{i1}\boldsymbol{v}_1 + \cdots + a_{ir}\boldsymbol{v}_r, \qquad \text{for } 1 \leqslant i \leqslant s.$$

Seek numbers $x_1, \ldots, x_s$, not all zero, such that

$$x_1\boldsymbol{u}_1 + \cdots + x_s\boldsymbol{u}_s = \boldsymbol{0},$$

i.e. $x_1(a_{11}\boldsymbol{v}_1 + \cdots + a_{1r}\boldsymbol{v}_r) + \cdots + x_s(a_{s1}\boldsymbol{v}_1 + \cdots + a_{sr}\boldsymbol{v}_r) = \boldsymbol{0}$,

i.e. (collecting terms in $\boldsymbol{v}_1$, $\boldsymbol{v}_2$, etc.)

$$(x_1a_{11} + \cdots + x_sa_{s1})\boldsymbol{v}_1 + \cdots + (x_1a_{1r} + \cdots + x_sa_{sr})\boldsymbol{v}_r = \boldsymbol{0}.$$

But $(\boldsymbol{v}_1, \ldots, \boldsymbol{v}_r)$ is LI, so we must have

$$\begin{aligned} x_1a_{11} + \cdots + x_sa_{s1} &= 0\\ x_1a_{12} + \cdots + x_sa_{s2} &= 0\\ &\vdots\\ x_1a_{1r} + \cdots + x_sa_{sr} &= 0. \end{aligned}$$

Rule

Let S be a subspace of $\mathbb{R}^n$ (for some n), let $(\boldsymbol{v}_1, \ldots, \boldsymbol{v}_r)$ be a basis for S, and let $\boldsymbol{u} \in S$. Then $\boldsymbol{u}$ may be written uniquely as

$$\boldsymbol{u} = a_1\boldsymbol{v}_1 + a_2\boldsymbol{v}_2 + \cdots + a_r\boldsymbol{v}_r.$$

Example 14.8 contains a demonstration of this.

We know that $\mathbb{R}^n$ has a basis consisting of n vectors (namely the standard basis), and we have seen other bases in the case of $\mathbb{R}^3$, each containing three vectors. There is a general rule about bases here.

Rule

Let S be a subspace of $\mathbb{R}^n$ (for some n). Every basis for S has the same number of elements.

Our justification for this rule depends on the result derived in Example 14.9: if S has a basis consisting of r vectors, then any list of vectors containing more than r vectors is LD. To demonstrate the rule, let $X = (\boldsymbol{v}_1, \ldots, \boldsymbol{v}_k)$ and $Y = (\boldsymbol{w}_1, \ldots, \boldsymbol{w}_l)$ be bases for S. If $l > k$ then, by Example 14.9, Y is LD. But Y is LI, since it is a basis. Thus we cannot have $l > k$. Similarly we cannot have $k > l$. Hence $k = l$, and all bases have the same number of elements.

This last rule is precisely what is needed to ensure that the next definition is sensible.

Definition

Let S be a subspace of $\mathbb{R}^n$, for some n. The *dimension* of S is the number of elements in a basis for S.

This is a set of r equations in the s unknowns $x_1, \ldots, x_s$. Since $s > r$ there are more unknowns than equations. They are consistent, since $x_1 = \cdots = x_s = 0$ satisfies them. By consideration of rank (see Chapter 8) this solution cannot be unique, so there do exist numbers $x_1, \ldots, x_s$, not all zero, such that

$$x_1\boldsymbol{u}_1 + \cdots + x_s\boldsymbol{u}_s = 0,$$

and the list Y is LD, as required.

14.10 Subspaces in geometry (see Examples 13.6 and 13.7).

(i) The origin. $\{\mathbf{0}\}$ is a subspace of $\mathbb{R}^3$ with dimension zero (by convention). It has no basis.

(ii) A straight line through the origin in the direction of the vector $\begin{bmatrix} l \\ m \\ n \end{bmatrix}$. Here $\left(\begin{bmatrix} l \\ m \\ n \end{bmatrix}\right)$ is a basis, since every element of the subspace has the form $\begin{bmatrix} lt \\ mt \\ nt \end{bmatrix}$, for some $t \in \mathbb{R}$.

(iii) A plane through the origin. Take any two points P_1 and P_2 in the plane so that O, P_1 and P_2 do not all lie on a straight line. Then $\overrightarrow{OP_1}$ and $\overrightarrow{OP_2}$ together constitute a basis. Thus the dimension is 2.

(iv) The whole of $\mathbb{R}^3$. The standard basis vectors $\begin{bmatrix} 1 \\ 0 \\ 0 \end{bmatrix}$, $\begin{bmatrix} 0 \\ 1 \\ 0 \end{bmatrix}$ and $\begin{bmatrix} 0 \\ 0 \\ 1 \end{bmatrix}$ represent, geometrically, unit vectors in the directions of the coordinate axes.

14.11 Proof that if S is a subspace of $\mathbb{R}^n$ with dimension k then every LI list containing k vectors from S is a basis for S.

Let S be a subspace of $\mathbb{R}^n$ with dimension k, and let X be a LI list of k vectors from S, say $X = (\boldsymbol{u}_1, \ldots, \boldsymbol{u}_k)$. We need to show that X spans S, so that X is a basis. Let $\boldsymbol{v} \in S$, with $\boldsymbol{v} \notin X$. Then $(\boldsymbol{u}_1, \ldots, \boldsymbol{u}_k, \boldsymbol{v})$ contains $k+1$ elements, and so is LD, by the result of Example 14.9 above. So

$$a_1\boldsymbol{u}_1 + \cdots + a_k\boldsymbol{u}_k + a_{k+1}\boldsymbol{v} = \mathbf{0},$$

for some $a_1, \ldots, a_{k+1} \in \mathbb{R}$, not all zero. Now $a_{k+1} \neq 0$, for otherwise X would be LD. Hence

$$\boldsymbol{v} = -(a_1/a_{k+1})\boldsymbol{u}_1 - \cdots - (a_k/a_{k+1})\boldsymbol{u}_k,$$

so $\boldsymbol{v}$ is in the subspace of $\mathbb{R}^n$ spanned by X. This depended on the assumption that $\boldsymbol{v}$ did not belong to X, but certainly all elements of X are also in the subspace spanned by X. Thus X spans S.

Now we can say that the dimension of $\mathbb{R}^n$ itself (under this algebraic definition) is n. We can also talk of the dimension of a subspace. This fits with geometrical ideas, in the case of subspaces of $\mathbb{R}^3$. We have seen that a subspace of $\mathbb{R}^3$ may correspond generally to one of:

(i) the set consisting of the origin itself,
(ii) a straight line through the origin,
(iii) a plane through the origin, or
(iv) the whole of $\mathbb{R}^3$.

Such subspaces have (respectively) dimensions 0, 1, 2 and 3, in both geometric and algebraic senses. See Examples 14.10.

To find the dimension of a subspace we normally have to find a basis for the subspace. We already have a routine for doing this (as used in Example 14.7). Conversely, if the dimension is known, then it is possible to shorten the task of finding a basis.

Rule
Let S be a subspace of $\mathbb{R}^n$ with dimension k. Then every LI list of vectors from S which consists of k vectors is a basis for S.

See Example 14.11 for a justification of this. Notice that it contains an argument similar to one used earlier in this chapter to show that a vector may be expressed as a linear combination of a given LI list of vectors.

This leads us on to a useful rule about subspaces. The term 'proper subspace' is used for a subspace of $\mathbb{R}^n$ other than $\mathbb{R}^n$ itself. As might be expected, a proper subspace of $\mathbb{R}^n$ has dimension less than n.

Rule
Every subspace of $\mathbb{R}^n$, other than $\mathbb{R}^n$ itself, has dimension strictly less than n. Equivalently, if S is known to be a subspace of $\mathbb{R}^n$, and the dimension of S is equal to n, then S must be $\mathbb{R}^n$ itself.

It is easier to show the second form of the rule. Let S be a subspace of $\mathbb{R}^n$, with dimension n. Then S has a basis, B say, which is a LI list containing n vectors. But every LI list of n vectors in $\mathbb{R}^n$ is a basis for $\mathbb{R}^n$. Hence B is also a basis for $\mathbb{R}^n$. It follows that $\mathbb{R}^n$ is the space spanned by B, and therefore that $\mathbb{R}^n$ is the same space as S.

14.12 Find a LI list which spans the subspace S of $\mathbb{R}^4$, where

$$S=\left\{\begin{bmatrix}x_1\\x_2\\x_3\\x_4\end{bmatrix}\in\mathbb{R}^4\colon x_1-2x_2+x_3+5x_4=0\right\}.$$

(The fact that S is a subspace of $\mathbb{R}^4$ we shall suppose to have been demonstrated.)

Pick out any $\boldsymbol{v}_1\in S$ other than $\mathbf{0}$, say $\boldsymbol{v}_1=\begin{bmatrix}2\\1\\0\\0\end{bmatrix}$. Next pick out $\boldsymbol{v}_2\in S$ such that $(\boldsymbol{v}_1,\boldsymbol{v}_2)$ is LI. Take $\boldsymbol{v}_2=\begin{bmatrix}1\\1\\1\\0\end{bmatrix}$, say. We can ensure that $(\boldsymbol{v}_1,\boldsymbol{v}_2)$ is LI by having more non-zero entries in $\boldsymbol{v}_2$ than in $\boldsymbol{v}_1$. Next pick out $\boldsymbol{v}_3\in S$ such that $(\boldsymbol{v}_1,\boldsymbol{v}_2,\boldsymbol{v}_3)$ is LI. Take $\boldsymbol{v}_3=\begin{bmatrix}2\\-1\\1\\-1\end{bmatrix}$, say. Here choosing a vector whose fourth entry is non-zero ensures that $(\boldsymbol{v}_1,\boldsymbol{v}_2,\boldsymbol{v}_3)$ is LI.

It is not possible to continue this process. There is no vector $\boldsymbol{v}_4$ in S such that $(\boldsymbol{v}_1,\boldsymbol{v}_2,\boldsymbol{v}_3,\boldsymbol{v}_4)$ is LI. We can see this as follows. First we are working in $\mathbb{R}^4$, whose dimension is 4. Second, S is a proper subspace of $\mathbb{R}^4$. (This is easy to justify: all we need to do is find a vector in $\mathbb{R}^4$ which is not in S.) Consequently, S will have dimension less than 4, and so we cannot have a LI list in S which contains four vectors. The list

$$\left(\begin{bmatrix}2\\1\\0\\0\end{bmatrix},\begin{bmatrix}1\\1\\1\\0\end{bmatrix},\begin{bmatrix}2\\-1\\1\\-1\end{bmatrix}\right)$$

will therefore serve as an answer. Of course we could have made different choices and arrived at a different list: as we have seen, there will always be many different bases for any subspace.

Finally let us return to consider the first rule of this chapter, which says that every subspace of $\mathbb{R}^n$ has a finite spanning list. Now that we know about bases and dimension, the problem of finding a spanning list for a given subspace can be thought of rather differently than at the start of this chapter. For one thing, a basis is a spanning list, so finding a basis will do, and our rules about bases can provide assistance. Here are some ideas. They are illustrated in Example 14.12.

The method of Example 14.4 can be generalised so that it works without a given spanning list. Instead, after making an arbitrary first choice, at each stage we attempt to extend the list of chosen vectors by finding any vector in the subspace such that the extended list of chosen vectors is LI. If we do this, it will eventually become impossible to find any more vectors with which to extend the LI list of chosen vectors. At that point we shall have a LI spanning list. Otherwise there would be some vector $\boldsymbol{x}$ in the subspace which was not a linear combination of the chosen vectors, in which case $\boldsymbol{x}$ could be used to extend the chosen list. The main difficulty with this process is knowing when to stop. How can we tell that a search for another vector to include in the chosen list will fail? Ideas concerning dimension can help here. The list being constructed will be a LI spanning list, i.e. a basis for the subspace. The number of elements it contains will be equal to the dimension of the subspace. Depending on how the subspace is presented, we may be able to estimate its dimension. At worst, we always know that a subspace of $\mathbb{R}^n$ necessarily has dimension $\leqslant n$.

Summary

Different lists of vectors may be spanning lists for the same subspace. For every subspace there is a linearly independent list which spans the subspace. An algorithm for finding such a spanning list is given. Various results are given concerning spanning lists and linear dependence, leading to the ideas of basis and dimension. Some properties of bases are derived and illustrated.

Exercises

1. In each case show that the given pairs of lists span the same subspace of $\mathbb{R}^3$.

   (i) $\left(\begin{bmatrix}1\\0\\0\end{bmatrix},\begin{bmatrix}0\\1\\0\end{bmatrix}\right)$, $\left(\begin{bmatrix}1\\1\\0\end{bmatrix},\begin{bmatrix}1\\-1\\0\end{bmatrix}\right)$.

   (ii) $\left(\begin{bmatrix}1\\0\\3\end{bmatrix},\begin{bmatrix}1\\1\\5\end{bmatrix},\begin{bmatrix}-1\\2\\1\end{bmatrix}\right)$, $\left(\begin{bmatrix}0\\1\\2\end{bmatrix},\begin{bmatrix}3\\-5\\-1\end{bmatrix},\begin{bmatrix}1\\-1\\1\end{bmatrix}\right)$.

   (iii) $\left(\begin{bmatrix}1\\0\\0\end{bmatrix},\begin{bmatrix}0\\1\\0\end{bmatrix},\begin{bmatrix}0\\0\\1\end{bmatrix}\right)$, $\left(\begin{bmatrix}2\\1\\0\end{bmatrix},\begin{bmatrix}-1\\1\\2\end{bmatrix},\begin{bmatrix}1\\3\\3\end{bmatrix}\right)$.

2. The list $\left(\begin{bmatrix}1\\-1\\-1\end{bmatrix},\begin{bmatrix}1\\3\\1\end{bmatrix}\right)$ is LI.

   Which of the following vectors may be taken along with these two to form a LI list of three vectors? Express each of the other vectors as a linear combination of these two vectors.

   $$\begin{bmatrix}0\\1\\0\end{bmatrix},\begin{bmatrix}1\\1\\0\end{bmatrix},\begin{bmatrix}1\\-3\\-2\end{bmatrix},\begin{bmatrix}3\\2\\1\end{bmatrix},\begin{bmatrix}5\\3\\-1\end{bmatrix}.$$

3. For each of the following lists of vectors, find a LI list which spans the same subspace of $\mathbb{R}^2$, $\mathbb{R}^3$ or $\mathbb{R}^4$, as appropriate.

   (i) $\left(\begin{bmatrix}2\\1\end{bmatrix},\begin{bmatrix}-4\\-2\end{bmatrix},\begin{bmatrix}1\\1\end{bmatrix}\right)$.

   (ii) $\left(\begin{bmatrix}1\\1\end{bmatrix},\begin{bmatrix}2\\2\end{bmatrix},\begin{bmatrix}3\\3\end{bmatrix}\right)$.

   (iii) $\left(\begin{bmatrix}2\\1\\5\end{bmatrix},\begin{bmatrix}-1\\2\\0\end{bmatrix},\begin{bmatrix}1\\1\\3\end{bmatrix}\right)$.

   (iv) $\left(\begin{bmatrix}3\\1\\0\end{bmatrix},\begin{bmatrix}1\\-3\\2\end{bmatrix},\begin{bmatrix}0\\5\\-3\end{bmatrix},\begin{bmatrix}1\\2\\-1\end{bmatrix}\right)$.

   (v) $\left(\begin{bmatrix}1\\1\\0\end{bmatrix},\begin{bmatrix}1\\0\\1\end{bmatrix},\begin{bmatrix}0\\1\\1\end{bmatrix},\begin{bmatrix}1\\2\\3\end{bmatrix}\right)$.

4. Show that the following lists of vectors are bases for the spaces indicated.

   (i) $\left(\begin{bmatrix}1\\1\end{bmatrix},\begin{bmatrix}1\\2\end{bmatrix}\right)$, for $\mathbb{R}^2$.

   (ii) $\left(\begin{bmatrix}9\\10\end{bmatrix},\begin{bmatrix}11\\12\end{bmatrix}\right)$, for $\mathbb{R}^2$.

(iii) $\left(\begin{bmatrix}0\\1\\2\end{bmatrix}, \begin{bmatrix}-1\\-2\\1\end{bmatrix}, \begin{bmatrix}1\\1\\1\end{bmatrix}\right)$, for $\mathbb{R}^3$.

(iv) $\left(\begin{bmatrix}1\\2\\3\end{bmatrix}, \begin{bmatrix}2\\3\\1\end{bmatrix}, \begin{bmatrix}3\\1\\2\end{bmatrix}\right)$, for $\mathbb{R}^3$.

(v) $\left(\begin{bmatrix}0\\1\\1\\1\end{bmatrix}, \begin{bmatrix}1\\0\\1\\1\end{bmatrix}, \begin{bmatrix}1\\1\\0\\1\end{bmatrix}, \begin{bmatrix}1\\1\\1\\0\end{bmatrix}\right)$, for $\mathbb{R}^4$.

5. In each case below, find a basis for the subspace (of $\mathbb{R}^2$, $\mathbb{R}^3$ or $\mathbb{R}^4$, as appropriate) spanned by the given list.

(i) $\left(\begin{bmatrix}-2\\3\end{bmatrix}, \begin{bmatrix}4\\-6\end{bmatrix}, \begin{bmatrix}1\\2\end{bmatrix}\right)$.

(ii) $\left(\begin{bmatrix}1\\1\end{bmatrix}, \begin{bmatrix}0\\0\end{bmatrix}, \begin{bmatrix}1\\-1\end{bmatrix}, \begin{bmatrix}2\\2\end{bmatrix}\right)$.

(iii) $\left(\begin{bmatrix}2\\-2\\2\end{bmatrix}, \begin{bmatrix}-3\\3\\3\end{bmatrix}, \begin{bmatrix}1\\-1\\2\end{bmatrix}, \begin{bmatrix}0\\0\\1\end{bmatrix}\right)$.

(iv) $\left(\begin{bmatrix}0\\1\\-1\end{bmatrix}, \begin{bmatrix}-1\\1\\0\end{bmatrix}, \begin{bmatrix}-2\\5\\-3\end{bmatrix}, \begin{bmatrix}1\\2\\-1\end{bmatrix}\right)$.

(v) $\left(\begin{bmatrix}1\\-1\\2\\1\end{bmatrix}, \begin{bmatrix}0\\1\\1\\3\end{bmatrix}, \begin{bmatrix}2\\-3\\3\\-1\end{bmatrix}, \begin{bmatrix}-3\\5\\-4\\3\end{bmatrix}, \begin{bmatrix}-1\\0\\1\\1\end{bmatrix}\right)$.

6. The following lists are all bases for $\mathbb{R}^3$. For each of them, express the vector $\begin{bmatrix}1\\0\\0\end{bmatrix}$ as a linear combination of the basis vectors.

(i) $\left(\begin{bmatrix}0\\1\\1\end{bmatrix}, \begin{bmatrix}1\\0\\1\end{bmatrix}, \begin{bmatrix}1\\1\\0\end{bmatrix}\right)$.

(ii) $\left(\begin{bmatrix}1\\-1\\1\end{bmatrix}, \begin{bmatrix}0\\1\\2\end{bmatrix}, \begin{bmatrix}1\\1\\1\end{bmatrix}\right)$.

(iii) $\left(\begin{bmatrix}2\\2\\-1\end{bmatrix}, \begin{bmatrix}2\\-1\\2\end{bmatrix}, \begin{bmatrix}-1\\2\\2\end{bmatrix}\right)$.

(iv) $\left(\begin{bmatrix}1\\1\\0\end{bmatrix}, \begin{bmatrix}-1\\1\\0\end{bmatrix}, \begin{bmatrix}0\\0\\1\end{bmatrix}\right)$.

7. Show that the following pairs of subspaces have the same dimension.

(i) $\left\{\begin{bmatrix} x \\ y \\ z \end{bmatrix} \in \mathbb{R}^3 \colon x = 0\right\}$,

$\left\{\begin{bmatrix} x \\ y \\ z \end{bmatrix} \in \mathbb{R}^3 \colon x - y + 2z = 0\right\}$.

(ii) $\left\{\begin{bmatrix} x \\ y \\ z \end{bmatrix} \in \mathbb{R}^3 \colon x = y = z\right\}$,

$\left\{\begin{bmatrix} x \\ y \\ z \end{bmatrix} \in \mathbb{R}^3 \colon 4x = y \text{ and } y + z = 0\right\}$.

(iii) $\left\{\begin{bmatrix} x \\ y \\ z \\ w \end{bmatrix} \in \mathbb{R}^4 \colon x = y\right\}$,

$\left\{\begin{bmatrix} x \\ y \\ z \\ w \end{bmatrix} \in \mathbb{R}^4 \colon z = w\right\}$.

8. Find LI lists of vectors which span the following subspaces of $\mathbb{R}^3$. (All of these have previously been shown to be subspaces.)

(i) $\left\{\begin{bmatrix} x \\ y \\ z \end{bmatrix} \in \mathbb{R}^3 \colon x - y = z\right\}$.

(ii) $\left\{\begin{bmatrix} x \\ y \\ z \end{bmatrix} \in \mathbb{R}^3 \colon x = 0\right\}$.

(iii) $\left\{\begin{bmatrix} x \\ y \\ z \end{bmatrix} \in \mathbb{R}^3 \colon 2x - y + 3z = 0\right\}$.

(iv) $\left\{\begin{bmatrix} x \\ y \\ z \end{bmatrix} \in \mathbb{R}^3 \colon y = 2x - z \text{ and } x = y + 2z\right\}$.

Examples

15.1 Algebraic operations in $\mathbb{R}_n$.

$$[a_1 \quad a_2 \quad \cdots \quad a_n] = [b_1 \quad b_2 \quad \cdots \quad b_n]$$

if and only if $a_1 = b_1, \ldots, a_n = b_n$.

$$[a_1 \quad a_2 \quad \cdots \quad a_n] + [b_1 \quad b_2 \quad \cdots \quad b_n]$$
$$= [a_1 + b_1 \quad a_2 + b_2 \quad \cdots \quad a_n + b_n].$$
$$k[a_1 \quad a_2 \quad \cdots \quad a_n] = [ka_1 \quad ka_2 \quad \cdots \quad ka_n], \qquad \text{for } k \in \mathbb{R}.$$

Associative and commutative laws hold for these operations just as they do for matrices in general and for column vectors.

15.2 Representation of simultaneous equations by a matrix equation involving vectors in $\mathbb{R}_n$.

(i) The equations

$$ax + by = h_1$$
$$cx + dy = h_2$$

may be represented as

$$[ax + by \quad cx + dy] = [h_1 \quad h_2],$$

or $$[ax \quad cx] + [by \quad dy] = [h_1 \quad h_2],$$

or $$[a \quad c]x + [b \quad d]y = [h_1 \quad h_2].$$

These may also be represented as

$$[x \quad y]\begin{bmatrix} a & c \\ b & d \end{bmatrix} = [h_1 \quad h_2].$$

Notice that the matrix of coefficients on the left-hand side appears transposed. This complication is one reason for preferring to use column vectors.

(ii) The equations

$$\begin{aligned} x_1 - 2x_2 + x_3 &= 4 \\ 3x_1 - x_2 &= 1 \\ -x_1 + x_2 - 3x_3 &= -2 \end{aligned}$$

may be written as

$$[1 \quad 3 \quad -1]x_1 + [-2 \quad -1 \quad 1]x_2 + [1 \quad 0 \quad -3]x_3 = [4 \quad 1 \quad -2],$$

or $$[x_1 \quad x_2 \quad x_3]\begin{bmatrix} 1 & 3 & -1 \\ -2 & -1 & 1 \\ 1 & 0 & -3 \end{bmatrix} = [4 \quad 1 \quad -2].$$

15.3 The space spanned by a list of vectors in $\mathbb{R}_3$.

Let $\boldsymbol{v}_1 = [1 \quad 2 \quad 3]$, $\boldsymbol{v}_2 = [4 \quad 5 \quad 6]$ and $\boldsymbol{v}_3 = [7 \quad 8 \quad 9]$. Then the subspace of $\mathbb{R}_3$ spanned by $(\boldsymbol{v}_1, \boldsymbol{v}_2, \boldsymbol{v}_3)$ consists of all vectors $\boldsymbol{v}$ which can be written as

$$a_1\boldsymbol{v}_1 + a_2\boldsymbol{v}_2 + a_3\boldsymbol{v}_3, \qquad \text{with } a_1,\ a_2 \text{ and } a_3 \in \mathbb{R},$$

i.e. $$a_1[1 \quad 2 \quad 3] + a_2[4 \quad 5 \quad 6] + a_3[7 \quad 8 \quad 9],$$

i.e. $$[a_1 + 4a_2 + 7a_3 \quad 2a_1 + 5a_2 + 8a_3 \quad 3a_1 + 6a_2 + 9a_3].$$

15

Rank

Now that we have the idea of dimension we can consider the idea of rank in a slightly different way. Recall that the rank of a matrix has been defined as the number of non-zero rows remaining after the standard Gaussian elimination process has been applied. Since rows are involved here we shall introduce the idea of a row vector, and show that row vectors can be treated in very much the same way that column vectors have been treated in earlier chapters.

Let $\mathbb{R}_n$ stand for the set of all $1 \times n$ matrices with real entries, i.e. the set of all row vectors with n components. Then $\mathbb{R}_n$ has substantially the same properties as $\mathbb{R}^n$. This is illustrated in Example 15.1. Linear dependence and independence have the same definitions. We can define subspaces, bases and dimension in exactly the same way. $\mathbb{R}_n$ and its algebraic operations can even be given the same geometrical interpretation as $\mathbb{R}^n$. The actual working of examples is rather different for $\mathbb{R}_n$, because it is not as clear intuitively how to represent a linear combination of vectors in $\mathbb{R}_n$ as a product of a matrix with a vector of coefficients. See Example 15.2.

What concerns us here is the idea of the space spanned by a list of vectors in $\mathbb{R}_n$, i.e. the set of all row vectors which may be written as linear combinations of the given list of row vectors. This is illustrated in Example 15.3. An example of this is the row space of a matrix, which is defined as follows.

15.4 The row space of a matrix.

Let $A = [a_{ij}]_{3 \times 3}$.

Then $[x \quad y \quad z]A$

$$= [xa_{11} + ya_{21} + za_{31} \quad xa_{12} + ya_{22} + za_{32} \quad xa_{13} + ya_{23} + za_{33}]$$
$$= x[a_{11} \quad a_{12} \quad a_{13}] + y[a_{21} \quad a_{22} \quad a_{23}] + z[a_{31} \quad a_{32} \quad a_{33}].$$

Thus the set of all products $\boldsymbol{x}A$, with $\boldsymbol{x} \in \mathbb{R}_3$, is the set of all linear combinations of the rows of A, i.e. the row space of A.

15.5 Proof that if A is a 3×3 matrix and T is a product of 3×3 elementary matrices, then A and TA have the same row space.

Let A have rows $\boldsymbol{v}_1$, $\boldsymbol{v}_2$ and $\boldsymbol{v}_3$. Pre-multiplying A by a single elementary matrix has one of three effects, none of which affect the row space. Take each in turn.

(a) Interchanging two rows. This cannot affect the set of all vectors expressible as linear combinations of the rows.

(b) Adding a multiple of one row to another. This gives (say) rows $\boldsymbol{v}_1$, $\boldsymbol{v}_2 + a\boldsymbol{v}_1$ and $\boldsymbol{v}_3$. Any linear combination of these is certainly a linear combination of $\boldsymbol{v}_1$, $\boldsymbol{v}_2$ and $\boldsymbol{v}_3$. Conversely, any linear combination of $\boldsymbol{v}_1$, $\boldsymbol{v}_2$ and $\boldsymbol{v}_3$ is a linear combination of the new rows thus:

$$b_1\boldsymbol{v}_1 + b_2\boldsymbol{v}_2 + b_3\boldsymbol{v}_3 = (b_1 - ab_2)\boldsymbol{v}_1 + b_2(\boldsymbol{v}_2 + a\boldsymbol{v}_1) + b_3\boldsymbol{v}_3.$$

(c) Multiplying one row by a non-zero scalar. This gives rows (say) $\boldsymbol{v}_1$, $\boldsymbol{v}_2$ and $a\boldsymbol{v}_3$. It is a straightforward exercise to show that the row space is unchanged by this.

The matrix TA is the result of applying a finite sequence of these row operations to A. Hence TA has the same row space as A.

The same argument can be generalised to apply to matrices of any size.

Definition
Given any $p \times q$ matrix A, the *row space* of A is the space spanned by the list of rows of A. The row space of A is a subspace of $\mathbb{R}_q$.

Though it is not quite so convenient to write down, the row space of a matrix A, like the column space, can be thought of as a set of products. If $\boldsymbol{x}$ is a $1 \times p$ matrix (an element of $\mathbb{R}_p$) and A is a $p \times q$ matrix, then the product $\boldsymbol{x}A$ is a $1 \times q$ matrix (an element of $\mathbb{R}_q$). In fact we have

$$\text{row space of } A = \{\boldsymbol{x}A : \boldsymbol{x} \in \mathbb{R}_q\}.$$

For more detail of this see Example 15.4.

The row space and the column space of a $p \times q$ matrix A are apparently unrelated (the former being a subspace of $\mathbb{R}_q$, the latter a subspace of $\mathbb{R}^p$). But there is a surprising connection. They both have the same dimension, and this dimension is the same as the rank of A. We can state this as a rule.

Rule
Let A be a $p \times q$ matrix. The rank of A, the dimension of the row space of A, and the dimension of the column space of A are all equal.

This rule is not easy to justify. We proceed in two stages, considering in turn the row space and the column space. First recall that the Gaussian elimination procedure amounts to the successive premultiplication of the given matrix by a sequence of elementary matrices (one corresponding to each elementary row operation). It follows that the result of applying the GE process to a $p \times q$ matrix A is a matrix TA which is in row-echelon form. Moreover, the matrix T, which is a product of $p \times p$ elementary matrices, is an invertible matrix. We are never normally concerned about what the matrix T actually is, but for our present argument we need to know that it exists. Example 15.5 shows that the matrices A and TA have the same row space. The non-zero rows of TA form a LI list of vectors in $\mathbb{R}_q$ (demonstration of this is left as an exercise), so the number of non-zero rows of TA is the dimension of the row space of TA. This means that the rank of A is equal to the dimension of the row space of TA, which in turn is equal to the dimension of the row space of A. So we have demonstrated the part of the rule concerning the row space.

15.6 Proof that if A is a $p \times q$ matrix and T is a product of elementary $p \times p$ matrices, then the column spaces of A and TA have the same dimension.

The column space of A is $\{A\boldsymbol{x}: \boldsymbol{x} \in \mathbb{R}^q\}$, and the column space of TA is $\{TA\boldsymbol{x}: \boldsymbol{x} \in \mathbb{R}^q\}$. Let $(\boldsymbol{v}_1, \ldots, \boldsymbol{v}_r)$ be a basis for the column space of A. We show that the list $(T\boldsymbol{v}_1, \ldots, T\boldsymbol{v}_r)$ is a basis for the column space of TA.

First, let $\boldsymbol{u} \in \{TA\boldsymbol{x}: \boldsymbol{x} \in \mathbb{R}^q\}$,

say $\quad \boldsymbol{u} = TA\boldsymbol{y} \qquad (\boldsymbol{y} \in \mathbb{R}^q)$.

Then $A\boldsymbol{y}$ belongs to the column space of A, so

$$A\boldsymbol{y} = a_1\boldsymbol{v}_1 + \cdots + a_r\boldsymbol{v}_r, \qquad \text{for some } a_1, \ldots, a_r \in \mathbb{R}.$$

Hence $\quad TA\boldsymbol{y} = a_1T\boldsymbol{v}_1 + \cdots + a_rT\boldsymbol{v}_r$,

so $\boldsymbol{u}$ belongs to the space spanned by $(T\boldsymbol{v}_1, \ldots, T\boldsymbol{v}_r)$, and we have shown that $(T\boldsymbol{v}_1, \ldots, T\boldsymbol{v}_r)$ spans the column space of TA.

Second, we show that $(T\boldsymbol{v}_1, \ldots, T\boldsymbol{v}_r)$ is LI. Suppose that

$$b_1T\boldsymbol{v}_1 + \cdots + b_rT\boldsymbol{v}_r = \boldsymbol{0}.$$

Then $\quad T(b_1\boldsymbol{v}_1 + \cdots + b_r\boldsymbol{v}_r) = \boldsymbol{0}$.

But T is invertible, being a product of elementary matrices, and so

$$b_1\boldsymbol{v}_1 + \cdots + b_r\boldsymbol{v}_r = T^{-1}\boldsymbol{0} = \boldsymbol{0}.$$

Now $(\boldsymbol{v}_1, \ldots, \boldsymbol{v}_r)$ is LI, so $b_1 = \cdots = b_r = 0$.

Hence $(T\boldsymbol{v}_1, \ldots, T\boldsymbol{v}_r)$ is LI, and so forms a basis for the column space of TA.

We have now shown that the column spaces of A and TA have bases with the same number of elements, so they have the same dimension.

15.7 A basis for the column space of a matrix in row-echelon form.

$$\begin{bmatrix} 1 & 2 & 3 & 1 & 1 & 5 & 0 \\ 0 & 1 & 4 & 2 & 0 & 1 & 1 \\ 0 & 0 & 0 & 1 & 1 & 2 & 1 \\ 0 & 0 & 0 & 0 & 0 & 1 & 3 \\ 0 & 0 & 0 & 0 & 0 & 0 & 0 \end{bmatrix}$$

is in row-echelon form. The columns whose last non-zero entry is a 1 which is itself the first non-zero entry in its own row are the first, second, fourth and sixth columns. A basis for the column space is

$$\left(\begin{bmatrix} 1 \\ 0 \\ 0 \\ 0 \\ 0 \end{bmatrix}, \begin{bmatrix} 2 \\ 1 \\ 0 \\ 0 \\ 0 \end{bmatrix}, \begin{bmatrix} 1 \\ 2 \\ 1 \\ 0 \\ 0 \end{bmatrix}, \begin{bmatrix} 5 \\ 1 \\ 2 \\ 1 \\ 0 \end{bmatrix} \right).$$

You should convince yourself that all of the columns can in fact be written as linear combinations of these.

Next we turn to column spaces. Let A and TA be as above, so that TA is in row-echelon form. Example 15.6 shows that the column spaces of A and of TA, though not necessarily the same, have the same dimension. Now what is the dimension of the column space of TA? It happens to be the number of non-zero rows in TA (which we know to be equal to the rank of A), because TA is in row-echelon form. One basis for the column space consists of those columns of TA whose last non-zero entry is a 1 which is itself the first non-zero entry in its own row. This sounds complicated, but see Example 15.7 for an illustration. We have therefore shown that the rank of A is equal to the dimension of the column space of TA, which as above is equal to the dimension of the column space of A. This yields the result promised.

The rank of a matrix is in many books defined as the dimension of the row space or column space. In such books the notions of row space, column space and dimension must be introduced before that of rank, whereas our definition allows rank to be defined at an earlier stage, in terms of more elementary notions.

Attention is drawn again to the Equivalence Theorem, given in Chapter 12. It may be valuable to give it again here, in a slightly different form.

Theorem (Equivalence Theorem, second version)
Let A be a $p \times p$ matrix. The following are equivalent.

(i) A is singular.
(ii) The rank of A is less than p.
(iii) The columns of A form a LD list in $\mathbb{R}^p$.
(iiia) The rows of A form a LD list in $\mathbb{R}_p$.
(iv) The equation $A\mathbf{x} = \mathbf{0}$ has non-trivial solutions.
(v) The determinant of A is equal to zero.

Summary
The set $\mathbb{R}_n$ of row vectors is described and discussed. The idea of the subspace of $\mathbb{R}_n$ spanned by a finite list of vectors is introduced and the example of the row space of a matrix is used in a demonstration that the row space and the column space of a matrix have the same dimension, equal to the rank of the matrix.

Exercises

1. In each case below, rewrite the given set of equations first as an equation connecting vectors in $\mathbb{R}_2$ and then as an equation of the form $\boldsymbol{x}A = \boldsymbol{h}$, with $\boldsymbol{x}, \boldsymbol{h} \in \mathbb{R}_2$.

   (i) $\begin{aligned} 3x + y &= 2 \\ x - 2y &= 1 \end{aligned}$ (ii) $\begin{aligned} x - y &= 1 \\ x + y &= 3 \end{aligned}$

   (iii) $\begin{aligned} 2x \quad &= 1 \\ x + 3y &= 2. \end{aligned}$

2. Repeat Exercise 1, but with the following sets of equations and with vectors in $\mathbb{R}_3$.

   (i) $\begin{aligned} x - y + z &= 1 \\ x + 2y - z &= 4 \\ -x - y - 2z &= -2 \end{aligned}$ (ii) $\begin{aligned} 2x + y - z &= 0 \\ x \quad - 2z &= 1 \\ 2y + z &= 3. \end{aligned}$

3. Find whether the following lists of vectors in $\mathbb{R}_3$ are LI or LD.

   (i) $([1 \;\; 1 \;\; 2], [-1 \;\; 2 \;\; 1], [1 \;\; 0 \;\; -1])$.
   (ii) $([0 \;\; 1 \;\; -1], [2 \;\; 1 \;\; 1], [2 \;\; 2 \;\; 0])$.
   (iii) $([1 \;\; -1 \;\; 1], [2 \;\; 1 \;\; 1], [1 \;\; -4 \;\; 2])$.
   (iv) $([2 \;\; 2 \;\; 1], [2 \;\; 1 \;\; 2], [2 \;\; 4 \;\; -1])$.

4. For each of the following matrices, find (by separate calculations) the dimensions of the row space and the column space. (They should of course be the same.)

   (i) $\begin{bmatrix} 0 & 2 & 3 & 1 \\ 3 & 1 & 3 & -1 \\ 0 & 2 & 3 & 1 \\ 4 & 2 & 5 & -1 \end{bmatrix}$. (ii) $\begin{bmatrix} 1 & -2 & 1 & -1 \\ 0 & 1 & 3 & 5 \\ 2 & 0 & 1 & 5 \\ -1 & 1 & 1 & 1 \end{bmatrix}$.

   (iii) $\begin{bmatrix} 2 & 1 & 1 \\ 1 & 1 & -1 \\ 1 & 0 & 2 \\ -2 & -1 & -1 \end{bmatrix}$. (iv) $\begin{bmatrix} 1 & 0 & 1 & 2 \\ 2 & 1 & 1 & -1 \\ 3 & 1 & 0 & 1 \end{bmatrix}$.

   (v) $\begin{bmatrix} 1 & 3 & 1 & 2 & 0 \\ -1 & 0 & -4 & 1 & 3 \\ 1 & 2 & 2 & 1 & -1 \end{bmatrix}$. (vi) $\begin{bmatrix} 1 & 0 & 0 & 1 \\ 0 & 1 & 0 & 2 \\ 0 & 0 & 1 & 3 \end{bmatrix}$.

Examples

16.1 Matrices representing functions.

(i) $\begin{bmatrix} 0 & 0 & 0 \\ 0 & 0 & 0 \\ 0 & 0 & 0 \end{bmatrix}$ represents the function $f: \mathbb{R}^3 \to \mathbb{R}^3$, where

$$f\left(\begin{bmatrix} x \\ y \\ z \end{bmatrix}\right) = \begin{bmatrix} 0 \\ 0 \\ 0 \end{bmatrix}.$$

(ii) $\begin{bmatrix} 1 & 0 & 0 \\ 0 & 1 & 0 \\ 0 & 0 & 1 \end{bmatrix}$ represents the function $f: \mathbb{R}^3 \to \mathbb{R}^3$, where

$$f\left(\begin{bmatrix} x \\ y \\ z \end{bmatrix}\right) = \begin{bmatrix} x \\ y \\ z \end{bmatrix}.$$

(iii) $\begin{bmatrix} 1 & 2 \\ -1 & 3 \\ 4 & 1 \end{bmatrix}$ represents a function $f: \mathbb{R}^2 \to \mathbb{R}^3$, where

$$f\left(\begin{bmatrix} x \\ y \end{bmatrix}\right) = \begin{bmatrix} 1 & 2 \\ -1 & 3 \\ 4 & 1 \end{bmatrix}\begin{bmatrix} x \\ y \end{bmatrix} = \begin{bmatrix} x + 2y \\ -x + 3y \\ 4x + y \end{bmatrix}.$$

(iv) $\begin{bmatrix} 3 & 1 & -1 \\ 1 & 2 & -3 \end{bmatrix}$ represents a function $f: \mathbb{R}^3 \to \mathbb{R}^2$, where

$$f\left(\begin{bmatrix} x_1 \\ x_2 \\ x_3 \end{bmatrix}\right) = \begin{bmatrix} 3 & 1 & -1 \\ 1 & 2 & -3 \end{bmatrix}\begin{bmatrix} x_1 \\ x_2 \\ x_3 \end{bmatrix}$$
$$= \begin{bmatrix} 3x_1 + x_2 - x_3 \\ x_1 + 2x_2 - 3x_3 \end{bmatrix}.$$

(v) $\begin{bmatrix} 1 & 2 & 0 \\ 0 & 2 & 1 \\ 1 & -1 & 3 \end{bmatrix}$ represents a function $f: \mathbb{R}^3 \to \mathbb{R}^3$, where

$$f\left(\begin{bmatrix} x_1 \\ x_2 \\ x_3 \end{bmatrix}\right) = \begin{bmatrix} 1 & 2 & 0 \\ 0 & 2 & 1 \\ 1 & -1 & 3 \end{bmatrix}\begin{bmatrix} x_1 \\ x_2 \\ x_3 \end{bmatrix}$$
$$= \begin{bmatrix} x_1 + 2x_2 \\ 2x_2 + x_3 \\ x_1 - x_2 + 3x_3 \end{bmatrix}.$$

16

Linear transformations

Now that we know about subspaces we can take note of an important application of the simple idea of multiplying a column vector by a matrix. Let A be some fixed $p \times q$ matrix. Then for any q-vector $\boldsymbol{x}$, the product $A\boldsymbol{x}$ is a p-vector, i.e. for any $\boldsymbol{x} \in \mathbb{R}^q$, $A\boldsymbol{x} \in \mathbb{R}^p$. In other words, this matrix A determines a function from $\mathbb{R}^q$ to $\mathbb{R}^p$. Every matrix can be thought of in this way. Some matrices determine simple functions in an obvious way, for example a zero matrix or an identity matrix. See Examples 16.1. For most matrices, however, the nature of the corresponding function will not be as clear. This is what we shall investigate in this chapter.

The first thing to notice is that functions which are determined in this way have some convenient algebraic properties, which are consequences of properties of matrix multiplication. Let A be a $p \times q$ matrix and let $\boldsymbol{x}$ and $\boldsymbol{y}$ be q-vectors. Then $\boldsymbol{x} + \boldsymbol{y}$ is also a q-vector, and

$$A(\boldsymbol{x} + \boldsymbol{y}) = A\boldsymbol{x} + A\boldsymbol{y}.$$

Putting it another way, if we denote by f_A the function which is determined as above by the matrix A, then

$$f_A(\boldsymbol{x} + \boldsymbol{y}) = f_A(\boldsymbol{x}) + f_A(\boldsymbol{y}).$$

Similarly, for any $k \in \mathbb{R}$,

$$f_A(k\boldsymbol{x}) = kf_A(\boldsymbol{x}).$$

Not all functions have such properties. Functions which do have them are significant mathematically. They are in a sense 'compatible' with the algebraic operations in $\mathbb{R}^q$. These two properties, then, are the essence of the following definition.

16.2 Proof that every linear transformation can be represented by a matrix. Let $f\colon \mathbb{R}^q \to \mathbb{R}^p$ be a linear transformation. Denote the standard basis for $\mathbb{R}^q$ by $(\boldsymbol{e}_1, \ldots, \boldsymbol{e}_q)$. Then for any $p \times q$ matrix H, $H\boldsymbol{e}_i$ is a p-vector, and in fact is the ith column of H. In order for a matrix A to have the property

$$f(\boldsymbol{x}) = A\boldsymbol{x} \qquad \text{for every } \boldsymbol{x} \in \mathbb{R}^q,$$

it will certainly be necessary to have, for each i,

$$f(\boldsymbol{e}_i) = A\boldsymbol{e}_i = \text{the } i\text{th column of } A.$$

We therefore take the columns of A to be the vectors $f(\boldsymbol{e}_1), \ldots, f(\boldsymbol{e}_q)$.
Of course we must now check that $f(\boldsymbol{x}) = A\boldsymbol{x}$ for every $\boldsymbol{x} \in \mathbb{R}^q$. Let $\boldsymbol{x} \in \mathbb{R}^q$, and suppose that

$$\boldsymbol{x} = a_1\boldsymbol{e}_1 + \cdots + a_q\boldsymbol{e}_q.$$

Then $\quad f(\boldsymbol{x}) = a_1 f(\boldsymbol{e}_1) + \cdots + a_q f(\boldsymbol{e}_q),$

using the fact that f is a linear transformation.

Also $\quad \boldsymbol{x} = \begin{bmatrix} a_1 \\ \vdots \\ a_q \end{bmatrix} \quad$ and $\quad A\boldsymbol{x} = A\begin{bmatrix} a_1 \\ \vdots \\ a_q \end{bmatrix}$

$$= a_1 f(\boldsymbol{e}_1) + \cdots + a_q f(\boldsymbol{e}_q),$$

since the columns of A are the vectors $f(\boldsymbol{e}_1), \ldots, f(\boldsymbol{e}_q)$.
Conclusion: f is represented by the matrix A whose columns are the vectors $f(\boldsymbol{e}_1), \ldots, f(\boldsymbol{e}_q)$.

16.3 Let f be the linear transformation from $\mathbb{R}^2$ to $\mathbb{R}^3$ for which

$$f\left(\begin{bmatrix} 1 \\ 0 \end{bmatrix}\right) = \begin{bmatrix} 1 \\ 1 \\ 1 \end{bmatrix} \quad \text{and} \quad f\left(\begin{bmatrix} 0 \\ 1 \end{bmatrix}\right) = \begin{bmatrix} 2 \\ 0 \\ -1 \end{bmatrix}.$$

Verify that

$$f(\boldsymbol{x}) = \begin{bmatrix} 1 & 2 \\ 1 & 0 \\ 1 & -1 \end{bmatrix}\boldsymbol{x}, \qquad \text{for each } \boldsymbol{x} \in \mathbb{R}^2.$$

Let $\quad \boldsymbol{x} \in \mathbb{R}^2, \quad$ say $\boldsymbol{x} = \begin{bmatrix} x_1 \\ x_2 \end{bmatrix}$.

Then $\quad \boldsymbol{x} = x_1\begin{bmatrix} 1 \\ 0 \end{bmatrix} + x_2\begin{bmatrix} 0 \\ 1 \end{bmatrix} = x_1\boldsymbol{e}_1 + x_2\boldsymbol{e}_2.$

And

$$\begin{aligned} f(\boldsymbol{x}) &= f(x_1\boldsymbol{e}_1 + x_2\boldsymbol{e}_2) \\ &= f(x_1\boldsymbol{e}_1) + f(x_2\boldsymbol{e}_2) \\ &= x_1 f(\boldsymbol{e}_1) + x_2 f(\boldsymbol{e}_2) \\ &= x_1\begin{bmatrix} 1 \\ 1 \\ 1 \end{bmatrix} + x_2\begin{bmatrix} 2 \\ 0 \\ -1 \end{bmatrix} = \begin{bmatrix} x_1 + 2x_2 \\ x_1 \\ x_1 - x_2 \end{bmatrix}. \end{aligned}$$

Also

$$\begin{bmatrix} 1 & 2 \\ 1 & 0 \\ 1 & -1 \end{bmatrix}\begin{bmatrix} x_1 \\ x_2 \end{bmatrix} = \begin{bmatrix} x_1 + 2x_2 \\ x_1 \\ x_1 - x_2 \end{bmatrix}.$$

Definition

Let f be a function from $\mathbb{R}^q$ to $\mathbb{R}^p$. Then f is a *linear transformation* if f has the properties

(i) $f(\boldsymbol{x}+\boldsymbol{y})=f(\boldsymbol{x})+f(\boldsymbol{y})$ for all $\boldsymbol{x}, \boldsymbol{y} \in \mathbb{R}^q$.
(ii) $f(k\boldsymbol{x})=kf(\boldsymbol{x})$ for all $\boldsymbol{x} \in \mathbb{R}^q$ and all $k \in \mathbb{R}$.

A function which is represented by a matrix as above has the properties required for a linear transformation. Can every linear transformation be represented by a matrix? Conveniently, the answer is 'yes'. A demonstration is given as Example 16.2, and an illustration is given as Example 16.3. It is worth noting here what form the matrix takes. Let f be a linear transformation from $\mathbb{R}^q$ to $\mathbb{R}^p$, and let $(\boldsymbol{e}_1, \ldots, \boldsymbol{e}_q)$ denote the standard basis for $\mathbb{R}^q$ ($\boldsymbol{e}_j$ has 1 in the jth position and 0s elsewhere). Then f is represented by the $p \times q$ matrix A whose columns are the vectors $f(\boldsymbol{e}_1), \ldots, f(\boldsymbol{e}_q)$.

16.4 Specification of linear transformations.

(i) Show that $f\colon \mathbb{R}^3 \to \mathbb{R}^1$ is a linear transformation, where

$$f\left(\begin{bmatrix} x_1 \\ x_2 \\ x_3 \end{bmatrix}\right) = 3x_1 + x_2 - x_3 \qquad (x_1, x_2, x_3 \in \mathbb{R}).$$

We have to verify the requirements of the definition of linear transformation.

$$\begin{aligned} f\left(\begin{bmatrix} x_1 \\ x_2 \\ x_3 \end{bmatrix} + \begin{bmatrix} y_1 \\ y_2 \\ y_3 \end{bmatrix}\right) &= f\left(\begin{bmatrix} x_1 + y_1 \\ x_2 + y_2 \\ x_3 + y_3 \end{bmatrix}\right) \\ &= 3(x_1 + y_1) + (x_2 + y_2) - (x_3 + y_3) \\ &= 3x_1 + x_2 - x_3 + 3y_1 + y_2 - y_3 \\ &= f\left(\begin{bmatrix} x_1 \\ x_2 \\ x_3 \end{bmatrix}\right) + f\left(\begin{bmatrix} y_1 \\ y_2 \\ y_3 \end{bmatrix}\right). \end{aligned}$$

Also
$$\begin{aligned} f\left(a\begin{bmatrix} x_1 \\ x_2 \\ x_3 \end{bmatrix}\right) &= f\left(\begin{bmatrix} ax_1 \\ ax_2 \\ ax_3 \end{bmatrix}\right) \\ &= 3(ax_1) + (ax_2) - (ax_3) \\ &= a(3x_1 + x_2 - x_3) \\ &= af\left(\begin{bmatrix} x_1 \\ x_2 \\ x_3 \end{bmatrix}\right). \end{aligned}$$

(ii) Let $f\colon \mathbb{R}^3 \to \mathbb{R}^2$ be given by

$$f(\boldsymbol{x}) = \begin{bmatrix} 1 & 2 & 1 \\ 1 & -1 & 3 \end{bmatrix}\boldsymbol{x}, \qquad \text{for each } \boldsymbol{x} \in \mathbb{R}^3.$$

It is a consequence of the discussion in the text on page 191 that f defined in this way is a linear transformation.

(iii) Let $f\colon \mathbb{R}^3 \to \mathbb{R}^3$ be a linear transformation such that

$$f\left(\begin{bmatrix} 1 \\ 0 \\ 0 \end{bmatrix}\right) = \begin{bmatrix} 0 \\ 1 \\ 1 \end{bmatrix}, \quad f\left(\begin{bmatrix} 0 \\ 1 \\ 0 \end{bmatrix}\right) = \begin{bmatrix} 1 \\ 2 \\ 1 \end{bmatrix}, \quad f\left(\begin{bmatrix} 0 \\ 0 \\ 1 \end{bmatrix}\right) = \begin{bmatrix} 2 \\ 3 \\ 1 \end{bmatrix}.$$

Then for all $\boldsymbol{x} \in \mathbb{R}^3$,

$$f(\boldsymbol{x}) = \begin{bmatrix} 0 & 1 & 2 \\ 1 & 2 & 3 \\ 1 & 1 & 1 \end{bmatrix}\boldsymbol{x}.$$

Notice that specifying the values of f on the basis vectors is sufficient to determine the value of f on every vector in $\mathbb{R}^3$ (given, of course, that f is presumed to be a linear transformation).

How can a linear transformation be specified? Here are some ways.

(a) A rule for calculating its values can be given. It must be such that properties (i) and (ii) hold.
(b) It can be specified as the linear transformation represented by a given matrix.
(c) The images of the standard basis vectors $\boldsymbol{e}_1, \ldots, \boldsymbol{e}_q$ may be given.

These are all illustrated in Examples 16.4.

16.5 Some geometrical transformations on $\mathbb{R}^2$.

$\begin{bmatrix} x \\ y \end{bmatrix} \mapsto \begin{bmatrix} x \\ -y \end{bmatrix}$: reflection in the x-axis.

$\begin{bmatrix} x \\ y \end{bmatrix} \mapsto \begin{bmatrix} -x \\ y \end{bmatrix}$: reflection in the y-axis.

$\begin{bmatrix} x \\ y \end{bmatrix} \mapsto \begin{bmatrix} y \\ x \end{bmatrix}$: reflection in the line $y = x$.

$\begin{bmatrix} x \\ y \end{bmatrix} \mapsto \begin{bmatrix} x + ay \\ y \end{bmatrix}$: shear parallel to the x-axis (each point is shifted parallel to the x-axis by a distance proportional to y).

$\begin{bmatrix} x \\ y \end{bmatrix} \mapsto \begin{bmatrix} x \\ y + ax \end{bmatrix}$: shear parallel to the y-axis.

$\begin{bmatrix} x \\ y \end{bmatrix} \mapsto \begin{bmatrix} x \\ 0 \end{bmatrix}$: projection on to the x-axis.

$\begin{bmatrix} x \\ y \end{bmatrix} \mapsto \begin{bmatrix} 0 \\ y \end{bmatrix}$: projection on to the y-axis.

These are represented, respectively, by the matrices

$$\begin{bmatrix} 1 & 0 \\ 0 & -1 \end{bmatrix}, \begin{bmatrix} -1 & 0 \\ 0 & 1 \end{bmatrix}, \begin{bmatrix} 0 & 1 \\ 1 & 0 \end{bmatrix}, \begin{bmatrix} 1 & a \\ 0 & 1 \end{bmatrix}, \begin{bmatrix} 1 & 0 \\ a & 1 \end{bmatrix}, \begin{bmatrix} 1 & 0 \\ 0 & 0 \end{bmatrix}, \begin{bmatrix} 0 & 0 \\ 0 & 1 \end{bmatrix}.$$

16.6 Rotations about the coordinate axes in $\mathbb{R}^3$.

(i) Rotation through an angle α about the x-axis is the linear transformation represented by the matrix

$$\begin{bmatrix} 1 & 0 & 0 \\ 0 & \cos\alpha & -\sin\alpha \\ 0 & \sin\alpha & \cos\alpha \end{bmatrix}.$$

(ii) A rotation about the y-axis is represented by a matrix

$$\begin{bmatrix} \cos\alpha & 0 & -\sin\alpha \\ 0 & 1 & 0 \\ \sin\alpha & 0 & \cos\alpha \end{bmatrix}.$$

(iii) A rotation about the z-axis is represented by a matrix

$$\begin{bmatrix} \cos\alpha & -\sin\alpha & 0 \\ \sin\alpha & \cos\alpha & 0 \\ 0 & 0 & 1 \end{bmatrix}.$$

In two and three dimensions, some linear transformations have simple geometrical effects. For example, in two dimensions, the linear transformation from $\mathbb{R}^2$ to $\mathbb{R}^2$ which takes

$$\begin{bmatrix} x \\ y \end{bmatrix} \text{ to } \begin{bmatrix} ax \\ ay \end{bmatrix} \quad \left(\text{i.e. to } \begin{bmatrix} a & 0 \\ 0 & a \end{bmatrix}\begin{bmatrix} x \\ y \end{bmatrix}\right),$$

where a is a non-zero constant, can be thought of as a dilatation of the plane. Each point (x, y) is taken to the point (ax, ay). Other illustrations are given in Examples 16.5.

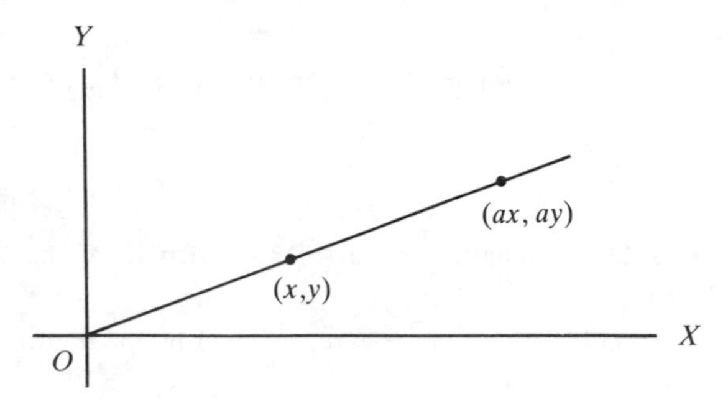

One important geometrical operation is rotation (in two or three dimensions) about an axis through the origin. Algebraically, such an operation is a linear transformation. In two dimensions, rotation through an angle α takes a point (x, y) to the point (x', y'), where

$$x' = x \cos \alpha - y \sin \alpha \qquad \text{and} \qquad y' = y \sin \alpha + x \cos \alpha.$$

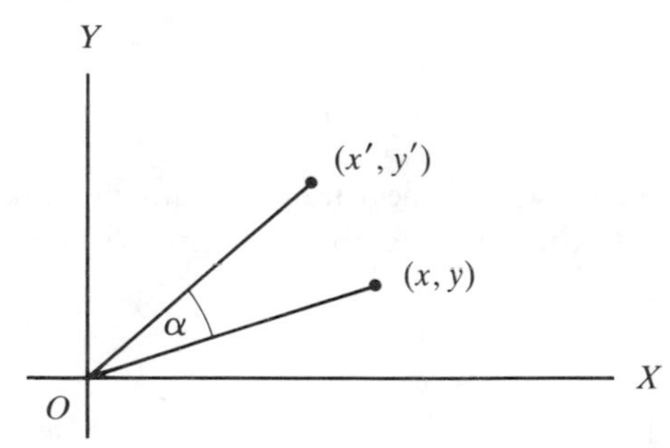

In algebraic terms, the image of $\begin{bmatrix} x \\ y \end{bmatrix}$ is $A\begin{bmatrix} x \\ y \end{bmatrix}$, where

$$A = \begin{bmatrix} \cos \alpha & -\sin \alpha \\ \sin \alpha & \cos \alpha \end{bmatrix}.$$

In three dimensions a matrix which represents a rotation is normally rather more complicated, but we give in Example 16.6 the matrices which represent rotations about the three coordinate axes. There is further discussion of rotations in Chapter 18.

16.7 One-to-one linear transformations.

(i) The linear transformation from $\mathbb{R}^3$ to $\mathbb{R}^3$ represented by the matrix

$$A=\begin{bmatrix}1 & -1 & 2\\ 0 & 1 & -1\\ 2 & 1 & 1\end{bmatrix}$$

is not one-to-one.

As shown opposite,

$$A\begin{bmatrix}0\\1\\2\end{bmatrix}=A\begin{bmatrix}1\\0\\1\end{bmatrix}=\begin{bmatrix}3\\-1\\3\end{bmatrix}.$$

Hence the function represented by A is not one-to-one. Notice that the difference

$$\begin{bmatrix}0\\1\\2\end{bmatrix}-\begin{bmatrix}1\\0\\1\end{bmatrix},\qquad \text{namely}\begin{bmatrix}-1\\1\\1\end{bmatrix}$$

is a solution to the equation $A\mathbf{x}=\mathbf{0}$. What is the rank of this matrix A? A standard calculation shows it to be 2.

(ii) The linear transformation from $\mathbb{R}^2$ to $\mathbb{R}^3$ represented by the matrix

$$B=\begin{bmatrix}2 & 1\\ 1 & 0\\ 1 & -1\end{bmatrix}$$

is one-to-one.

The rank of B is 2 (standard calculation), so it follows, by the argument in the text, that whenever $\mathbf{y}\neq\mathbf{z}$ in $\mathbb{R}^2$, $B\mathbf{y}\neq B\mathbf{z}$ in $\mathbb{R}^3$, so B represents a one-to-one function.

(iii) Let $C=\begin{bmatrix}1 & 1 & 0\\ 1 & 0 & 1\\ 0 & 1 & 1\end{bmatrix}$.

This 3×3 matrix has rank 3, so it represents a linear transformation from $\mathbb{R}^3$ to $\mathbb{R}$ which is one-to-one. There do not exist distinct vectors $\mathbf{y}$ and $\mathbf{z}$ in $\mathbb{R}^3$ for which $C\mathbf{y}=C\mathbf{z}$.

(iv) Let $D=\begin{bmatrix}-1 & 2 & 1\\ 2 & -4 & 2\\ 1 & -2 & -1\end{bmatrix}$.

This 3×3 matrix has rank 1, so it represents a linear transformation which is not one-to-one. A convenient way to find distinct vectors which are taken to the same vector is to solve the equation $D\mathbf{x}=\mathbf{0}$, to obtain $\mathbf{x}$ different from $\mathbf{0}$. Then $\mathbf{x}$ and $\mathbf{0}$ are distinct vectors which are both taken to $\mathbf{0}$ by multiplication by D. The usual process of solving equations yields such an $\mathbf{x}$, say

$$\mathbf{x}=\begin{bmatrix}2\\1\\0\end{bmatrix}.$$

When we consider the linear transformation represented by a matrix A, we may note the possibility that vectors which are different may be taken to the same vector. For example

$$\begin{bmatrix} 1 & -1 & 2 \\ 0 & 1 & -1 \\ 2 & 1 & 1 \end{bmatrix}\begin{bmatrix} 0 \\ 1 \\ 2 \end{bmatrix}=\begin{bmatrix} 3 \\ -1 \\ 3 \end{bmatrix}$$

and

$$\begin{bmatrix} 1 & -1 & 2 \\ 0 & 1 & -1 \\ 2 & 1 & 1 \end{bmatrix}\begin{bmatrix} 1 \\ 0 \\ 1 \end{bmatrix}=\begin{bmatrix} 3 \\ -1 \\ 3 \end{bmatrix}.$$

On the other hand, it may happen that different vectors are always taken to different vectors by the process of multiplying by a particular matrix. Let us pursue this a little, because it is important, and it links in with previous work.

Let A be a $p \times q$ matrix and let $\mathbf{y}$ and $\mathbf{z}$ be two p-vectors which are different, but for which

$$A\mathbf{y} = A\mathbf{z}$$

(as in the example above). Then it follows that

$$A(\mathbf{y} - \mathbf{z}) = \mathbf{0}.$$

Since $\mathbf{y} - \mathbf{z} \neq \mathbf{0}$, it follows that the equation $A\mathbf{x} = \mathbf{0}$ has a solution other than $\mathbf{x} = \mathbf{0}$. Using a rule from Chapter 8, this must mean that the rank of A is less than q. Examples 16.7 illustrate this, and the converse case also, which we now describe.

Suppose that A is a $p \times q$ matrix such that whenever $\mathbf{y} \neq \mathbf{z}$ (in $\mathbb{R}^q$), we have $A\mathbf{y} \neq A\mathbf{z}$. Then it follows as a particular case of this that if $\mathbf{x} \neq \mathbf{0}$ then

$$A\mathbf{x} \neq A\mathbf{0},$$

i.e. $$A\mathbf{x} \neq \mathbf{0}.$$

Thus the equation $A\mathbf{x} = \mathbf{0}$ has a unique solution $\mathbf{x} = \mathbf{0}$. Again from Chapter 8, it follows that the rank of A must be equal to q.

In this chapter we are dealing with functions (linear transformations) represented by matrices. What has been figuring in the above discussion is an important property that functions in general may have. This is the property of being one-to-one.

Definition

A function f from a set S to a set T is *one-to-one* if, whenever x and y are different elements of S, then $f(x)$ and $f(y)$ are different elements of T.

16.8 Image and column space.

Let $$A=\begin{bmatrix}3 & -1 & 1\\ 0 & 2 & 1\\ 1 & -1 & 1\end{bmatrix}.$$

The column space of A is the set of all 3-vectors which are linear combinations of the columns of A, i.e. all vectors

$$\begin{bmatrix}3\\0\\1\end{bmatrix}x_1+\begin{bmatrix}-1\\2\\-1\end{bmatrix}x_2+\begin{bmatrix}1\\1\\1\end{bmatrix}x_3,$$

where x_1, x_2 and x_3 are real numbers.

The image of the linear transformation represented by A is the set of all 3-vectors $A\mathbf{x}$, for all possible values of $\mathbf{x}$. This is the set of all products

$$\begin{bmatrix}3 & -1 & 1\\ 0 & 2 & 1\\ 1 & -1 & 1\end{bmatrix}\begin{bmatrix}x_1\\x_2\\x_3\end{bmatrix},$$

where x_1, x_2 and x_3 are real numbers. This is the same set as described above. Both sets may be written

$$\left\{\begin{bmatrix}3x_1- & x_2+x_3\\ & 2x_2+x_3\\ x_1- & x_2+x_3\end{bmatrix}: x_1, x_2, x_3\in\mathbb{R}\right\}.$$

16.9 Diagrammatic representation of a linear transformation from $\mathbb{R}^q$ to $\mathbb{R}^p$.

$\ker(f)=\{\mathbf{x}\in\mathbb{R}^q\colon f(\mathbf{x})=\mathbf{0}\}$.

$\operatorname{im}(f)=\{\mathbf{y}\in\mathbb{R}^p\colon \mathbf{y}=f(\mathbf{x})$ for some $\mathbf{x}\in\mathbb{R}^q\}$.

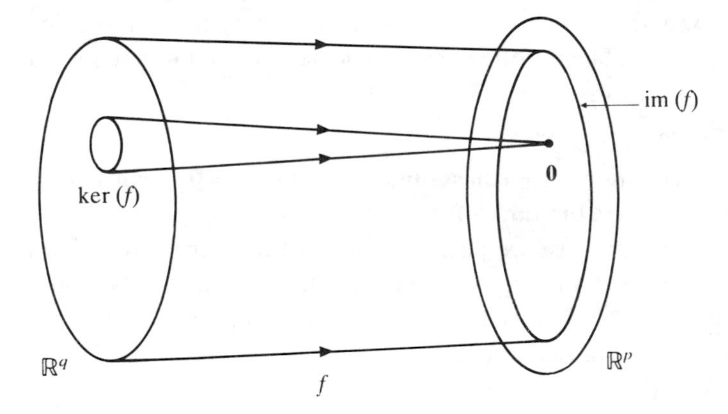

The conclusion which can be reached from the arguments above is that a $p \times q$ matrix A represents a function f from $\mathbb{R}^q$ to $\mathbb{R}^p$ which is one-to-one if and only if the rank of A is equal to q.

The term 'one-to-one' is indicative of the property it describes. A function which is one-to-one provides a correspondence between elements of two sets: each x corresponds with $f(x)$, and each value $f(x)$ corresponds with x only. If f is a linear transformation from $\mathbb{R}^q$ to $\mathbb{R}^p$, the image of f (denoted by $\mathrm{im}(f)$) is the set of all values of the function f, i.e. the set

$$\{\boldsymbol{y} \in \mathbb{R}^p : \boldsymbol{y} = f(\boldsymbol{x}) \text{ for some } \boldsymbol{x} \in \mathbb{R}^q\}.$$

If f is one-to-one, then the elements of $\mathbb{R}^q$ and the elements of $\mathrm{im}(f)$ can be matched off in pairs, and in a sense, $\mathrm{im}(f)$ is a 'copy' of $\mathbb{R}^q$. We can make this more concrete by considering matrices. If f is represented by the matrix A, the image of f is a set we already know about, namely the column space of A. This was implicit in Chapter 12, but is made explicit in Example 16.8. From Chapter 13 we know that the column space of a $p \times q$ matrix is a subspace of $\mathbb{R}^p$. We also know (from the definition of rank) that the column space of such a matrix has dimension equal to the rank of the matrix. Putting all this together, the image of a linear transformation from $\mathbb{R}^q$ to $\mathbb{R}^p$ is a subspace of $\mathbb{R}^p$, whose dimension is equal to the rank of a matrix which represents the linear transformation.

Next we consider another subspace which can be associated with a linear transformation f. This is the kernel of f. Again, matrices will help to give a clear idea. We know that if the equation $A\boldsymbol{x} = \boldsymbol{0}$ has any solution other than $\boldsymbol{x} = \boldsymbol{0}$, then it has infinitely many solutions. Indeed we already know (see Chapter 13) that the set

$$\{\boldsymbol{x} \in \mathbb{R}^q : A\boldsymbol{x} = \boldsymbol{0}\}$$

is a subspace of $\mathbb{R}^q$. If we think of A as representing a linear transformation f then this set is the set of all vectors which are taken to the zero vector by the linear transformation. This set is the kernel of f.

Definition

Let f be a linear transformation from $\mathbb{R}^q$ to $\mathbb{R}^p$. The *kernel* of f, denoted by $\ker(f)$, is the set

$$\{\boldsymbol{x} \in \mathbb{R}^q : f(\boldsymbol{x}) = \boldsymbol{0}\}.$$

This is represented diagrammatically in Example 16.9. The diagram also gives an indication of the image of f.

16.10 Kernels and images of linear transformations. In each case let f be the linear transformation represented by the given matrix.

(i) $A=\begin{bmatrix}1 & 0\\ 2 & 1\\ 1 & -1\end{bmatrix}$. Here f is from $\mathbb{R}^2$ to $\mathbb{R}^3$.

The rank of A is equal to 2, the number of columns of A, so here f is one-to-one. It follows that $\ker(f)=\{\mathbf{0}\}$, and so $\dim(\ker(f))=0$. Consequently, $\dim(\operatorname{im}(f))=2$. Notice that it is not possible for a linear transformation from $\mathbb{R}^q$ to $\mathbb{R}^p$ to have image with dimension greater than q.

(ii) $A=\begin{bmatrix}1 & 3\\ 2 & 6\\ 1 & 3\end{bmatrix}$. Here f is from $\mathbb{R}^2$ to $\mathbb{R}^3$.

In this case, f is not one-to-one. The image of f is the column space of A, and it is easy to see that this has dimension 1 (the columns are multiples of each other). Hence

$$\dim(\ker(f))=1 \quad \text{and} \quad \dim(\operatorname{im}(f))=1.$$

(iii) $A=\begin{bmatrix}1 & 2 & 0\\ 0 & -1 & 1\\ 1 & 0 & 1\end{bmatrix}$, so that f is from $\mathbb{R}^3$ to $\mathbb{R}^3$.

The rank of A is equal to 3 (check this). Hence the column space of A has dimension 3, so $\dim(\operatorname{im}(f))=3$. Consequently, $\dim(\ker(f))=0$ (f is one-to-one).

(iv) $A=\begin{bmatrix}1 & -2 & 4\\ 1 & 1 & 1\\ 2 & 3 & 1\end{bmatrix}$, so that f is from $\mathbb{R}^3$ to $\mathbb{R}^3$.

Here the rank of A is equal to 2, so

$$\dim(\operatorname{im}(f))=2 \quad \text{and} \quad \dim(\ker(f))=1.$$

(v) $A=\begin{bmatrix}1 & 2 & 3\\ 2 & 4 & 6\\ 3 & 6 & 9\end{bmatrix}$, so that f is from $\mathbb{R}^3$ to $\mathbb{R}^3$.

Here the rank of A is equal to 1, so

$$\dim(\operatorname{im}(f))=1 \quad \text{and} \quad \dim(\ker(f))=2.$$

If these examples are not clear, then it would be a useful exercise to find the kernel in each case explicitly by solving the equation $A\mathbf{x}=\mathbf{0}$ in the standard way.

Although the kernel of a linear transformation is familiar as the solution space of an equation of the form $A\boldsymbol{x}=\mathbf{0}$, our previous work has not given us a convenient way of finding the dimension of such a solution space. There is a simple and important relationship between the dimensions of the kernel and image of a linear transformation, which we state here without proof.

Rule

Let f be a linear transformation from $\mathbb{R}^q$ to $\mathbb{R}^p$. Then the sum of the dimensions of the kernel of f and the image of f is equal to q.

Examples 16.10 give some specific matrices which represent linear transformations, with the dimensions of the corresponding kernels and images. Notice the particular case of a one-to-one linear transformation. The kernel of such a linear transformation is the subspace $\{\mathbf{0}\}$, which has dimension 0.

Linear transformations from $\mathbb{R}^p$ to $\mathbb{R}^p$ are an important special case. As noted in Chapter 14, a subspace of $\mathbb{R}^p$ which has dimension equal to p must necessarily be the whole of $\mathbb{R}^p$. The image of a one-to-one linear transformation from $\mathbb{R}^p$ to $\mathbb{R}^p$ is such a subspace, because its kernel has dimension 0, and the sum of the dimensions of the kernel and image is equal to p. So in this situation a one-to-one linear transformation f provides a pairing of elements of $\mathbb{R}^p$. Each $\boldsymbol{x}$ is paired with $f(\boldsymbol{x})$, and each $\boldsymbol{y}\in\mathbb{R}^p$ is actually equal to $f(\boldsymbol{x})$ for one particular $\boldsymbol{x}$.

Let us end this chapter by drawing together all the strands we have considered into a rule.

Rule

Let f be a linear transformation from $\mathbb{R}^p$ to $\mathbb{R}^p$, represented by the $p\times p$ matrix A. The following are equivalent.

(i) f is one-to-one.
(ii) $\ker(f)=\{\mathbf{0}\}$.
(iii) $\operatorname{im}(f)=\mathbb{R}^p$.
(iv) A is invertible.
(v) The rank of A is equal to p.

This is very similar to the Equivalence Theorem from Chapter 12, especially if we note that (ii) may be re-written as: the equation $A\boldsymbol{x}=\mathbf{0}$ has no solution other than $\boldsymbol{x}=\mathbf{0}$. Thus partial justification for the rule is contained in the body of this chapter. Details of the proof are omitted.

Summary

The idea of a linear transformation is introduced by noting some properties held by functions represented by matrices. It is shown that any linear transformation is represented by the matrix whose columns are the images of the standard basis vectors. Some geometrical examples are given. A criterion is found, involving the notion of rank, for deciding whether a given linear transformation is one-to-one. Kernels and images of linear transformations are introduced and the relationship between their dimensions is stated as a rule.

Exercises

1. (i) Let f be a linear transformation from $\mathbb{R}^2$ to $\mathbb{R}^3$ such that

$$f\left(\begin{bmatrix}1\\0\end{bmatrix}\right)=\begin{bmatrix}-1\\1\\2\end{bmatrix} \quad \text{and} \quad f\left(\begin{bmatrix}0\\1\end{bmatrix}\right)=\begin{bmatrix}3\\0\\1\end{bmatrix}.$$

Show that for each vector $\begin{bmatrix}x\\y\end{bmatrix}$ in $\mathbb{R}^2$,

$$f\left(\begin{bmatrix}x\\y\end{bmatrix}\right)=\begin{bmatrix}-1&3\\1&0\\2&1\end{bmatrix}\begin{bmatrix}x\\y\end{bmatrix}.$$

(ii) Let g be a linear transformation from $\mathbb{R}^3$ to $\mathbb{R}^3$ such that

$$g\left(\begin{bmatrix}1\\0\\0\end{bmatrix}\right)=\begin{bmatrix}2\\1\\0\end{bmatrix},\quad g\left(\begin{bmatrix}0\\1\\0\end{bmatrix}\right)=\begin{bmatrix}1\\0\\2\end{bmatrix} \quad \text{and} \quad g\left(\begin{bmatrix}0\\0\\1\end{bmatrix}\right)=\begin{bmatrix}0\\2\\1\end{bmatrix}.$$

Show that for each vector $\begin{bmatrix}x\\y\\z\end{bmatrix}$ in $\mathbb{R}^3$,

$$g\left(\begin{bmatrix}x\\y\\z\end{bmatrix}\right)=\begin{bmatrix}2&1&0\\1&0&2\\0&2&1\end{bmatrix}\begin{bmatrix}x\\y\\z\end{bmatrix}.$$

(iii) Let h be a linear transformation from $\mathbb{R}^2$ to $\mathbb{R}^3$ such that

$$h\left(\begin{bmatrix}1\\1\end{bmatrix}\right)=\begin{bmatrix}1\\2\\1\end{bmatrix},\quad h\left(\begin{bmatrix}-1\\1\end{bmatrix}\right)=\begin{bmatrix}-1\\0\\3\end{bmatrix}.$$

Calculate $h\left(\begin{bmatrix}1\\0\end{bmatrix}\right)$ and $h\left(\begin{bmatrix}0\\1\end{bmatrix}\right)$, and hence find a matrix A such that for each vector $\boldsymbol{x}\in\mathbb{R}^2$, $h(\boldsymbol{x})=A\boldsymbol{x}$.

2. Let f be a function from $\mathbb{R}^4$ to $\mathbb{R}^3$ given by

$$f\left(\begin{bmatrix} x \\ y \\ z \\ w \end{bmatrix}\right) = \begin{bmatrix} x+y+z \\ y+z+w \\ z+w+x \end{bmatrix}.$$

Show that f is a linear transformation, and find a matrix A which represents it.

3. Let f be the linear transformation from $\mathbb{R}^3$ to $\mathbb{R}^3$ for which

$$f\left(\begin{bmatrix} 1 \\ 0 \\ 0 \end{bmatrix}\right) = \begin{bmatrix} 2 \\ 1 \\ 1 \end{bmatrix}, \quad f\left(\begin{bmatrix} 0 \\ 1 \\ 0 \end{bmatrix}\right) = \begin{bmatrix} 0 \\ -3 \\ 3 \end{bmatrix} \quad \text{and} \quad f\left(\begin{bmatrix} 0 \\ 0 \\ 1 \end{bmatrix}\right) = \begin{bmatrix} 2 \\ 0 \\ 2 \end{bmatrix}.$$

Show that

$$f\left(\begin{bmatrix} 4 \\ 3 \\ -2 \end{bmatrix}\right) = f\left(\begin{bmatrix} -2 \\ 1 \\ 5 \end{bmatrix}\right).$$

Is f one-to-one? Is the image of f equal to $\mathbb{R}^3$?

4. In each case below determine whether the linear transformation represented by the given matrix is one-to-one.

(i) $\begin{bmatrix} 1 & -1 \\ 1 & 1 \end{bmatrix}$ $(\mathbb{R}^2 \to \mathbb{R}^2)$

(ii) $\begin{bmatrix} 1 & -1 & 0 \\ 1 & 1 & 1 \end{bmatrix}$ $(\mathbb{R}^3 \to \mathbb{R}^2)$

(iii) $\begin{bmatrix} 1 & 2 \\ 2 & 1 \\ 1 & 1 \end{bmatrix}$ $(\mathbb{R}^2 \to \mathbb{R}^3)$

(iv) $\begin{bmatrix} 3 & 2 & 1 \\ 1 & 1 & -1 \\ 1 & 0 & 3 \end{bmatrix}$ $(\mathbb{R}^3 \to \mathbb{R}^3)$

(v) $\begin{bmatrix} 0 & 1 & -1 \\ 1 & 2 & 1 \\ 2 & 3 & 3 \\ 2 & 1 & 5 \end{bmatrix}$ $(\mathbb{R}^3 \to \mathbb{R}^4)$

(vi) $\begin{bmatrix} 1 & 2 \\ 3 & 4 \\ 5 & 6 \\ 7 & 8 \end{bmatrix}$ $(\mathbb{R}^2 \to \mathbb{R}^4)$

(vii) $\begin{bmatrix} 2 & 1 & 1 & 3 \\ 1 & 2 & -1 & 3 \\ -1 & 1 & -2 & 1 \end{bmatrix}$ $(\mathbb{R}^4 \to \mathbb{R}^3)$

(viii) $\begin{bmatrix} 1 & 2 & -3 & 3 \\ -1 & 1 & -3 & 0 \\ 3 & 1 & 1 & 4 \end{bmatrix}$ $(\mathbb{R}^4 \to \mathbb{R}^3)$

5. Verify that the rotations whose matrices are given in Example 16.6 are one-to-one functions.

6. Find the kernels and images of all of the linear transformations whose matrices are listed in Exercise 4 above.

Examples

17.1 Components with respect to bases other than the standard basis.

(i) $B = \left(\begin{bmatrix} 1 \\ 1 \end{bmatrix}, \begin{bmatrix} 1 \\ -1 \end{bmatrix} \right)$ is a basis for $\mathbb{R}^2$. Let $\boldsymbol{v} = \begin{bmatrix} 1 \\ 3 \end{bmatrix}$. Then $\boldsymbol{v} = \begin{bmatrix} 1 \\ 0 \end{bmatrix} + 3\begin{bmatrix} 0 \\ 1 \end{bmatrix}$ in terms of the standard basis. Also

$$\boldsymbol{v} = 2\begin{bmatrix} 1 \\ 1 \end{bmatrix} - \begin{bmatrix} 1 \\ -1 \end{bmatrix},$$

so $\boldsymbol{v}$ may be represented by the 2-vector $\begin{bmatrix} 2 \\ -1 \end{bmatrix}$ of components with respect to the basis B.

(ii) $B = \left(\begin{bmatrix} 1 \\ 1 \\ 0 \end{bmatrix}, \begin{bmatrix} 1 \\ 0 \\ 1 \end{bmatrix}, \begin{bmatrix} 0 \\ 1 \\ 1 \end{bmatrix} \right)$ is a basis for $\mathbb{R}^3$. Let $\boldsymbol{v} = \begin{bmatrix} 1 \\ 5 \\ 2 \end{bmatrix}$.

Then $\quad \boldsymbol{v} = \begin{bmatrix} 1 \\ 0 \\ 0 \end{bmatrix} + 5\begin{bmatrix} 0 \\ 1 \\ 0 \end{bmatrix} + 2\begin{bmatrix} 0 \\ 0 \\ 1 \end{bmatrix}.$

Also $\quad \boldsymbol{v} = 2\begin{bmatrix} 1 \\ 1 \\ 0 \end{bmatrix} - \begin{bmatrix} 1 \\ 0 \\ 1 \end{bmatrix} + 3\begin{bmatrix} 0 \\ 1 \\ 1 \end{bmatrix},$

so $\boldsymbol{v}$ may be represented by the 3-vector $\begin{bmatrix} 2 \\ -1 \\ 3 \end{bmatrix}$ of components with respect to the basis B.

(iii) $B = \left(\begin{bmatrix} 1 \\ 2 \\ 2 \end{bmatrix}, \begin{bmatrix} 0 \\ 1 \\ -1 \end{bmatrix}, \begin{bmatrix} 2 \\ 1 \\ -2 \end{bmatrix} \right)$ is a basis for $\mathbb{R}^3$. Find the components of $\boldsymbol{v} = \begin{bmatrix} 3 \\ 7 \\ -4 \end{bmatrix}$ with respect to B. To do this we require to solve the equation

$$\begin{bmatrix} 1 \\ 2 \\ 2 \end{bmatrix} x_1 + \begin{bmatrix} 0 \\ 1 \\ -1 \end{bmatrix} x_2 + \begin{bmatrix} 2 \\ 1 \\ -2 \end{bmatrix} x_3 = \begin{bmatrix} 3 \\ 7 \\ -4 \end{bmatrix}.$$

The GE process gives

$$\begin{bmatrix} 1 & 0 & 2 & 3 \\ 2 & 1 & 1 & 7 \\ 2 & -1 & -2 & -4 \end{bmatrix} \rightarrow \begin{bmatrix} 1 & 0 & 2 & 3 \\ 0 & 1 & -3 & 1 \\ 0 & 0 & 1 & 1 \end{bmatrix},$$

yielding the (unique) solution $x_1 = 1$, $x_2 = 4$, $x_3 = 1$.

Hence $\begin{bmatrix} 3 \\ 7 \\ -4 \end{bmatrix}$ has components $\begin{bmatrix} 1 \\ 4 \\ 1 \end{bmatrix}$ with respect to the basis B.

17

Change of basis

The elements of $\mathbb{R}^n$ are $n \times 1$ matrices, and we have seen how the entries in such a matrix can be regarded as components of a vector. When we express a given n-vector as a linear combination of the standard basis vectors, the coefficients which appear are the components of the vector, and they are in fact just the entries in the original $n \times 1$ matrix.

There are, of course, other bases for $\mathbb{R}^n$ besides the standard basis, and a property which every basis has is that every vector in $\mathbb{R}^n$ may be expressed (uniquely) as a linear combination of the basis vectors. Let $B = (\mathbf{v}_1, \ldots, \mathbf{v}_n)$ be a basis for $\mathbb{R}^n$ and let $\mathbf{v} \in \mathbb{R}^n$. Then we may write

$$\mathbf{v} = a_1\mathbf{v}_1 + \cdots + a_n\mathbf{v}_n,$$

where the coefficients $a_1, \ldots, a_n$ are uniquely determined. See Chapter 14. These coefficients are called the components of $\mathbf{v}$ with respect to the basis B. Indeed $\mathbf{v}$ may be represented as the n-vector $\begin{bmatrix} a_1 \\ \vdots \\ a_n \end{bmatrix}$ with respect to B. Examples 17.1 give some particular cases of this.

17.2 Components with respect to a basis in a subspace of $\mathbb{R}^3$.

Let $S = \left\{ \begin{bmatrix} x \\ y \\ z \end{bmatrix} \in \mathbb{R}^3 : 2x - y - 3z = 0 \right\}$.

S may be shown to be a subspace of $\mathbb{R}^3$ by the methods of Chapter 13. Also by our earlier methods it can be shown that the list $B = \left(\begin{bmatrix} 1 \\ 2 \\ 0 \end{bmatrix}, \begin{bmatrix} 1 \\ -1 \\ 1 \end{bmatrix} \right)$ is a basis for S (so S has dimension 2). Every vector in S may be represented uniquely as a linear combination of these vectors. For example,

$$\begin{bmatrix} 7 \\ -1 \\ 5 \end{bmatrix} \in S, \qquad \text{and} \qquad \begin{bmatrix} 7 \\ -1 \\ 5 \end{bmatrix} = 2 \begin{bmatrix} 1 \\ 2 \\ 0 \end{bmatrix} + 5 \begin{bmatrix} 1 \\ -1 \\ 1 \end{bmatrix}.$$

This expression is found by solving equations, as in Example 17.1(iii) above. Thus $\begin{bmatrix} 7 \\ -1 \\ 5 \end{bmatrix}$ may be represented by the 2-vector $\begin{bmatrix} 2 \\ 5 \end{bmatrix}$ with respect to the basis B for S.

17.3 Geometrical representation for a change of basis in $\mathbb{R}^2$.

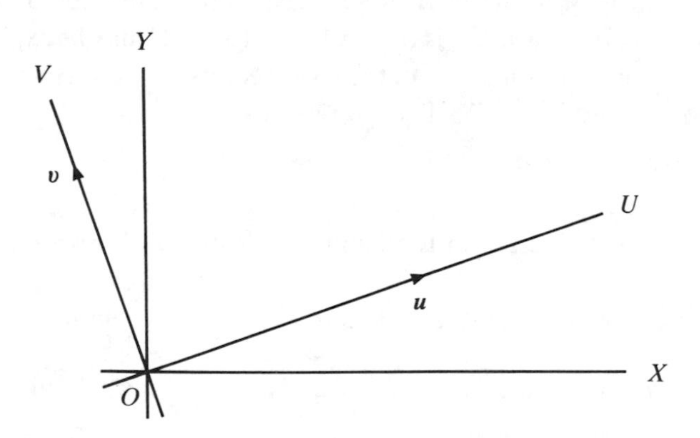

OU and OV are new axes. In geometry they would perhaps normally be chosen to be perpendicular, but for our algebraic purposes they need not be (they must of course not be in the same or opposite directions). Let $\boldsymbol{u}$ and $\boldsymbol{v}$ be vectors in the directions of OU and OV. These may be, but need not be, unit vectors. Then $(\boldsymbol{u}, \boldsymbol{v})$ is LI, and so is a basis for $\mathbb{R}^2$. Suppose that

$$\boldsymbol{u} = \begin{bmatrix} u_1 \\ u_2 \end{bmatrix} \quad \text{and} \quad \boldsymbol{v} = \begin{bmatrix} v_1 \\ v_2 \end{bmatrix}, \quad \text{and let } \boldsymbol{x} = \begin{bmatrix} x \\ y \end{bmatrix}.$$

How can we find the components of $\boldsymbol{x}$ with respect to the basis $(\boldsymbol{u}, \boldsymbol{v})$? Just solve the equation

$$\boldsymbol{x} = a\boldsymbol{u} + b\boldsymbol{v}$$

for a and b. This equation may be written

This is a very general notion which applies not just to the spaces $\mathbb{R}^n$ and their various bases, but also to subspaces of $\mathbb{R}^n$. If S is a subspace of $\mathbb{R}^n$ and B is a basis for S, then every vector in S has components with respect to B and may be represented by a column vector of components. The number of components, of course, is the number of vectors in the basis, i.e. the dimension of S. See Example 17.2.

It is important to remember that a basis is a list of vectors, rather than a set. The elements of a basis are to be taken in a particular order, and consequently the components of any vector with respect to a given basis will be in the corresponding order.

It may be helpful to bring in some geometrical ideas. Points in three-dimensional space may be represented by ordered triples of coordinates with respect to fixed coordinate axes. The connection between algebra and geometry is made by associating a point $P(x, y, z)$ with its position vector $\overrightarrow{OP} = \begin{bmatrix} x \\ y \\ z \end{bmatrix}$. The standard basis vectors in $\mathbb{R}^3$ are usually denoted by $\boldsymbol{i}$, $\boldsymbol{j}$ and $\boldsymbol{k}$ respectively in the directions of the x-axis, the y-axis and the z-axis. The coordinates x, y and z of the point P are the components of the vector $\overrightarrow{OP}$ with respect to the standard basis. It is quite a common geometrical operation to change the axes, i.e. to consider a different set of axes (for simplicity here we assume that the origin remains fixed), and to try to find formulas which relate coordinates of any given point with respect to the different sets of axes. This can be done using geometrical methods, but it is much better dealt with algebraically. Example 17.3 shows the procedures involved in $\mathbb{R}^2$.

$$x = au_1 + bv_1,$$
$$y = au_2 + bv_2$$

which in turn may be written

$$\begin{bmatrix} x \\ y \end{bmatrix} = P\begin{bmatrix} a \\ b \end{bmatrix}, \qquad \text{where } P = \begin{bmatrix} u_1 & v_1 \\ u_2 & v_2 \end{bmatrix}.$$

P is invertible because its columns form a LI list, so we may write

$$\begin{bmatrix} a \\ b \end{bmatrix} = P^{-1}\begin{bmatrix} x \\ y \end{bmatrix},$$

which yields the components a and b which we seek. As an illustration, consider the case where

$$\boldsymbol{u} = \begin{bmatrix} 1 \\ 1 \end{bmatrix}, \quad \boldsymbol{v} = \begin{bmatrix} -1 \\ 1 \end{bmatrix} \quad \text{and} \quad P = \begin{bmatrix} 1 & -1 \\ 1 & 1 \end{bmatrix}.$$

Then $\quad P^{-1} = \begin{bmatrix} \frac{1}{2} & \frac{1}{2} \\ -\frac{1}{2} & \frac{1}{2} \end{bmatrix} \quad$ and $\quad \begin{bmatrix} a \\ b \end{bmatrix} = \begin{bmatrix} \frac{1}{2} & \frac{1}{2} \\ -\frac{1}{2} & \frac{1}{2} \end{bmatrix}\begin{bmatrix} x \\ y \end{bmatrix},$

i.e. $\quad a = \frac{1}{2}x + \frac{1}{2}y,$

and $\quad b = -\frac{1}{2}x + \frac{1}{2}y.$

17.4 Change of coordinates in three dimensions.

Let $\quad \boldsymbol{u} = \begin{bmatrix} 1 \\ 1 \\ 0 \end{bmatrix}, \quad \boldsymbol{v} = \begin{bmatrix} 1 \\ 0 \\ 1 \end{bmatrix}, \quad \boldsymbol{w} = \begin{bmatrix} 0 \\ 1 \\ 1 \end{bmatrix}.$

Then $(\boldsymbol{u}, \boldsymbol{v}, \boldsymbol{w})$ is a basis for $\mathbb{R}^3$. We show how to find the components of any given vector in terms of this basis.

Let $\quad \boldsymbol{x} = \begin{bmatrix} x \\ y \\ z \end{bmatrix}, \quad$ and let $\boldsymbol{x} = a\boldsymbol{u} + b\boldsymbol{v} + c\boldsymbol{w}$. Then (from the text opposite)

$$\begin{bmatrix} a \\ b \\ c \end{bmatrix} = P^{-1}\begin{bmatrix} x \\ y \\ z \end{bmatrix}, \qquad \text{where } P = \begin{bmatrix} 1 & 1 & 0 \\ 1 & 0 & 1 \\ 0 & 1 & 1 \end{bmatrix}.$$

By computing P^{-1} in the normal way, we may derive

$$a = \tfrac{1}{2}x + \tfrac{1}{2}y - \tfrac{1}{2}z,$$
$$b = \tfrac{1}{2}x - \tfrac{1}{2}y + \tfrac{1}{2}z,$$
$$c = -\tfrac{1}{2}x + \tfrac{1}{2}y + \tfrac{1}{2}z.$$

These are the equations which yield the new components. In Example 17.1(ii) we dealt with the single case of $\begin{bmatrix} x \\ y \\ z \end{bmatrix} = \begin{bmatrix} 1 \\ 5 \\ 2 \end{bmatrix}$. We now have general equations which we could use to obtain new components without solving equations every time (but of course the work of inverting a matrix remains). Let us check the answer obtained in Example 17.1:

$$a = \tfrac{1}{2} + \tfrac{5}{2} - \tfrac{2}{2} = 2$$
$$b = \tfrac{1}{2} - \tfrac{5}{2} + \tfrac{2}{2} = -1$$
$$c = -\tfrac{1}{2} + \tfrac{5}{2} + \tfrac{2}{2} = 3,$$

which agrees with the previous result.

Let us here deal with $\mathbb{R}^3$. See Example 17.4.

Let OU, OV and OW be new coordinate axes in three dimensions. It may be helpful to imagine that they are rectangular and right-handed, as they would normally be in a geometrical situation, but from the point of view of the algebra they need not be (and in Example 17.4 they are not). All we require is that OU, OV and OW are not coplanar, so the vectors in these directions are linearly independent. We shall find a way of converting between coordinates with respect to OU, OV and OW, and coordinates with respect to the standard axes OX, OY and OZ. Let $\mathbf{u}$, $\mathbf{v}$ and $\mathbf{w}$ be vectors in the directions of OU, OV and OW respectively. Then the list $(\mathbf{u}, \mathbf{v}, \mathbf{w})$ is a basis for $\mathbb{R}^3$ (being a LI list in $\mathbb{R}^3$ with three members). Suppose that

$$\mathbf{u} = \begin{bmatrix} u_1 \\ u_2 \\ u_3 \end{bmatrix}, \qquad \mathbf{v} = \begin{bmatrix} v_1 \\ v_2 \\ v_3 \end{bmatrix}, \qquad \mathbf{w} = \begin{bmatrix} w_1 \\ w_2 \\ w_3 \end{bmatrix}.$$

Let $\mathbf{x}$ be any vector in $\mathbb{R}^3$, say $\mathbf{x} = \begin{bmatrix} x \\ y \\ z \end{bmatrix}$. Then $\mathbf{x}$ is the position vector of the point which has coordinates (x, y, z) with respect to the axes OX, OY and OZ. But we may write

$$\mathbf{x} = a\mathbf{u} + b\mathbf{v} + c\mathbf{w},$$

where a, b and c are the components of $\mathbf{x}$ with respect to the basis $(\mathbf{u}, \mathbf{v}, \mathbf{w})$. This equation may be written

$$\begin{aligned} x &= au_1 + bv_1 + cw_1, \\ y &= au_2 + bv_2 + cw_2, \\ z &= au_3 + bv_3 + cw_3, \end{aligned}$$

and hence as a matrix equation

$$\begin{bmatrix} x \\ y \\ z \end{bmatrix} = P \begin{bmatrix} a \\ b \\ c \end{bmatrix}, \qquad \text{where } P = \begin{bmatrix} u_1 & v_1 & w_1 \\ u_2 & v_2 & w_2 \\ u_3 & v_3 & w_3 \end{bmatrix}.$$

This matrix equation is what we are seeking. It relates the vector $\begin{bmatrix} x \\ y \\ z \end{bmatrix}$ of components with respect to the standard basis with the vector $\begin{bmatrix} a \\ b \\ c \end{bmatrix}$ of components with respect to the new basis. The columns of the matrix P are the vectors $\mathbf{u}$, $\mathbf{v}$ and $\mathbf{w}$, so they form a LI list. Consequently (by the Equivalence Theorem, Chapter 12), P is invertible and we can write also

$$\begin{bmatrix} a \\ b \\ c \end{bmatrix} = P^{-1} \begin{bmatrix} x \\ y \\ z \end{bmatrix}.$$

17.5 Change of basis when neither basis is the standard basis.

Let $$X=\left(\begin{bmatrix}1\\-1\\2\end{bmatrix},\begin{bmatrix}0\\1\\1\end{bmatrix},\begin{bmatrix}3\\1\\-1\end{bmatrix}\right)$$

and $$Y=\left(\begin{bmatrix}0\\2\\1\end{bmatrix},\begin{bmatrix}1\\2\\0\end{bmatrix},\begin{bmatrix}1\\1\\1\end{bmatrix}\right).$$

X and Y are both bases for $\mathbb{R}^3$. We find a matrix Q such that for any vector $\boldsymbol{x}\in\mathbb{R}^3$, if $\boldsymbol{x}$ has components $\begin{bmatrix}a\\b\\c\end{bmatrix}$ with respect to X and $\begin{bmatrix}l\\m\\n\end{bmatrix}$ with respect to Y, then

$$\begin{bmatrix}l\\m\\n\end{bmatrix}=Q\begin{bmatrix}a\\b\\c\end{bmatrix}.$$

Suppose that $\boldsymbol{x}=\begin{bmatrix}x\\y\\z\end{bmatrix}$ with respect to the standard basis. Then

$$\begin{bmatrix}x\\y\\z\end{bmatrix}=P_1\begin{bmatrix}a\\b\\c\end{bmatrix},\qquad \text{where } P_1=\begin{bmatrix}1&0&3\\-1&1&1\\2&1&-1\end{bmatrix}.$$

Also $$\begin{bmatrix}x\\y\\z\end{bmatrix}=P_2\begin{bmatrix}l\\m\\n\end{bmatrix},\qquad \text{where } P_2=\begin{bmatrix}0&1&1\\2&2&1\\1&0&1\end{bmatrix}.$$

Consequently

$$\begin{bmatrix}l\\m\\n\end{bmatrix}=P_2^{-1}\begin{bmatrix}x\\y\\z\end{bmatrix}=P_2^{-1}P_1\begin{bmatrix}a\\b\\c\end{bmatrix}.$$

The required matrix Q representing the change of basis from X to Y is just $P_2^{-1}P_1$. Standard calculation gives

$$Q=\tfrac{1}{3}\begin{bmatrix}-1&2&-6\\-4&-1&6\\7&1&3\end{bmatrix}$$

in this example.

Note that the change of basis in the opposite direction is given by the matrix $P_1^{-1}P_2$, which of course is the inverse of $P_2^{-1}P_1$.

This form is more apparently useful, as it gives the new components in terms of the old.

The above is an algebraic process which may be generalised. The same argument may be carried through without the geometrical thread, for vectors in $\mathbb{R}^n$, for any value of n. The result is a general rule.

Rule

Let B be any basis for $\mathbb{R}^n$, and let $\boldsymbol{x} \in \mathbb{R}^n$. If the column vector of components of $\boldsymbol{x}$ with respect to the basis B is denoted by $\boldsymbol{x}'$, then

$$\boldsymbol{x} = P\boldsymbol{x}' \quad \text{and} \quad \boldsymbol{x}' = P^{-1}\boldsymbol{x},$$

where P is the $n \times n$ matrix whose columns consist of the vectors in the basis B in order.

It is important to remember here again that components form an ordered list, and that the order of the components, the order of the basis vectors and the order of the columns in the matrix P must all be consistent.

The above procedure can be further generalised to cover a change from components with respect to any basis to components with respect to any other basis. The columns of the matrix in that case are just the components of the new basis vectors with respect to the old basis. There is a more convenient way, however. We may proceed in two stages. First convert to components with respect to the standard basis, and then convert from there to components with respect to the desired basis, each time using the process detailed above. Example 17.5 illustrates this. It deals with a change from one basis to another in $\mathbb{R}^3$, when neither basis is the standard basis.

Summary

An element of a subspace of $\mathbb{R}^n$ may be represented uniquely by a column vector of its components with respect to a given basis. In the case of $\mathbb{R}^n$ itself, it is shown how such a column of components is related to the components with respect to the standard basis, via a matrix equation. Change of axes in $\mathbb{R}^2$ and $\mathbb{R}^3$ are given as examples.

Exercises

1. In each case below, a basis for $\mathbb{R}^2$ or $\mathbb{R}^3$ is given, together with a vector from the same space. Find the components of the vector with respect to the basis.

   (i) $\left(\begin{bmatrix}1\\2\end{bmatrix},\begin{bmatrix}2\\1\end{bmatrix}\right), \quad \begin{bmatrix}1\\5\end{bmatrix}.$

   (ii) $\left(\begin{bmatrix}-1\\1\end{bmatrix},\begin{bmatrix}1\\1\end{bmatrix}\right), \quad \begin{bmatrix}5\\1\end{bmatrix}.$

   (iii) $\left(\begin{bmatrix}2\\1\\1\end{bmatrix},\begin{bmatrix}1\\2\\1\end{bmatrix},\begin{bmatrix}1\\1\\2\end{bmatrix}\right), \quad \begin{bmatrix}0\\5\\3\end{bmatrix}.$

   (iv) $\left(\begin{bmatrix}-1\\1\\1\end{bmatrix},\begin{bmatrix}0\\1\\-1\end{bmatrix},\begin{bmatrix}2\\1\\2\end{bmatrix}\right), \quad \begin{bmatrix}0\\2\\5\end{bmatrix}.$

2. Let $\boldsymbol{u}=\begin{bmatrix}1\\2\end{bmatrix}$ and $\boldsymbol{v}=\begin{bmatrix}2\\1\end{bmatrix}$. Find a matrix P such that $P^{-1}\begin{bmatrix}x\\y\end{bmatrix}=\begin{bmatrix}a\\b\end{bmatrix}$, for every $\begin{bmatrix}x\\y\end{bmatrix}\in\mathbb{R}^2$, where a and b are the components of the vector $\begin{bmatrix}x\\y\end{bmatrix}$ with respect to the basis $(\boldsymbol{u}, \boldsymbol{v})$.

3. As in Exercise 2 above, find the matrix representing the change of basis in $\mathbb{R}^3$, where the new basis is

$$\left(\begin{bmatrix}1\\2\\1\end{bmatrix},\begin{bmatrix}2\\5\\0\end{bmatrix},\begin{bmatrix}1\\1\\4\end{bmatrix}\right).$$

4. Repeat Exercise 3, but with the new basis

$$\left(\begin{bmatrix}1\\1\\0\end{bmatrix},\begin{bmatrix}1\\2\\1\end{bmatrix},\begin{bmatrix}0\\1\\0\end{bmatrix}\right).$$

5. Repeat Exercise 3, but with the new basis

$$\left(\begin{bmatrix}1\\0\\1\end{bmatrix},\begin{bmatrix}2\\1\\0\end{bmatrix},\begin{bmatrix}3\\2\\3\end{bmatrix}\right).$$

6. Let X be a basis for $\mathbb{R}^3$ such that components $\begin{bmatrix}a\\b\\c\end{bmatrix}$ with respect to X are related to components $\begin{bmatrix}x\\y\\z\end{bmatrix}$ with respect to the standard basis

according to the equation

$$\begin{bmatrix} a \\ b \\ c \end{bmatrix} = \begin{bmatrix} 1 & 3 & -1 \\ -1 & -2 & 2 \\ 2 & 4 & -3 \end{bmatrix} \begin{bmatrix} x \\ y \\ z \end{bmatrix}.$$

Find such a basis X.

Examples

18.1 Given that the matrix

$$A=\begin{bmatrix}\frac{2}{3} & -\frac{2}{3} & \frac{1}{3}\\ \frac{2}{3} & \frac{1}{3} & -\frac{2}{3}\\ \frac{1}{3} & \frac{2}{3} & \frac{2}{3}\end{bmatrix}$$

represents a rotation about an axis through the origin, find the direction of this axis.

Suppose that $\boldsymbol{v}$ is in the direction of the axis of rotation. Then

$$A\boldsymbol{v}=\boldsymbol{v}.$$

Hence $A\boldsymbol{v}-\boldsymbol{v}=\boldsymbol{0}$,

i.e. $(A-I)\boldsymbol{v}=\boldsymbol{0}$.

Now $A-I=\begin{bmatrix}-\frac{1}{3} & -\frac{2}{3} & \frac{1}{3}\\ \frac{2}{3} & -\frac{2}{3} & -\frac{2}{3}\\ \frac{1}{3} & \frac{2}{3} & -\frac{1}{3}\end{bmatrix}$. Let $\boldsymbol{v}=\begin{bmatrix}x\\ y\\ z\end{bmatrix}$.

The GE process can be used to solve $(A-I)\boldsymbol{v}=\boldsymbol{0}$:

$$\begin{bmatrix}-\frac{1}{3} & -\frac{2}{3} & \frac{1}{3}\\ \frac{2}{3} & -\frac{2}{3} & -\frac{2}{3}\\ \frac{1}{3} & \frac{2}{3} & -\frac{1}{3}\end{bmatrix}\to\begin{bmatrix}1 & 2 & -1\\ 0 & 1 & 0\\ 0 & 0 & 0\end{bmatrix},$$

so the solution is $z=t$, $y=0$, $x=t$ $(t\in\mathbb{R})$. In other words, if $\boldsymbol{v}$ is any multiple of $\begin{bmatrix}1\\ 0\\ 1\end{bmatrix}$ then $A\boldsymbol{v}=\boldsymbol{v}$ and, furthermore, there are no other such vectors. It follows that the axis of rotation is in the direction of the vector $\begin{bmatrix}1\\ 0\\ 1\end{bmatrix}$.

18

Eigenvalues and eigenvectors

Recall (from Chapter 16) that a rotation in three dimensions about an axis through the origin may be represented algebraically as a linear transformation from $\mathbb{R}^3$ to $\mathbb{R}^3$, and that a linear transformation may be represented by multiplication by a matrix. This situation will serve to illustrate the procedures which are to be discussed in this chapter. Example 18.1 gives the details of the following. Suppose that we are told that the linear transformation f from $\mathbb{R}^4$ to $\mathbb{R}^3$ represents a rotation about an axis through the origin and that f is represented by the 3×3 matrix A. How can we find the direction of the axis of rotation? We can use the fact that any vector in the direction of this axis (and only such vectors) will be left unchanged by the rotation, so if $\mathbf{v}$ is such a vector we must have

$$A\mathbf{v} = \mathbf{v}.$$

Elementary methods will serve to find such vectors $\mathbf{v}$, in the following way. Rewrite the equation as

$$A\mathbf{v} - \mathbf{v} = \mathbf{0}, \text{ then as } (A - I)\mathbf{v} = \mathbf{0},$$

where I is the 3×3 identity matrix. Now solve in the normal way, using the Gaussian elimination process. Of course we are interested only in non-trivial solutions (i.e. other than $\mathbf{v} = \mathbf{0}$), and we know that such solutions will exist if and only if the matrix $A - I$ is singular. This is bound to happen if A represents a rotation as above, for there must be an axis of rotation, so there must be such a vector $\mathbf{v}$. And of course there will be infinitely many solutions, all multiples of each other, all solutions being in the direction of the axis of rotation.

There is a more general idea in the background, here, though, and it turns out to be a remarkably useful one which crops up wherever linear algebra is applied. Given any $n \times n$ matrix A, there may be some n-vectors $\mathbf{x}$ for which $A\mathbf{x}$ and $\mathbf{x}$ are multiples of each other (which in geometrical terms means having the same or opposite directions). Such vectors are called eigenvectors, and the property above is made precise in the definition below.

18.2 Examples of eigenvalues and eigenvectors.

(i) $A = \begin{bmatrix} 2 & 1 \\ 2 & 3 \end{bmatrix}$.

$$\begin{bmatrix} 2 & 1 \\ 2 & 3 \end{bmatrix}\begin{bmatrix} 1 \\ -1 \end{bmatrix} = \begin{bmatrix} 1 \\ -1 \end{bmatrix}$$

and $$\begin{bmatrix} 2 & 1 \\ 2 & 3 \end{bmatrix}\begin{bmatrix} 1 \\ 2 \end{bmatrix} = \begin{bmatrix} 4 \\ 8 \end{bmatrix} = 4\begin{bmatrix} 1 \\ 2 \end{bmatrix}.$$

So $\begin{bmatrix} 1 \\ -1 \end{bmatrix}$ and $\begin{bmatrix} 1 \\ 2 \end{bmatrix}$ are eigenvectors of A, with corresponding eigenvalues 1 and 4 respectively.

(ii) $A = \begin{bmatrix} 1 & 1 & 0 \\ 1 & 0 & 1 \\ 0 & 1 & 1 \end{bmatrix}$.

Here $$A\begin{bmatrix} 1 \\ 1 \\ 1 \end{bmatrix} = \begin{bmatrix} 2 \\ 2 \\ 2 \end{bmatrix}, \quad A\begin{bmatrix} 1 \\ 0 \\ -1 \end{bmatrix} = \begin{bmatrix} 1 \\ 0 \\ -1 \end{bmatrix} \quad \text{and} \quad A\begin{bmatrix} 1 \\ -2 \\ 1 \end{bmatrix} = \begin{bmatrix} -1 \\ 2 \\ -1 \end{bmatrix},$$

so $\begin{bmatrix} 1 \\ 1 \\ 1 \end{bmatrix}$, $\begin{bmatrix} 1 \\ 0 \\ -1 \end{bmatrix}$ and $\begin{bmatrix} 1 \\ -2 \\ 1 \end{bmatrix}$ are eigenvectors of A with corresponding eigenvalues 2, 1 and -1 respectively.

(iii) $A = \begin{bmatrix} 1 & -3 \\ -2 & 6 \end{bmatrix}$.

Here $$A\begin{bmatrix} 1 \\ -2 \end{bmatrix} = \begin{bmatrix} 7 \\ -14 \end{bmatrix}, \quad \text{and} \quad A\begin{bmatrix} 3 \\ 1 \end{bmatrix} = \begin{bmatrix} 0 \\ 0 \end{bmatrix},$$

so $\begin{bmatrix} 1 \\ -2 \end{bmatrix}$ and $\begin{bmatrix} 3 \\ 1 \end{bmatrix}$ are eigenvectors of A with corresponding eigenvalues 7 and 0 respectively.

In all of the above cases the matrix has other eigenvectors also. Can you find some?

Definition

Let A be a $n \times n$ matrix. A non-zero n-vector $\boldsymbol{x}$ for which $A\boldsymbol{x}$ is a multiple of $\boldsymbol{x}$ is called an *eigenvector* of A.

Notice that we exclude the zero vector from consideration. $A\boldsymbol{0}$ is always a multiple of $\boldsymbol{0}$ in a trivial way. We are interested in non-trivial ways in which $A\boldsymbol{x}$ is a multiple of $\boldsymbol{x}$.

The numbers which may appear as multipliers in this context have a special significance. They are called eigenvalues of the matrix A. Each eigenvector has a corresponding eigenvalue.

Definition

Let A be a $n \times n$ matrix and let $\boldsymbol{x}$ be an eigenvector of A. If $A\boldsymbol{x} = k\boldsymbol{x}$ then k is said to be the *eigenvalue* of A corresponding to $\boldsymbol{x}$.

Some examples of eigenvalues and eigenvectors are given in Examples 18.2. Notice that in the case of a matrix A representing a rotation, we have $A\boldsymbol{v} = \boldsymbol{v}$ when $\boldsymbol{v}$ is in the direction of the axis of rotation, so such vectors $\boldsymbol{v}$ are eigenvectors of this A, and 1 is the corresponding eigenvalue for each of them.

Among Examples 18.2 there is one matrix for which 0 is an eigenvalue. This is an important special case. 0 is an eigenvalue of A if there is a non-zero vector $\boldsymbol{x}$ for which

$$A\boldsymbol{x} = 0\boldsymbol{x}, \qquad \text{i.e. } A\boldsymbol{x} = \boldsymbol{0}.$$

Of course this is equivalent to saying that A is singular.

Rule

A square matrix A has 0 as an eigenvalue if and only if A is singular.

18.3 Finding eigenvalues and eigenvectors.

(i) Let $A=\begin{bmatrix}2 & 1\\ 2 & 3\end{bmatrix}$.

Seek values of k for which $(A-kI)\mathbf{x}=\mathbf{0}$ has non-trivial solutions. So seek values of k for which $\det(A-kI)=0$.

$$\begin{aligned}\det(A-kI)&=\begin{vmatrix}2-k & 1\\ 2 & 3-k\end{vmatrix}\\ &=(2-k)(3-k)-2\\ &=4-5k+k^2\\ &=(4-k)(1-k).\end{aligned}$$

Hence the values we seek are 4 and 1. These are the only eigenvalues of A. To find corresponding eigenvectors, take each eigenvalue in turn, and solve the equation $(A-kI)\mathbf{x}=0$.

$k=4$: Solve $\begin{bmatrix}-2 & 1\\ 2 & -1\end{bmatrix}\begin{bmatrix}x\\ y\end{bmatrix}=\begin{bmatrix}0\\ 0\end{bmatrix}$.

The solution is $y=t$, $x=\frac{1}{2}t$ $(t\in\mathbb{R})$. So all vectors $\begin{bmatrix}\frac{1}{2}t\\ t\end{bmatrix}$ $(t\in\mathbb{R},\ t\neq 0)$ are eigenvectors of A corresponding to $k=4$. Note that they all lie in the same direction. One such vector is $\begin{bmatrix}1\\ 2\end{bmatrix}$ (see Example 18.2(i)).

$k=1$: Solve $\begin{bmatrix}1 & 1\\ 2 & 2\end{bmatrix}\begin{bmatrix}x\\ y\end{bmatrix}=\begin{bmatrix}0\\ 0\end{bmatrix}$.

The solution is $y=t$, $x=-t$ $(t\in\mathbb{R})$. So all vectors $\begin{bmatrix}t\\ -t\end{bmatrix}$ $(t\in\mathbb{R},\ t\neq 0)$ are eigenvectors of A corresponding to $k=1$. One such vector is the one discovered in Example 18.2(i), namely $\begin{bmatrix}1\\ -1\end{bmatrix}$.

(ii) $A=\begin{bmatrix}1 & 1 & 0\\ 1 & 0 & 1\\ 0 & 1 & 1\end{bmatrix}$.

Solve the equation

$$\det(A-kI)=0,$$

i.e. $$\begin{vmatrix}1-k & 1 & 0\\ 1 & -k & 1\\ 0 & 1 & 1-k\end{vmatrix}=0$$

i.e. $(1-k)[-k(1-k)-1]-1(1-k-0)=0$

i.e. $(1-k)(k^2-k-1)-(1-k)=0$

i.e. $(1-k)(k^2-k-2)=0$

i.e. $(1-k)(k-2)(k+1)=0$.

This is satisfied by $k=1$, $k=2$ and $k=-1$. These are the eigenvalues of A. To find the eigenvectors, take each eigenvalue in turn.

There is a routine process for finding eigenvalues and eigenvectors. It is as follows (illustrated by Example 18.3). Given a square matrix A, we seek numbers k and non-zero vectors $\mathbf{x}$ which satisfy the equation

$$A\mathbf{x} = k\mathbf{x}.$$

The first thing is to find all numbers k for which corresponding non-zero vectors can exist (i.e. all eigenvalues of the matrix). Write the equation as

$$A\mathbf{x} - k\mathbf{x} = \mathbf{0}, \text{ then as } (A - kI)\mathbf{x} = \mathbf{0},$$

where I is an appropriately sized identity matrix. This equation has non-trivial solutions for $\mathbf{x}$ if and only if $A - kI$ is a singular matrix, i.e. if and only if

$$\det(A - kI) = 0.$$

So we seek values of k which satisfy this equation. If A is a $n \times n$ matrix then this equation will turn out (when the determinant is evaluated) to be a polynomial equation in k of degree n. We may therefore expect n values for k as roots of the equation, some of which may be complex numbers, of course, and some of which may be repeated roots. In any event there will be finitely many real eigenvalues k (possibly none).

Definition

The equation $\det(A - kI) = 0$ is called the *characteristic equation* of A.

$k=1$: Solve $\begin{bmatrix} 0 & 1 & 0 \\ 1 & -1 & 1 \\ 0 & 1 & 0 \end{bmatrix}\begin{bmatrix} x \\ y \\ z \end{bmatrix}=\begin{bmatrix} 0 \\ 0 \\ 0 \end{bmatrix}$.

The solution is $z=t$, $y=0$, $x=-t$ $(t\in\mathbb{R})$.

$k=2$: Solve $\begin{bmatrix} -1 & 1 & 0 \\ 1 & -2 & 1 \\ 0 & 1 & -1 \end{bmatrix}\begin{bmatrix} x \\ y \\ z \end{bmatrix}=\begin{bmatrix} 0 \\ 0 \\ 0 \end{bmatrix}$.

The solution is $z=t$, $y=t$, $x=t$ $(t\in\mathbb{R})$.

$k=-1$: Solve $\begin{bmatrix} 2 & 1 & 0 \\ 1 & 1 & 1 \\ 0 & 1 & 2 \end{bmatrix}\begin{bmatrix} x \\ y \\ z \end{bmatrix}=\begin{bmatrix} 0 \\ 0 \\ 0 \end{bmatrix}$.

The solution is $z=t$, $y=-2t$, $x=t$ $(t\in\mathbb{R})$.

Thus all eigenvalues and eigenvectors are:

$k=1$ with eigenvectors $\begin{bmatrix} -t \\ 0 \\ t \end{bmatrix}$ $(t\in\mathbb{R},\ t\neq 0)$,

$k=2$ with eigenvectors $\begin{bmatrix} t \\ t \\ t \end{bmatrix}$ $(t\in\mathbb{R},\ t\neq 0)$,

$k=-1$ with eigenvectors $\begin{bmatrix} t \\ -2t \\ t \end{bmatrix}$ $(t\in\mathbb{R},\ t\neq 0)$.

18.4 What is the eigenspace of A corresponding to the eigenvalue 2, where A is the matrix

$$\begin{bmatrix} 1 & 1 & 0 \\ 1 & 0 & 1 \\ 0 & 1 & 1 \end{bmatrix}?$$

The work has been done in Example 18.3(ii). The eigenvalue 2 corresponds to the set

$$\left\{\begin{bmatrix} t \\ t \\ t \end{bmatrix} : t\in\mathbb{R},\ t\neq 0\right\}$$

of eigenvectors. The eigenspace is just

$$\left\{\begin{bmatrix} t \\ t \\ t \end{bmatrix} : t\in\mathbb{R}\right\}.$$

Notice that the eigenspace contains the zero vector, which is not an eigenvector.

Once we have found an eigenvalue of A, the problem of finding corresponding eigenvectors is just a matter of solving equations. This was done in Example 18.1, where the eigenvalue concerned was 1, and it was done in Example 18.3 for each of the eigenvalues separately. The eigenvectors are solutions $\mathbf{x}$ (other than $\mathbf{0}$) of the equation

$$(A - kI)\mathbf{x} = \mathbf{0},$$

where k is the eigenvalue in question. Of course we know that there must be infinitely many solutions to this equation, so to each eigenvalue there always corresponds an infinite set of eigenvectors. Indeed it is not hard to see that this set, when taken together with $\mathbf{0}$, constitutes a subspace of $\mathbb{R}^n$. See Exercise 6 in Chapter 13. It is called the *eigenspace* of A corresponding to the eigenvalue k. See Example 18.4.

18.5 Given that -3 is an eigenvalue of the matrix

$$\begin{bmatrix} -2 & 2 & 2 \\ 2 & 1 & 4 \\ 2 & 4 & 1 \end{bmatrix},$$

find the eigenspace corresponding to it.

Solve the equation $(A+3I)\mathbf{x}=\mathbf{0}$,

i.e. $$\begin{bmatrix} 1 & 2 & 2 \\ 2 & 4 & 4 \\ 2 & 4 & 4 \end{bmatrix}\begin{bmatrix} x \\ y \\ z \end{bmatrix}=\begin{bmatrix} 0 \\ 0 \\ 0 \end{bmatrix}.$$

Obtain, by the standard process, the solution

$$z=t, \qquad y=u, \qquad x=-2t-2u \qquad (t, u \in \mathbb{R}).$$

Hence $$\left\{\begin{bmatrix} -2t-2u \\ u \\ t \end{bmatrix} : t, u \in \mathbb{R}\right\}$$

is the required eigenspace. Geometrically, this is the plane with equation $x+2y+2z=0$.

18.6 Let $A=\begin{bmatrix} 1 & 1 \\ -1 & 1 \end{bmatrix}$. Find all eigenvalues and eigenvectors of A.

Solve the characteristic equation $\det(A-kI)=0$,

i.e. $$\begin{vmatrix} 1-k & 1 \\ -1 & 1-k \end{vmatrix}=0$$

i.e. $$(1-k)^2+1=0$$

i.e. $$k^2-2k+2=0.$$

This equation has no real roots. Consequently this matrix A, as a real matrix, has no eigenvalues and no eigenvectors. However, we can regard A as a complex matrix, since real numbers can be regarded as complex numbers with no imaginary part. The characteristic equation above does have complex roots, namely $k=1+i$ and $k=1-i$, and these are complex eigenvalues of A. We can go on to find complex eigenvectors, by the standard method.

$k=1+i$: Solve $$\begin{aligned} -ix+\ y&=0, \\ -x-iy&=0 \end{aligned}$$

obtaining the solution $y=t$, $x=-it$ $(t \in \mathbb{C})$. So all vectors $\begin{bmatrix} -it \\ t \end{bmatrix}$ $(t \in \mathbb{C},$ $t \neq 0)$ are complex eigenvectors of A corresponding to the eigenvalue $1+i$.

$k=1-i$: Solve $$\begin{aligned} ix+\ y&=0, \\ -x+iy&=0 \end{aligned}$$

obtaining the solution $y=t$, $x=it$ $(t \in \mathbb{C})$. So all vectors $\begin{bmatrix} it \\ t \end{bmatrix}$ $(t \in \mathbb{C}, t \neq 0)$ are complex eigenvectors of A corresponding to the eigenvalue $1-i$.

Example 18.5 shows a case where an eigenspace has dimension 2. In most of the situations we shall come across, eigenspaces will have dimension 1.

Eigenvalues which are complex numbers have no significance in relation to the algebra of $\mathbb{R}^n$. But the ideas extend – the process of finding eigenvectors can be carried out – if we treat complex eigenvalues in just the same way as real ones. We may then expect to find complex eigenvectors. These will of course not be vectors in $\mathbb{R}^n$, but instead will belong to $\mathbb{C}^n$, the set of all $n \times 1$ matrices with complex numbers as entries. We shall not pursue this, but in what follows we shall need to be aware of these possibilities. A simple case is given in Example 18.6.

For some matrices, all of the eigenvalues are real. We have seen some examples like this already. But there is a class of matrices for which we can be sure in advance that all eigenvalues are real. This is given in the next rule.

Rule

A real symmetric matrix always has only real eigenvalues.

We give a demonstration of this here, but it is quite difficult, and it may be skipped without loss of subsequent understanding. Let k be an eigenvalue of the real symmetric matrix A, and suppose only that $k \in \mathbb{C}$. We show that $k = \bar{k}$ (its complex conjugate) so that k is real. Let $\mathbf{x}$ be an eigenvector corresponding to k (with entries which may be complex), so that

$$A\mathbf{x} = k\mathbf{x}.$$

Transposing both sides, we obtain

$$\mathbf{x}^{\mathrm{T}}A^{\mathrm{T}} = k\mathbf{x}^{\mathrm{T}},$$

i.e. $\mathbf{x}^{\mathrm{T}}A = k\mathbf{x}^{\mathrm{T}}$, since A is symmetric.

Now take complex conjugates of both sides:

$$\bar{\mathbf{x}}^{\mathrm{T}}\bar{A} = \bar{k}\bar{x}^{\mathrm{T}},$$

i.e. $\bar{\mathbf{x}}^{\mathrm{T}}A = \bar{k}\bar{x}^{\mathrm{T}}$, since A has only real entries.

Consider the matrix product $\bar{\mathbf{x}}^{\mathrm{T}}A\mathbf{x}$.

First, $\bar{\mathbf{x}}^{\mathrm{T}}A\mathbf{x} = (\bar{\mathbf{x}}^{\mathrm{T}}A)\mathbf{x} = \bar{k}\bar{\mathbf{x}}^{\mathrm{T}}\mathbf{x}$, from above.

Second, $\bar{\mathbf{x}}^{\mathrm{T}}A\mathbf{x} = \bar{\mathbf{x}}^{\mathrm{T}}(A\mathbf{x}) = \bar{\mathbf{x}}^{\mathrm{T}}k\mathbf{x} = k\bar{\mathbf{x}}^{\mathrm{T}}\mathbf{x}$.

Hence $\bar{k}\bar{\mathbf{x}}^{\mathrm{T}}\mathbf{x} = k\bar{\mathbf{x}}^{\mathrm{T}}\mathbf{x}$.

It follows that $k = \bar{k}$, as required, because $\bar{\mathbf{x}}^{\mathrm{T}}\mathbf{x}$ cannot be equal to 0. To

18.7 Properties of eigenvalues. Let $A = \begin{bmatrix} 1 & 2 \\ 2 & 1 \end{bmatrix}$. Then $\begin{bmatrix} 1 \\ 1 \end{bmatrix}$ is an eigenvector of A corresponding to the eigenvalue 3, because

$$\begin{bmatrix} 1 & 2 \\ 2 & 1 \end{bmatrix}\begin{bmatrix} 1 \\ 1 \end{bmatrix} = \begin{bmatrix} 3 \\ 3 \end{bmatrix} = 3\begin{bmatrix} 1 \\ 1 \end{bmatrix}.$$

(i) Let $a \in \mathbb{R}$. Then $aA = \begin{bmatrix} a & 2a \\ 2a & a \end{bmatrix}$.

Now $$(aA)\begin{bmatrix} 1 \\ 1 \end{bmatrix} = \begin{bmatrix} a & 2a \\ 2a & a \end{bmatrix}\begin{bmatrix} 1 \\ 1 \end{bmatrix} = \begin{bmatrix} 3a \\ 3a \end{bmatrix} = 3a\begin{bmatrix} 1 \\ 1 \end{bmatrix},$$

so $\begin{bmatrix} 1 \\ 1 \end{bmatrix}$ is an eigenvector of aA, with corresponding eigenvalue $3a$.

(ii) $$A^2 = \begin{bmatrix} 5 & 4 \\ 4 & 5 \end{bmatrix}.$$

$$A^2\begin{bmatrix} 1 \\ 1 \end{bmatrix} = \begin{bmatrix} 5 & 4 \\ 4 & 5 \end{bmatrix}\begin{bmatrix} 1 \\ 1 \end{bmatrix} = \begin{bmatrix} 9 \\ 9 \end{bmatrix} = 9\begin{bmatrix} 1 \\ 1 \end{bmatrix},$$

so $\begin{bmatrix} 1 \\ 1 \end{bmatrix}$ is an eigenvector of A^2, with corresponding eigenvalue 9.

(iii) $$A^{-1} = \begin{bmatrix} -\frac{1}{3} & \frac{2}{3} \\ \frac{2}{3} & -\frac{1}{3} \end{bmatrix}.$$

$$A^{-1}\begin{bmatrix} 1 \\ 1 \end{bmatrix} = \begin{bmatrix} -\frac{1}{3} & \frac{2}{3} \\ \frac{2}{3} & -\frac{1}{3} \end{bmatrix}\begin{bmatrix} 1 \\ 1 \end{bmatrix} = \begin{bmatrix} \frac{1}{3} \\ \frac{1}{3} \end{bmatrix} = \frac{1}{3}\begin{bmatrix} 1 \\ 1 \end{bmatrix},$$

so $\begin{bmatrix} 1 \\ 1 \end{bmatrix}$ is an eigenvector of A, with corresponding eigenvalue $\frac{1}{3}$.

(iv) Let $c \in \mathbb{R}$. Then $c\begin{bmatrix} 1 \\ 1 \end{bmatrix} = \begin{bmatrix} c \\ c \end{bmatrix}$,

and $$A\begin{bmatrix} c \\ c \end{bmatrix} = \begin{bmatrix} 1 & 2 \\ 2 & 1 \end{bmatrix}\begin{bmatrix} c \\ c \end{bmatrix} = \begin{bmatrix} c + 2c \\ 2c + c \end{bmatrix} = 3\begin{bmatrix} c \\ c \end{bmatrix}.$$

see this, let $\boldsymbol{x} = \begin{bmatrix} x_1 \\ \vdots \\ x_n \end{bmatrix}$. Then $\bar{\boldsymbol{x}}^{\mathrm{T}} = [\bar{x}_1, \ldots, \bar{x}_n]$, and

$$\begin{aligned} \bar{\boldsymbol{x}}^{\mathrm{T}}\boldsymbol{x} &= \bar{x}_1 x_1 + \cdots + \bar{x}_n x_n \\ &= |x_1|^2 + \cdots + |x_n|^2. \end{aligned}$$

This last is a sum of non-negative real numbers. This sum can be zero only if each of the terms is zero, i.e. only if $\boldsymbol{x} = \mathbf{0}$. But $\boldsymbol{x} \neq \mathbf{0}$, since $\boldsymbol{x}$ is an eigenvector. This completes the demonstration that a real symmetric matrix has only real eigenvalues.

We end this chapter with some properties of eigenvalues. These are all illustrated in Examples 18.7. Let A be a square matrix, and let $\boldsymbol{x}$ be an eigenvector of A, with corresponding eigenvalue k.

(a) For any $b \in \mathbb{R}$, $\boldsymbol{x}$ is an eigenvector of bA, with corresponding eigenvalue bk.
(b) For any natural number r, $\boldsymbol{x}$ is an eigenvector of A^r, with corresponding eigenvalue k^r.
(c) If A is invertible, then $\boldsymbol{x}$ is an eigenvector of A^{-1}, with corresponding eigenvalue k^{-1}.
(d) For any $c \in \mathbb{R}$, $c\boldsymbol{x}$ is an eigenvector of A, with corresponding eigenvalue k.

These can be easily demonstrated as follows.

For (a) $(bA)\boldsymbol{x} = b(A\boldsymbol{x}) = b(k\boldsymbol{x}) = (bk)\boldsymbol{x}$.

For (b) $(A^r)\boldsymbol{x} = (A^{r-1})A\boldsymbol{x} = A^{r-1}(k\boldsymbol{x}) = k(A^{r-1}\boldsymbol{x}) = k(A^{r-2})A\boldsymbol{x}$
$= k(A^{r-2})k\boldsymbol{x} = k^2(A^{r-2}\boldsymbol{x}) = \cdots = k^r\boldsymbol{x}$.

For (c) $A\boldsymbol{x} = k\boldsymbol{x} \;\Rightarrow\; A^{-1}A\boldsymbol{x} = A^{-1}(k\boldsymbol{x})$
$\Rightarrow\; \boldsymbol{x} = kA^{-1}\boldsymbol{x}$
$\Rightarrow\; (1/k)\boldsymbol{x} = A^{-1}\boldsymbol{x}$.

(Think about why k cannot be 0.)

For (d) $A(c\boldsymbol{x}) = c(A\boldsymbol{x}) = c(k\boldsymbol{x}) = k(c\boldsymbol{x})$.

Summary

The idea of an eigenvector of a matrix and its corresponding eigenvalue are introduced and defined. A routine process for finding eigenvalues and eigenvectors, involving solution of the characteristic equation, is described. It is shown that a real symmetric matrix always has only real eigenvalues. Some properties of eigenvalues and eigenvectors are listed and illustrated.

Exercises

1. Find all eigenvalues and corresponding eigenvectors of the following matrices.

(i) $\begin{bmatrix} 1 & 4 \\ 1 & 1 \end{bmatrix}$. (ii) $\begin{bmatrix} 1 & 1 \\ 1 & 1 \end{bmatrix}$.

(iii) $\begin{bmatrix} 1 & -1 \\ 2 & 4 \end{bmatrix}$. (iv) $\begin{bmatrix} 1 & 4 \\ 1 & -2 \end{bmatrix}$.

(v) $\begin{bmatrix} 1 & -1 & -1 \\ 0 & 3 & 2 \\ 0 & 2 & 0 \end{bmatrix}$. (vi) $\begin{bmatrix} 0 & -2 & 2 \\ 2 & -1 & 0 \\ 2 & -2 & 1 \end{bmatrix}$.

(vii) $\begin{bmatrix} 3 & 1 & 0 \\ 0 & -3 & 1 \\ 0 & 0 & 1 \end{bmatrix}$. (viii) $\begin{bmatrix} -1 & 0 & 0 \\ 2 & 1 & 0 \\ 1 & 2 & 3 \end{bmatrix}$.

2. Each of the following matrices has a repeated eigenvalue (i.e. one which is a repeated root of the characteristic equation). In each case find this eigenvalue and find its corresponding eigenspace.

(i) $\begin{bmatrix} 1 & 0 & 0 \\ 3 & 3 & -1 \\ -1 & 2 & 0 \end{bmatrix}$. (ii) $\begin{bmatrix} 0 & 2 & 3 \\ 2 & 3 & 6 \\ 3 & 6 & 8 \end{bmatrix}$.

(iii) $\begin{bmatrix} -1 & 2 \\ -2 & 3 \end{bmatrix}$. (iv) $\begin{bmatrix} -2 & 2 & 2 \\ 2 & 1 & 4 \\ 2 & 4 & 1 \end{bmatrix}$.

(v) $\begin{bmatrix} 1 & -2 & 1 & -1 \\ 1 & 0 & -1 & 1 \\ -1 & 2 & 3 & -1 \\ -2 & -4 & 2 & 0 \end{bmatrix}$.

3. Find all real eigenvalues and corresponding eigenvectors of the following matrices.

(i) $\begin{bmatrix} 0 & -1 & 1 \\ 0 & 1 & 0 \\ -1 & 0 & 0 \end{bmatrix}$. (ii) $\begin{bmatrix} 2 & -2 & 1 \\ 2 & 1 & -2 \\ 1 & 2 & 2 \end{bmatrix}$.

(iii) $\begin{bmatrix} -1 & 0 \\ 0 & -1 \end{bmatrix}$. (iv) $\begin{bmatrix} 1 & 3 \\ -2 & 1 \end{bmatrix}$.

4. For each of the matrices in Exercise 3 above, find all complex eigenvalues, and corresponding complex eigenvectors.
5. Find an inequality which must be satisfied by the numbers a, b, c and d if the matrix

$$\begin{bmatrix} a & b \\ c & d \end{bmatrix}$$

is to have only real eigenvalues.

6. Show that the matrix

$$\begin{bmatrix} 1/\sqrt{2} & 1/\sqrt{2} \\ -1/\sqrt{2} & 1/\sqrt{2} \end{bmatrix}$$

has no real eigenvalues and eigenvectors. Give a geometrical interpretation for this. (Hint: see Chapter 16.)

7. (i) A square matrix A is said to be nilpotent if $A^r = 0$ for some natural number r. Show that a nilpotent matrix can have no eigenvalues other than 0.

   (ii) Let A be a $n \times n$ matrix and let P be an invertible $n \times n$ matrix. Show that if $\boldsymbol{x}$ is an eigenvector of A with corresponding eigenvalue k, then $P^{-1}\boldsymbol{x}$ is an eigenvector of the matrix $P^{-1}AP$, also with corresponding eigenvalue k. Show further that A and $P^{-1}AP$ have the same eigenvalues.

Examples

19.1 LI lists of eigenvectors.

(i) Let $A = \begin{bmatrix} 1 & 1 & 0 \\ 1 & 0 & 1 \\ 0 & 1 & 1 \end{bmatrix}$.

See Example 18.3(ii). There are three distinct eigenvalues: 1, 2 and -1. Pick one eigenvector corresponding to each, say

$$\begin{bmatrix} -1 \\ 0 \\ 1 \end{bmatrix}, \begin{bmatrix} 1 \\ 1 \\ 1 \end{bmatrix}, \begin{bmatrix} 1 \\ -2 \\ 1 \end{bmatrix} \text{ respectively.}$$

These form a LI list.

(ii) Let $A = \begin{bmatrix} -2 & 2 & 2 \\ 2 & 1 & 4 \\ 2 & 4 & 1 \end{bmatrix}$. See Example 18.5. There is an eigenvalue -3, with corresponding set of eigenvectors

$$\left\{ \begin{bmatrix} -2t-2u \\ u \\ t \end{bmatrix} : t, u \in \mathbb{R},\ t, u \text{ not both } 0 \right\}.$$

There is also another eigenvalue 6, with corresponding set of eigenvectors

$$\left\{ \begin{bmatrix} \frac{1}{2}t \\ t \\ t \end{bmatrix} : t \in \mathbb{R},\ t \neq 0 \right\}.$$

Pick one eigenvector corresponding to each eigenvalue, say $\begin{bmatrix} -6 \\ 1 \\ 2 \end{bmatrix}, \begin{bmatrix} 1 \\ 2 \\ 2 \end{bmatrix}$ respectively.

These form a LI list, irrespective of the choice made.

(iii) Let $A = \begin{bmatrix} 2 & 2 & 3 \\ 1 & 2 & 1 \\ 2 & -2 & 1 \end{bmatrix}$.

Solving the characteristic equation $\det(A - kI) = 0$ gives eigenvalues $-1, 2, 4$. Note that this involves solving a cubic equation. The best way of doing this (in all of the examples in this book, at least) is by factorisation. If there are no obvious factors, then try to spot a solution by testing small integer values. When one solution is known, a factorisation can be produced. Here the characteristic equation is (when simplified)

$$k^3 - 5k^2 + 2k + 8 = 0.$$

Trial and error shows $k = -1$ to be a solution, and so $(k+1)$ must be a factor of the left-hand side. Hence the equation becomes

$$(k+1)(k^2 - 6k + 8) = 0,$$

i.e. $$(k+1)(k-2)(k-4) = 0.$$

19

Diagonalisation 1

Our ideas concerning linear dependence and independence are relevant in the context of eigenvalues and eigenvectors. In a way which is made precise in the rule below, a list of eigenvectors of a matrix can be expected to be linearly independent.

Rule

Let A be a square matrix, let $k_1, \ldots, k_r$ be eigenvalues of A which are all different, and let $\boldsymbol{x}_1, \ldots, \boldsymbol{x}_r$ (respectively) be corresponding eigenvectors. Then the list $(\boldsymbol{x}_1, \ldots, \boldsymbol{x}_r)$ is linearly independent.

Examples 19.1 give some matrices and some LI lists of eigenvectors, to illustrate this. The rule can be demonstrated by the following argument.

Suppose that $(\boldsymbol{x}_1, \ldots, \boldsymbol{x}_r)$ is LD. Then let l be the smallest number $(\leqslant r)$ such that $(\boldsymbol{x}_1, \ldots, \boldsymbol{x}_l)$ is LD. It is then the case that $(\boldsymbol{x}_1, \ldots, \boldsymbol{x}_{l-1})$ is LI, and that

$$\boldsymbol{x}_l = a_1\boldsymbol{x}_1 + \cdots + a_{l-1}\boldsymbol{x}_{l-1},$$

for some real numbers $a_1, \ldots, a_{l-1}$, not all of which are 0. But then

$$\begin{aligned} A\boldsymbol{x}_l &= A(a_1\boldsymbol{x}_1 + \cdots + a_{l-1}\boldsymbol{x}_{l-1}) \\ &= A(a_1\boldsymbol{x}_1) + \cdots + A(a_{l-1}\boldsymbol{x}_{l-1}) \\ &= a_1 A\boldsymbol{x}_1 + \cdots + a_{l-1}A\boldsymbol{x}_{l-1} \\ &= a_1 k_1 \boldsymbol{x}_1 + \cdots + a_{l-1}k_{l-1}\boldsymbol{x}_{l-1}. \end{aligned}$$

So $\quad k_l\boldsymbol{x}_l = a_1 k_1 \boldsymbol{x}_1 + \cdots + a_{l-1}k_{l-1}\boldsymbol{x}_{l-1}.$

But $\quad$ $$\begin{aligned} k_l\boldsymbol{x}_l &= k_l(a_1\boldsymbol{x}_1 + \cdots + a_{l-1}\boldsymbol{x}_{l-1}) \\ &= a_1 k_l \boldsymbol{x}_1 + \cdots + a_{l-1}k_l\boldsymbol{x}_{l-1}. \end{aligned}$$

Subtracting then gives us

$$\boldsymbol{0} = a_1(k_l - k_1)\boldsymbol{x}_1 + \cdots + a_{l-1}(k_l - k_{l-1})\boldsymbol{x}_{l-1}.$$

But none of the numbers $k_l - k_i$ is zero, because the chosen eigenvalues

Now find eigenvectors.

$k=-1$: $\begin{bmatrix} -t \\ 0 \\ t \end{bmatrix}$ $(t \in \mathbb{R},\ t \neq 0)$.

$k=2$: $\begin{bmatrix} -t \\ -\frac{3}{2}t \\ t \end{bmatrix}$ $(t \in \mathbb{R},\ t \neq 0)$.

$k=4$: $\begin{bmatrix} 4t \\ \frac{5}{2}t \\ t \end{bmatrix}$ $(t \in \mathbb{R},\ t \neq 0)$.

Pick one corresponding to each eigenvalue, say $\begin{bmatrix} 2 \\ 0 \\ -2 \end{bmatrix}, \begin{bmatrix} 4 \\ 6 \\ -4 \end{bmatrix}, \begin{bmatrix} 8 \\ 5 \\ 2 \end{bmatrix}$.

These form a LI list.

19.2 In each case below a matrix P is formed by taking a LI list of eigenvectors for the given matrix A as columns. The product $P^{-1}AP$ is given also, showing that the entries on the main diagonal are just the eigenvalues of A.

(i) See Example 18.2(i).

Let $A = \begin{bmatrix} 2 & 1 \\ 2 & 3 \end{bmatrix}$ and $P = \begin{bmatrix} 1 & 1 \\ -1 & 2 \end{bmatrix}$.

Then $P^{-1}AP = \begin{bmatrix} 1 & 0 \\ 0 & 4 \end{bmatrix}$.

(ii) See Example 18.2(ii).

Let $A = \begin{bmatrix} 1 & 1 & 0 \\ 1 & 0 & 1 \\ 0 & 1 & 1 \end{bmatrix}$ and $P = \begin{bmatrix} -1 & 1 & 1 \\ 0 & 1 & -2 \\ 1 & 1 & 1 \end{bmatrix}$.

Then $P^{-1}AP = \begin{bmatrix} 1 & 0 & 0 \\ 0 & 2 & 0 \\ 0 & 0 & -1 \end{bmatrix}$.

(iii) See Example 18.2(iii).

Let $A = \begin{bmatrix} 1 & -3 \\ -2 & 6 \end{bmatrix}$ and $P = \begin{bmatrix} 1 & 3 \\ -2 & 1 \end{bmatrix}$.

Then $P^{-1}AP = \begin{bmatrix} 7 & 0 \\ 0 & 0 \end{bmatrix}$.

(iv) See Example 19.1.

Let $A = \begin{bmatrix} 2 & 2 & 3 \\ 1 & 2 & 1 \\ 2 & -2 & 1 \end{bmatrix}$ and $P = \begin{bmatrix} 2 & 4 & 8 \\ 0 & 6 & 5 \\ -2 & -4 & 2 \end{bmatrix}$.

Then $P^{-1}AP = \begin{bmatrix} -1 & 0 & 0 \\ 0 & 2 & 0 \\ 0 & 0 & 4 \end{bmatrix}$.

are all different. Also, not all of $a_1, \ldots, a_{l-1}$ are zero. The above equation therefore contradicts the fact that $(\boldsymbol{x}_1, \ldots, \boldsymbol{x}_{l-1})$ is LI. Thus the demonstration of the rule is complete.

We know that a $n \times n$ matrix may have as many as n distinct eigenvalues (they are the roots of a polynomial equation of degree n). In that situation there must be a LI list consisting of n eigenvectors. When this happens, for a particular matrix A, there is an interesting and important consequence. This is illustrated in Examples 19.2. Given a $n \times n$ matrix A, if we can find a LI list of n eigenvectors, we may take these eigenvectors as the columns of a $n \times n$ matrix, say P. Because its columns form a LI list, this matrix P is invertible, so we may calculate the product matrix $P^{-1}AP$. The result of this process will always be a diagonal matrix, as happens in Examples 19.2. Furthermore, the entries on the main diagonal of this diagonal matrix are precisely the eigenvalues of the original matrix A. The remainder of this chapter is devoted to why and how this comes about, because it is a very important aspect of linear algebra.

19.3 Find an invertible matrix P such that $P^{-1}AP$ is a diagonal matrix, where

$$A=\begin{bmatrix}1&1&0\\1&0&1\\0&1&1\end{bmatrix}.$$

The way to proceed is as follows. Find all eigenvalues of A (done in Example 18.2(ii)). These are 1, 2 and -1. Next find eigenvectors $\boldsymbol{x}_1$, $\boldsymbol{x}_2$ and $\boldsymbol{x}_3$ corresponding to these. This was done also in Example 18.2(ii), but here we need just any three particular eigenvectors, say

$$\boldsymbol{x}_1=\begin{bmatrix}-1\\0\\1\end{bmatrix},\qquad \boldsymbol{x}_2=\begin{bmatrix}1\\1\\1\end{bmatrix},\qquad \boldsymbol{x}_3=\begin{bmatrix}1\\2\\-1\end{bmatrix}.$$

Other choices would be equally appropriate. Now form the matrix P with these vectors as columns

$$P=\begin{bmatrix}-1&1&1\\0&1&2\\1&1&-1\end{bmatrix}.$$

Then $P^{-1}AP$ is a diagonal matrix. To see this in detail, first note that P is invertible because its columns form a LI list. Next, observe that for $i=1, 2, 3$,

$$\boldsymbol{x}_i=P\boldsymbol{e}_i,$$

where $(\boldsymbol{e}_1, \boldsymbol{e}_2, \boldsymbol{e}_3)$ is the standard basis for $\mathbb{R}^3$. Consequently

$$\boldsymbol{e}_i=P^{-1}\boldsymbol{x}_i \qquad \text{for each } i.$$

Hence

$$\begin{aligned}P^{-1}AP\boldsymbol{e}_i&=P^{-1}A\boldsymbol{x}_i\\&=P^{-1}k_i\boldsymbol{x}_i\\&=k_iP^{-1}\boldsymbol{x}_i\\&=k_i\boldsymbol{e}_i,\end{aligned}$$

for each i, where k_i is the eigenvalue (1, 2 or -1) corresponding to $\boldsymbol{x}_i$. This tells us what the columns of $P^{-1}AP$ are:

first column is $k_1\boldsymbol{e}_1$, i.e. $\boldsymbol{e}_1$,
second column is $k_2\boldsymbol{e}_2$, i.e. $2\boldsymbol{e}_2$,
and third column is $k_3\boldsymbol{e}_3$, i.e. $-\boldsymbol{e}_3$.

So

$$P^{-1}AP=\begin{bmatrix}1&0&0\\0&2&0\\0&0&-1\end{bmatrix}.$$

The following is a general argument which explains this phenomenon. Example 19.3 illustrates its application in a particular case.

Let A be a $n \times n$ matrix, let $(\boldsymbol{x}_1, \ldots, \boldsymbol{x}_n)$ be a LI list consisting of eigenvectors of A, and let $k_1, \ldots, k_n$ (respectively) be corresponding eigenvalues. (These may be all different, but, as we shall see in Chapter 22, this requirement is not necessary.) For $1 \leqslant i \leqslant n$, let $\boldsymbol{e}_i$ denote the n-vector which has 1 in the ith position and 0s elsewhere. Construct a $n \times n$ matrix P having the vectors $\boldsymbol{x}_1, \ldots, \boldsymbol{x}_n$ as columns. Then P is invertible, since its columns form a LI list. We have, for $1 \leqslant i \leqslant n$,

$$P\boldsymbol{e}_i = \boldsymbol{x}_i \qquad \text{and} \qquad \boldsymbol{e}_i = P^{-1}\boldsymbol{x}_i.$$

Multiplying P by $\boldsymbol{e}_i$ just picks out the ith column. Indeed, the ith column of any matrix T may be obtained by evaluating the product $T\boldsymbol{e}_i$. In particular, the ith column of $P^{-1}AP$ is obtained thus:

$$\begin{aligned} P^{-1}AP\boldsymbol{e}_i &= P^{-1}A\boldsymbol{x}_i \\ &= P^{-1}(k_i\boldsymbol{x}_i) \\ &= k_iP^{-1}\boldsymbol{x}_i \\ &= k_i\boldsymbol{e}_i. \end{aligned}$$

Hence the ith column of $P^{-1}AP$ has k_i in the ith position and 0s elsewhere. $P^{-1}AP$ is therefore a diagonal matrix, with the numbers $k_1, \ldots, k_n$ down the main diagonal.

This leads us to an important result, which we shall give as the next rule, but we need a definition first, to collect together the ideas involved.

Definition

A square matrix A is *diagonalisable* if there is an invertible matrix P such that the matrix $P^{-1}AP$ is a diagonal matrix.

In the event that A is diagonalisable, it will necessarily be the case that the columns of the matrix P will be eigenvectors of A, and the matrix $P^{-1}AP$ will have the eigenvalues of A as its diagonal entries (think about it!).

Rule

Let A be a $n \times n$ matrix. If there is a LI list of n eigenvectors of A then A is diagonalisable.

19.4 Let $A=\begin{bmatrix} 1 & 0 & 1 \\ 0 & 1 & 1 \\ -1 & 1 & 2 \end{bmatrix}$.

Show that A is not diagonalisable.

The characteristic equation of A is

$$(1-k)^2(2-k)=0,$$

so the eigenvalues of A are 1 and 2. The set of eigenvectors of A corresponding to the eigenvalue 2 is

$$\left\{\begin{bmatrix} t \\ t \\ t \end{bmatrix}: t \in \mathbb{R},\, t \neq 0\right\}.$$

This is found in the usual way. Now consider the eigenvalue 1 in the usual way, i.e. try to solve the equation $(A-I)\mathbf{x}=\mathbf{0}$.

$$\begin{bmatrix} 0 & 0 & 1 \\ 0 & 0 & 1 \\ -1 & 1 & 1 \end{bmatrix}\begin{bmatrix} x \\ y \\ z \end{bmatrix}=\begin{bmatrix} 0 \\ 0 \\ 0 \end{bmatrix}.$$

The solution is $z=0$, $y=t$, $x=t$ $(t \in \mathbb{R})$, and so the set of eigenvectors corresponding to the eigenvalue 1 is

$$\left\{\begin{bmatrix} t \\ t \\ 0 \end{bmatrix}: t \in \mathbb{R},\, t \neq 0\right\}.$$

It is now apparent that there is no LI list of three eigenvectors of A. In any list of three eigenvectors, at least two would have to correspond to the same eigenvalue, and so in this case would be multiples of one another.

19.5 Matrices A and B for which there is an invertible matrix P such that $B=P^{-1}AP$ are said to be *similar*. Show that

(i) similar matrices have the same determinant,
(ii) similar matrices have the same rank,
(iii) similar matrices have the same eigenvalues.

(i)
$$\begin{aligned} \det(P^{-1}AP) &= \det(P^{-1})\det(A)\det(P) \\ &= \det(A)\det(P^{-1})\det(P) \\ &= \det(A)\det(P^{-1}P) \\ &= \det(A)\det(I) \\ &= \det(A). \end{aligned}$$

(ii) Suppose that $B=P^{-1}AP$. P is invertible, so P may be written as a product of elementary matrices (see Chapter 5). Likewise P^{-1}. By the method of Example 15.5, then, A and $P^{-1}A$ have the same row space, and so have the same rank.

By a similar argument relating to columns rather than rows, the matrices $P^{-1}A$ and $P^{-1}AP$ have the same column space, and so have the same rank. Thus A and $P^{-1}AP$ have the same rank.

The justification of this is contained in the preceding discussion. In fact, the converse of this result is also true. The existence of such a list of eigenvectors is a necessary condition for the matrix to be diagonalisable (but we shall not prove this). Consequently, the non-existence of such a list of eigenvectors demonstrates that a matrix is not diagonalisable. See Example 19.4.

The practical process of diagonalisation involves nothing new. The matrix P is made up from eigenvectors, and the diagonal matrix is determined by the eigenvalues. Of course, the order of the eigenvectors as columns of P corresponds to the order in which the eigenvalues occur in the diagonal matrix. Notice also that because any LI list of n vectors in $\mathbb{R}^n$ is a basis for $\mathbb{R}^n$, the columns of the matrix P will necessarily constitute a basis for $\mathbb{R}^n$.

Diagonalisation is very useful in applications of linear algebra. This is because diagonal matrices are very simple. If, for a given matrix A, we can find a matrix P such that $P^{-1}AP$ is a diagonal matrix, then properties or uses of the matrix A may be simplified by consideration of this diagonal matrix. Some illustration of this is given by Example 19.5. Other examples, showing practical uses, will be given in later chapters.

The rule above describes certain circumstances in which a matrix is diagonalisable. The rule is not easy to apply directly. We have seen some matrices for which there is a LI list of n eigenvectors, but in general it will be a substantial task to find out whether such a list exists or not. But it turns out that there is a whole class of matrices which we can show in general to have this property, and which are consequently diagonalisable. This is expressed in our next rule.

Rule

Let A be a real symmetric matrix with distinct eigenvalues (i.e. the characteristic equation of A has no repeated roots). Then A is diagonalisable.

This result is an immediate consequence of our earlier rules.

Take care not to fall into the trap of believing that the requirements of this rule are *necessary* for a matrix to be diagonalisable. We have already seen some diagonalisable matrices which are not symmetric, and we shall see more in later chapters. Indeed, in Chapter 22 we shall develop methods for diagonalising some matrices which do not satisfy these requirements.

(iii)
$$\begin{aligned}\det(P^{-1}AP - kI) &= \det(P^{-1}AP - kP^{-1}P)\\ &= \det(P^{-1}(A - kI)P)\\ &= \det(A - kI), \qquad \text{by (i) above.}\end{aligned}$$

This shows that the characteristic equation of $P^{-1}AP$ is the same as the characteristic equation of A, so $P^{-1}AP$ and A have the same eigenvalues, where A is any square matrix and P is any invertible matrix of the same size.

Summary
It is shown that eigenvectors corresponding to distinct eigenvalues constitute a linearly independent list. A routine procedure is described for finding, given a $n \times n$ matrix A which has a linearly independent list of n eigenvectors, an invertible matrix P such that $P^{-1}AP$ is a diagonal matrix. It is noted that this is always possible if A is symmetric.

Exercises

1. For each matrix A given in Exercise 1 of Chapter 18, write down an invertible matrix P such that $P^{-1}AP$ is a diagonal matrix. Verify by explicit calculation that the diagonal entries in $P^{-1}AP$ are the eigenvalues of A.
2. Which of the following matrices are diagonalisable? For those which are, find an invertible matrix P as in Exercise 1.

(i) $\begin{bmatrix} 2 & -1 \\ 1 & 0 \end{bmatrix}$. (ii) $\begin{bmatrix} 3 & 4 \\ -1 & 2 \end{bmatrix}$.

(iii) $\begin{bmatrix} 3 & 6 \\ -1 & -2 \end{bmatrix}$. (iv) $\begin{bmatrix} 1 & 0 \\ 1 & 1 \end{bmatrix}$.

(v) $\begin{bmatrix} 1 & 0 & 0 \\ 1 & 1 & 0 \\ 1 & 1 & 1 \end{bmatrix}$. (vi) $\begin{bmatrix} 0 & 2 & -1 \\ 1 & 1 & 1 \\ -1 & 1 & 1 \end{bmatrix}$.

(vii) $\begin{bmatrix} 1 & 0 & 1 \\ 2 & -1 & 1 \\ -1 & 2 & 1 \end{bmatrix}$. (viii) $\begin{bmatrix} -1 & 2 & 2 \\ 2 & 1 & 0 \\ 2 & 0 & 1 \end{bmatrix}$.

(ix) $\begin{bmatrix} 1 & 0 & 0 \\ 1 & 2 & 0 \\ 1 & 2 & 3 \end{bmatrix}$. (x) $\begin{bmatrix} 1 & 1 & -2 \\ 4 & 0 & 4 \\ 1 & -1 & 4 \end{bmatrix}$.

(xi) $\begin{bmatrix} 4 & 2 & 3 \\ 2 & 1 & 2 \\ -1 & -2 & 0 \end{bmatrix}$.

3. Let A be a square matrix, and suppose that P is a matrix such that $P^{-1}AP$ is a diagonal matrix. Show that for any real number c, $P^{-1}(cA)P$ is a diagonal matrix. Show further that, for any natural number r, $P^{-1}(A^r)P$ is a diagonal matrix.

Examples

20.1 Examples of calculation of dot products.

(i) In $\mathbb{R}^2$, let $x=\begin{bmatrix}1\\3\end{bmatrix}$, $y=\begin{bmatrix}2\\4\end{bmatrix}$.

Then $x\cdot y = 2+12=14$.

(ii) In $\mathbb{R}^3$, let $x=\begin{bmatrix}1\\2\\3\end{bmatrix}$, $y=\begin{bmatrix}2\\3\\4\end{bmatrix}$.

Then $x\cdot y = 2+6+12=20$.

(iii) In $\mathbb{R}^4$, let $x=\begin{bmatrix}1\\1\\2\\2\end{bmatrix}$, $y=\begin{bmatrix}-3\\1\\2\\1\end{bmatrix}$.

Then $x\cdot y = -3+1+4+2=4$.

(iv) In $\mathbb{R}^4$, let $x=\begin{bmatrix}2\\-1\\0\\1\end{bmatrix}$, $y=\begin{bmatrix}1\\1\\3\\-1\end{bmatrix}$.

Then $x\cdot y = 2-1+0-1=0$.

20.2 Properties of the dot product.

Let $x=\begin{bmatrix}x_1\\ \vdots\\ x_n\end{bmatrix}$, $y=\begin{bmatrix}y_1\\ \vdots\\ y_n\end{bmatrix}$, $z=\begin{bmatrix}z_1\\ \vdots\\ z_n\end{bmatrix}$ belong to $\mathbb{R}^n$.

(ii) $x\cdot(y+z)=x\cdot y+x\cdot z$.

Proof:
$$\begin{aligned}x\cdot(y+z) &= x_1(y_1+z_1)+\cdots+x_n(y_n+z_n)\\ &= x_1y_1+x_1z_1+\cdots+x_ny_n+x_nz_n\\ &= x_1y_1+\cdots+x_ny_n+x_1z_1+\cdots+x_nz_n\\ &= x\cdot y+x\cdot z.\end{aligned}$$

(iii) Let $a\in\mathbb{R}$. Then $(ax)\cdot y=a(x\cdot y)$.

Proof:
$$\begin{aligned}(ax)\cdot y &= (ax_1)y_1+\cdots+(ax_n)y_n\\ &= a(x_1y_1+\cdots+x_ny_n)\\ &= a(x\cdot y).\end{aligned}$$

Similarly,
$$x\cdot(ay)=a(x\cdot y).$$

20

The dot product

Chapter 10 dealt with the dot product of vectors in $\mathbb{R}^3$, and discussed how use can be made of this notion in geometry. We shall now extend those ideas and methods to the context of $\mathbb{R}^n$ for arbitrary n. In doing so we shall make use of geometrical intuition and geometrical terminology, even though they are not strictly applicable. It is certainly convenient to do so, and we shall find that there are general algebraic notions which correspond quite closely with geometric ones.

Definition

Let $\boldsymbol{x}, \boldsymbol{y} \in \mathbb{R}^n$. The *dot product* of $\boldsymbol{x}$ and $\boldsymbol{y}$ is defined by

$$\boldsymbol{x} \cdot \boldsymbol{y} = \boldsymbol{x}^{\mathrm{T}} \boldsymbol{y}.$$

More explicitly, if $\boldsymbol{x} = \begin{bmatrix} x_1 \\ \vdots \\ x_n \end{bmatrix}$ and $\boldsymbol{y} = \begin{bmatrix} y_1 \\ \vdots \\ y_n \end{bmatrix}$, then

$$\boldsymbol{x} \cdot \boldsymbol{y} = x_1 y_1 + \cdots + x_n y_n.$$

Here we do not distinguish between the 1×1 matrix $\boldsymbol{x}^{\mathrm{T}} \boldsymbol{y}$ and the number which is its single entry. So remember to note that $\boldsymbol{x} \cdot \boldsymbol{y}$ is always a scalar (that is to say, a number), not a vector. Some examples are given in Example 20.1.

Listed below are some properties of the dot product. Justifications where necessary are given in Examples 20.2.

(i) $\boldsymbol{x} \cdot \boldsymbol{y} = \boldsymbol{y} \cdot \boldsymbol{x}$

(ii) $\boldsymbol{x} \cdot (\boldsymbol{y} + \boldsymbol{z}) = \boldsymbol{x} \cdot \boldsymbol{y} + \boldsymbol{x} \cdot \boldsymbol{z}$

(iii) $(a\boldsymbol{x}) \cdot \boldsymbol{y} = \boldsymbol{x} \cdot (a\boldsymbol{y}) = a(\boldsymbol{x} \cdot \boldsymbol{y})$

(iv) If $\boldsymbol{x} = \begin{bmatrix} x_1 \\ \vdots \\ x_n \end{bmatrix}$ then $\boldsymbol{x} \cdot \boldsymbol{x} = x_1^2 + \cdots + x_n^2$

(v) $\boldsymbol{x} \cdot \boldsymbol{x} \geqslant 0$,

where $\boldsymbol{x}, \boldsymbol{y}, \boldsymbol{z} \in \mathbb{R}^n$, and $a \in \mathbb{R}$.

20.3 Proof of the Cauchy–Schwarz inequality:

$$|\boldsymbol{x}\cdot\boldsymbol{y}| \leqslant |\boldsymbol{x}||\boldsymbol{y}| \qquad \text{for all } \boldsymbol{x}, \boldsymbol{y}\in\mathbb{R}^n.$$

The result is obvious if $\boldsymbol{x}$ or $\boldsymbol{y}$ is $\boldsymbol{0}$. So suppose that $\boldsymbol{x}\neq\boldsymbol{0}$ and $\boldsymbol{y}\neq\boldsymbol{0}$. By property (iv) of the dot product,

$$((1/|\boldsymbol{x}|)\boldsymbol{x}-(1/|\boldsymbol{y}|)\boldsymbol{y})\cdot((1/|\boldsymbol{x}|)\boldsymbol{x}-(1/|\boldsymbol{y}|)\boldsymbol{y})\geqslant 0.$$

Hence $(1/|\boldsymbol{x}|)^2(\boldsymbol{x}\cdot\boldsymbol{x})+(1/|\boldsymbol{y}|)^2(\boldsymbol{y}\cdot\boldsymbol{y})-(2/(|\boldsymbol{x}||\boldsymbol{y}|))(\boldsymbol{x}\cdot\boldsymbol{y})\geqslant 0,$

i.e. $1+1-2(\boldsymbol{x}\cdot\boldsymbol{y})/(|\boldsymbol{x}||\boldsymbol{y}|)\geqslant 0.$

So $(\boldsymbol{x}\cdot\boldsymbol{y})/(|\boldsymbol{x}||\boldsymbol{y}|)\leqslant 1,$

giving $\boldsymbol{x}\cdot\boldsymbol{y}\leqslant|\boldsymbol{x}||\boldsymbol{y}|.$

Of course, $\boldsymbol{x}\cdot\boldsymbol{y}$ may be negative, so we need a lower bound also in order to bound the absolute value $|\boldsymbol{x}\cdot\boldsymbol{y}|$. Apply a similar procedure to

$$((1/|\boldsymbol{x}|)\boldsymbol{x}+(1/|\boldsymbol{y}|)\boldsymbol{y})\cdot((1/|\boldsymbol{x}|)\boldsymbol{x}+(1/|\boldsymbol{y}|)\boldsymbol{y})\geqslant 0,$$

obtaining (details omitted)

$$(\boldsymbol{x}\cdot\boldsymbol{y})/(|\boldsymbol{x}||\boldsymbol{y}|)\geqslant -1,$$

so $\boldsymbol{x}\cdot\boldsymbol{y}\geqslant -|\boldsymbol{x}||\boldsymbol{y}|.$

Putting these together yields

$$|\boldsymbol{x}\cdot\boldsymbol{y}|\leqslant|\boldsymbol{x}||\boldsymbol{y}|$$

as required.

20.4 Find the length of the given vector in each case.

(i) In $\mathbb{R}^2$, $\boldsymbol{x}=\begin{bmatrix}1\\2\end{bmatrix}$, $|\boldsymbol{x}|^2=1+4=5.$

(ii) In $\mathbb{R}^3$, $\boldsymbol{x}=\begin{bmatrix}1\\2\\3\end{bmatrix}$, $|\boldsymbol{x}|^2=1+4+9=14.$

$\boldsymbol{y}=\begin{bmatrix}-1\\-2\\2\end{bmatrix}$, $|\boldsymbol{y}|^2=1+4+4=9.$

(iii) In $\mathbb{R}^4$, $\boldsymbol{x}=\begin{bmatrix}1\\2\\3\\4\end{bmatrix}$, $|\boldsymbol{x}|^2=1+4+9+16=30.$

$\boldsymbol{y}=\begin{bmatrix}0\\-2\\1\\3\end{bmatrix}$, $|\boldsymbol{y}|^2=0+4+1+9=14.$

$\boldsymbol{z}=\begin{bmatrix}2\\1\\0\\3\end{bmatrix}$, $|\boldsymbol{z}|^2=2+1+0+9=12.$

Geometrical notions such as lengths and angles clearly have no meaning in $\mathbb{R}^n$ for $n > 3$. Nevertheless they have formal counterparts, as follows.

Definition

If $\boldsymbol{x} \in \mathbb{R}^n$ then the *length* of $\boldsymbol{x}$ is $(\boldsymbol{x} \cdot \boldsymbol{x})^{\frac{1}{2}}$. It is denoted by $|\boldsymbol{x}|$. If $\boldsymbol{x}, \boldsymbol{y} \in \mathbb{R}^n$ and $\boldsymbol{x}$ and $\boldsymbol{y}$ are non-zero, then we say that the *angle between* $\boldsymbol{x}$ and $\boldsymbol{y}$ is θ, where

$$\cos \theta = \frac{\boldsymbol{x} \cdot \boldsymbol{y}}{|\boldsymbol{x}||\boldsymbol{y}|}.$$

The definition of the angle between two vectors will make sense only if we know that the expression given for $\cos \theta$ always has a value which lies in the interval $[-1, 1]$. In fact it always does, by virtue of what is known as the Cauchy–Schwarz inequality:

$$|\boldsymbol{x} \cdot \boldsymbol{y}| \leqslant |\boldsymbol{x}||\boldsymbol{y}|, \qquad \text{for all } \boldsymbol{x}, \boldsymbol{y} \in \mathbb{R}^n.$$

A proof of this is given as Example 20.3. It can be omitted without prejudice to what follows.

Sample calculations of lengths and angles are given in Examples 20.4 and 20.5.

20.5 Find the cosine of the angle between given vectors.

(i) In $\mathbb{R}^2$, let $x = \begin{bmatrix} 1 \\ 2 \end{bmatrix}$, $y = \begin{bmatrix} 4 \\ 1 \end{bmatrix}$.

Then $\cos\theta = \dfrac{4+2}{\sqrt{1+4}\sqrt{16+1}} = 6/\sqrt{85}$.

(ii) In $\mathbb{R}^3$, let $x = \begin{bmatrix} 1 \\ 1 \\ 3 \end{bmatrix}$, $y = \begin{bmatrix} -1 \\ 1 \\ 1 \end{bmatrix}$.

Then $\cos\theta = \dfrac{-1+1+3}{\sqrt{1+1+9}\sqrt{1+1+1}} = 3/\sqrt{33}$.

(iii) In $\mathbb{R}^3$, let $x = \begin{bmatrix} 2 \\ -1 \\ 2 \end{bmatrix}$, $y = \begin{bmatrix} -1 \\ -1 \\ -1 \end{bmatrix}$.

Then $\cos\theta = \dfrac{-2+1-2}{\sqrt{9}\sqrt{3}} = -1/\sqrt{3}$.

(The negative sign indicates an obtuse angle.)

(iv) In $\mathbb{R}^4$, let $x = \begin{bmatrix} 2 \\ 1 \\ -1 \\ 2 \end{bmatrix}$, $y = \begin{bmatrix} 0 \\ 3 \\ 1 \\ 1 \end{bmatrix}$.

Then $\cos\theta = \dfrac{0+3-1+2}{\sqrt{4+1+1+4}\sqrt{0+9+1+1}} = 4/\sqrt{110}$.

20.6 Illustrations of orthogonal pairs of vectors.

(i) In $\mathbb{R}^2$, $x = \begin{bmatrix} 1 \\ 1 \end{bmatrix}$, $y = \begin{bmatrix} 1 \\ -1 \end{bmatrix}$, $x \cdot y = 0$.

$x = \begin{bmatrix} -1 \\ 3 \end{bmatrix}$, $y = \begin{bmatrix} 1 \\ \frac{1}{3} \end{bmatrix}$, $x \cdot y = 0$.

(ii) In $\mathbb{R}^3$, $x = \begin{bmatrix} 1 \\ 2 \\ 3 \end{bmatrix}$, $y = \begin{bmatrix} 1 \\ 1 \\ -1 \end{bmatrix}$, $x \cdot y = 0$.

$x = \begin{bmatrix} 2 \\ 3 \\ -1 \end{bmatrix}$, $y = \begin{bmatrix} 5 \\ -3 \\ 1 \end{bmatrix}$, $x \cdot y = 0$.

(iii) In $\mathbb{R}^4$, $x = \begin{bmatrix} 1 \\ 2 \\ -1 \\ 1 \end{bmatrix}$, $y = \begin{bmatrix} 3 \\ -1 \\ 1 \\ 0 \end{bmatrix}$, $x \cdot y = 0$.

$x = \begin{bmatrix} 1 \\ 2 \\ -1 \\ 1 \end{bmatrix}$, $y = \begin{bmatrix} 0 \\ 1 \\ 1 \\ -1 \end{bmatrix}$, $x \cdot y = 0$.

An important special case of the angle between two vectors is a right angle.

Definition
Two non-zero vectors in $\mathbb{R}^n$ are said to be *orthogonal* to each other if the angle between them is a right angle (equivalently, if their dot product is zero).

'Orthogonal' is the word normally used in the context of $\mathbb{R}^n$ to extend the notion of perpendicularity which strictly speaking applies only in the real world of geometry.

The vectors in the standard basis for $\mathbb{R}^n$ are orthogonal to each other. This is easy to check. Some other illustrations are given in Examples 20.6.

We shall say that a set or list of non-zero vectors is an orthogonal set (or list) if every vector in the set (or list) is orthogonal to every other vector in the set (or list). There is an important connection between orthogonality and linear independence.

Rule
Any orthogonal list of (non-zero) vectors in $\mathbb{R}^n$ is linearly independent.

This rule is justified as follows. Let $(\boldsymbol{x}_1, \ldots, \boldsymbol{x}_s)$ be an orthogonal list of vectors in $\mathbb{R}^n$, and suppose that

$$a_1\boldsymbol{x}_1 + \cdots + a_s\boldsymbol{x}_s = \boldsymbol{0}.$$

Then, for each i $(1 \leqslant i \leqslant s)$,

$$\boldsymbol{x}_i \cdot (a_1\boldsymbol{x}_1 + \cdots + a_s\boldsymbol{x}_s) = 0,$$

so $\quad a_1(\boldsymbol{x}_i \cdot \boldsymbol{x}_1) + \cdots + a_i(\boldsymbol{x}_i \cdot \boldsymbol{x}_i) + \cdots + a_s(\boldsymbol{x}_i \cdot \boldsymbol{x}_s) = 0,$

giving $\quad a_i|\boldsymbol{x}_i|^2 = 0,$

since $\quad \boldsymbol{x}_i \cdot \boldsymbol{x}_j = 0 \quad$ for every $j \neq i$.

It follows that, for each i $(1 \leqslant i \leqslant s)$, $a_i = 0$, since $\boldsymbol{x}_i$ is a non-zero vector. Thus all of the coefficients a_i must be 0, and so the list $(\boldsymbol{x}_1, \ldots, \boldsymbol{x}_s)$ is LI.

In the geometry of three dimensions, we may visualise the following situation. Given any non-zero vector $\boldsymbol{v}$ in $\mathbb{R}^3$, the set of all vectors which are perpendicular to $\boldsymbol{v}$ can be thought of as representing the plane (through the origin) which has $\boldsymbol{v}$ as its normal vector. This was dealt with in Chapter 10, and we have seen earlier (in Chapter 13) that such a plane is a subspace of $\mathbb{R}^3$. So we have a special case of a notion of a 'subspace orthogonal to a vector'. The general situation is given by the next rule.

20.7 Proof that if $\boldsymbol{x} \in \mathbb{R}^n$ ($\boldsymbol{x} \neq \boldsymbol{0}$) then the set

$$\{\boldsymbol{y} \in \mathbb{R}^n\colon \boldsymbol{y} \cdot \boldsymbol{x} = 0\}$$

is a subspace of $\mathbb{R}^n$.

Denote this set by S. We apply the rule given in Chapter 13.

First, $\boldsymbol{0} \in S$, so S is not empty.

Second, let $\boldsymbol{u}, \boldsymbol{v} \in S$, and let $k, l \in \mathbb{R}$.

Then we know that $\boldsymbol{u} \cdot \boldsymbol{x} = \boldsymbol{v} \cdot \boldsymbol{x} = 0$, and

$$\begin{aligned}(k\boldsymbol{u} + l\boldsymbol{v}) \cdot \boldsymbol{x} &= (k\boldsymbol{u}) \cdot \boldsymbol{x} + (l\boldsymbol{v}) \cdot \boldsymbol{x} \\ &= k(\boldsymbol{u} \cdot \boldsymbol{x}) + l(\boldsymbol{v} \cdot \boldsymbol{x}) \\ &= 0 + 0 = 0.\end{aligned}$$

Hence $k\boldsymbol{u} + l\boldsymbol{v} \in S$, and it follows that S is a subspace.

20.8 Orthogonal eigenvectors.

Let $$A = \begin{bmatrix} 2 & 1 & 0 \\ 1 & 3 & 1 \\ 0 & 1 & 2 \end{bmatrix}.$$

By the usual process, we find eigenvalues and eigenvectors of A:

$k = 1$: Eigenvectors $\begin{bmatrix} t \\ -t \\ t \end{bmatrix}$ $(t \in \mathbb{R},\ t \neq 0)$,

$k = 2$: Eigenvectors $\begin{bmatrix} -t \\ 0 \\ t \end{bmatrix}$ $(t \in \mathbb{R},\ t \neq 0)$,

$k = 4$: Eigenvectors $\begin{bmatrix} t \\ 2t \\ t \end{bmatrix}$ $(t \in \mathbb{R},\ t \neq 0)$.

Now pick any three eigenvectors, one for each eigenvalue,

say $$\boldsymbol{x}_1 = \begin{bmatrix} a \\ -a \\ a \end{bmatrix}, \quad \boldsymbol{x}_2 = \begin{bmatrix} -b \\ 0 \\ b \end{bmatrix}, \quad \boldsymbol{x}_3 = \begin{bmatrix} c \\ 2c \\ c \end{bmatrix}.$$

Each pair of these vectors is an orthogonal pair, irrespective of the choice of a, b and c (non-zero, of course). This is easily verified:

$$\begin{aligned}\boldsymbol{x}_1 \cdot \boldsymbol{x}_2 &= -ab + 0 + ab = 0, \\ \boldsymbol{x}_2 \cdot \boldsymbol{x}_3 &= -bc + 0 + bc = 0, \\ \boldsymbol{x}_3 \cdot \boldsymbol{x}_1 &= ac - 2ac + ac = 0.\end{aligned}$$

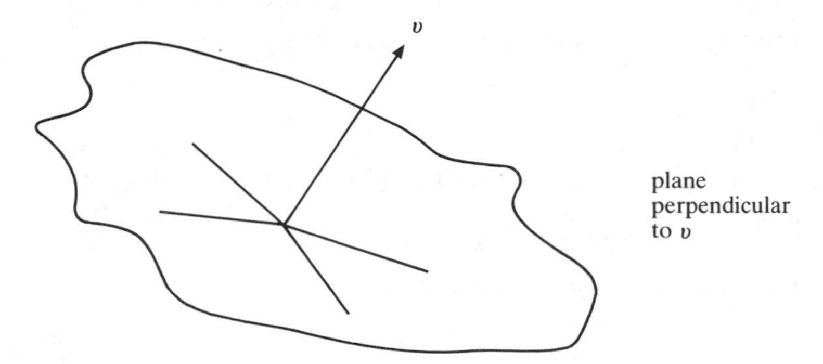

Rule

Let $\boldsymbol{x} \in \mathbb{R}^n$, with $\boldsymbol{x} \neq \mathbf{0}$. Then the set which consists of all vectors in $\mathbb{R}^n$ which are orthogonal to $\boldsymbol{x}$, together with $\mathbf{0}$, constitutes a subspace of $\mathbb{R}^n$.

A demonstration of this is given in Example 20.7.

Orthogonality of vectors crops up also in the context of eigenvectors. Example 20.8 gives a 3×3 matrix, its eigenvalues and its eigenvectors. It happens that the eigenvectors form an orthogonal set, as is verified in the example. There is a general result lying behind this.

Rule

Let A be a real symmetric matrix, and let k_1 and k_2 be distinct eigenvalues of A, with corresponding eigenvectors $\boldsymbol{x}_1$ and $\boldsymbol{x}_2$. Then $\boldsymbol{x}_1$ and $\boldsymbol{x}_2$ are orthogonal to each other.

To prove this, suppose that

$$A\boldsymbol{x}_1 = k_1\boldsymbol{x}_1 \qquad \text{and} \qquad A\boldsymbol{x}_2 = k_2\boldsymbol{x}_2.$$

Consider the product $\boldsymbol{x}_1^{\mathrm{T}} A\boldsymbol{x}_2$.

First, $\boldsymbol{x}_1^{\mathrm{T}} A\boldsymbol{x}_2 = \boldsymbol{x}_1^{\mathrm{T}} k_2\boldsymbol{x}_2 = k_2\boldsymbol{x}_1^{\mathrm{T}}\boldsymbol{x}_2$.

Second,
$$\begin{aligned} \boldsymbol{x}_1^{\mathrm{T}} A\boldsymbol{x}_2 &= (\boldsymbol{x}_1^{\mathrm{T}} A^{\mathrm{T}})\boldsymbol{x}_2 \qquad \text{(since } A \text{ is symmetric)} \\ &= (A\boldsymbol{x}_1)^{\mathrm{T}}\boldsymbol{x}_2 \\ &= (k_1\boldsymbol{x}_1)^{\mathrm{T}}\boldsymbol{x}_2 \\ &= k_1\boldsymbol{x}_1^{\mathrm{T}}\boldsymbol{x}_2. \end{aligned}$$

Hence $k_1\boldsymbol{x}_1^{\mathrm{T}}\boldsymbol{x}_2 = k_2\boldsymbol{x}_1^{\mathrm{T}}\boldsymbol{x}_2$,

so $(k_1 - k_2)\boldsymbol{x}_1^{\mathrm{T}}\boldsymbol{x}_2 = 0$.

But $k_1 \neq k_2$, so we must have

$$\boldsymbol{x}_1^{\mathrm{T}}\boldsymbol{x}_2 = 0, \qquad \text{i.e. } \boldsymbol{x}_1 \cdot \boldsymbol{x}_2 = 0,$$

so $\boldsymbol{x}_1$ and $\boldsymbol{x}_2$ are orthogonal. (Neither vector can be $\mathbf{0}$ because both are eigenvectors.)

20.9 Eigenvalues of a matrix which is not symmetric.

Let $A = \begin{bmatrix} 2 & 2 & 3 \\ 1 & 2 & 1 \\ 2 & -2 & 1 \end{bmatrix}$.

By the usual process (see Example 19.1(iii)), we find eigenvalues -1, 2 and 4 and corresponding eigenvectors:

$k=1$: Eigenvectors $\begin{bmatrix} -t \\ 0 \\ t \end{bmatrix}$ $(t \in \mathbb{R},\ t \neq 0)$,

$k=2$: Eigenvectors $\begin{bmatrix} -t \\ -\frac{3}{2}t \\ t \end{bmatrix}$ $(t \in \mathbb{R},\ t \neq 0)$,

$k=4$: Eigenvectors $\begin{bmatrix} 4t \\ \frac{5}{2}t \\ t \end{bmatrix}$ $(t \in \mathbb{R},\ t \neq 0)$.

Pick three eigenvectors, one for each eigenvalue, say

$$\boldsymbol{x}_1 = \begin{bmatrix} -a \\ 0 \\ a \end{bmatrix}, \quad \boldsymbol{x}_2 = \begin{bmatrix} -b \\ -\frac{3}{2}b \\ b \end{bmatrix}, \quad \boldsymbol{x}_3 = \begin{bmatrix} 4c \\ \frac{5}{2}c \\ c \end{bmatrix}.$$

No pair of these is orthogonal, for any values of a, b and c (non-zero, of course). Again, this is easily verified:

$$\begin{aligned} \boldsymbol{x}_1 \cdot \boldsymbol{x}_2 &= ab + 0 + ab &&= 2ab, \\ \boldsymbol{x}_2 \cdot \boldsymbol{x}_3 &= -4bc - \tfrac{15}{4}bc + bc &&= -\tfrac{27}{4}bc, \\ \boldsymbol{x}_3 \cdot \boldsymbol{x}_1 &= -4ac + 0 + ac &&= -3ac. \end{aligned}$$

Be sure to note the requirement in this rule that A be symmetric. See Example 20.9. We shall return to these ideas in Chapter 22 when we reconsider diagonalisation.

Summary

The dot product in $\mathbb{R}^n$ is introduced and compared with the dot product in $\mathbb{R}^3$. Geometrical ideas of lengths and angles are extended to $\mathbb{R}^n$. In particular, orthogonality of vectors in $\mathbb{R}^n$ is discussed, including the relationship between orthogonality and linear independence and including the orthogonality of eigenvectors corresponding to distinct eigenvalues of a matrix.

Exercises

1. In each case below, evaluate the dot product of the two given vectors.

(i) $\begin{bmatrix}2\\1\end{bmatrix}, \begin{bmatrix}3\\-1\end{bmatrix}$. (ii) $\begin{bmatrix}1\\0\end{bmatrix}, \begin{bmatrix}-1\\4\end{bmatrix}$.

(iii) $\begin{bmatrix}1\\2\end{bmatrix}, \begin{bmatrix}-4\\2\end{bmatrix}$. (iv) $\begin{bmatrix}-1\\2\end{bmatrix}, \begin{bmatrix}0\\0\end{bmatrix}$.

(v) $\begin{bmatrix}-1\\2\\1\end{bmatrix}, \begin{bmatrix}3\\1\\1\end{bmatrix}$. (vi) $\begin{bmatrix}1\\0\\0\end{bmatrix}, \begin{bmatrix}1\\0\\0\end{bmatrix}$.

(vii) $\begin{bmatrix}1\\2\\3\end{bmatrix}, \begin{bmatrix}1\\2\\3\end{bmatrix}$. (viii) $\begin{bmatrix}0\\1\\-3\end{bmatrix}, \begin{bmatrix}3\\-4\\1\end{bmatrix}$.

(ix) $\begin{bmatrix}1\\0\\2\\2\end{bmatrix}, \begin{bmatrix}3\\-2\\-2\\0\end{bmatrix}$. (x) $\begin{bmatrix}6\\-2\\1\\4\end{bmatrix}, \begin{bmatrix}2\\5\\-3\\1\end{bmatrix}$.

2. Find the length of each of the following vectors.

$$\begin{bmatrix}1\\1\end{bmatrix}, \begin{bmatrix}3\\-1\end{bmatrix}, \begin{bmatrix}3\\4\end{bmatrix},$$

$$\begin{bmatrix}1\\1\\1\end{bmatrix}, \begin{bmatrix}2\\3\\-1\end{bmatrix}, \begin{bmatrix}3\\0\\4\end{bmatrix}, \begin{bmatrix}-1\\1\\2\end{bmatrix}, \begin{bmatrix}7\\4\\4\end{bmatrix},$$

$$\begin{bmatrix}1\\1\\1\\1\end{bmatrix}, \begin{bmatrix}2\\-1\\1\\-2\end{bmatrix}, \begin{bmatrix}1\\3\\-2\\1\end{bmatrix}, \begin{bmatrix}-1\\4\\5\\1\end{bmatrix}.$$

3. Show that if $\boldsymbol{x}$ is an eigenvector of the matrix A corresponding to the eigenvalue k, then the unit vector $\boldsymbol{x}/|\boldsymbol{x}|$ is also an eigenvector of A corresponding to the eigenvalue k. Find three eigenvectors which have unit length for the matrix

$$\begin{bmatrix} 1 & 1 & 0 \\ 1 & 0 & 1 \\ 0 & 1 & 0 \end{bmatrix}.$$

(See Examples 18.2(ii) and 19.3.)

4. In each case below, find the cosine of the angle between the given vectors.

(i) $\begin{bmatrix} -1 \\ 2 \end{bmatrix}, \begin{bmatrix} 1 \\ 2 \end{bmatrix}$. (ii) $\begin{bmatrix} 2 \\ 3 \end{bmatrix}, \begin{bmatrix} 3 \\ 2 \end{bmatrix}$.

(iii) $\begin{bmatrix} 1 \\ 0 \\ 1 \end{bmatrix}, \begin{bmatrix} 2 \\ 1 \\ -1 \end{bmatrix}$. (iv) $\begin{bmatrix} -3 \\ 2 \\ 1 \end{bmatrix}, \begin{bmatrix} 1 \\ 2 \\ -3 \end{bmatrix}$.

(v) $\begin{bmatrix} 1 \\ -1 \\ 1 \\ -1 \end{bmatrix}, \begin{bmatrix} 2 \\ 0 \\ 2 \\ 0 \end{bmatrix}$. (vi) $\begin{bmatrix} 0 \\ 1 \\ -1 \\ 3 \end{bmatrix}, \begin{bmatrix} 2 \\ 0 \\ 2 \\ -1 \end{bmatrix}$.

(vii) $\begin{bmatrix} 2 \\ 1 \\ -1 \\ -1 \end{bmatrix}, \begin{bmatrix} 1 \\ 0 \\ 1 \\ 1 \end{bmatrix}$. (viii) $\begin{bmatrix} 1 \\ 1 \\ 1 \\ 1 \\ 1 \end{bmatrix}, \begin{bmatrix} 1 \\ -1 \\ 1 \\ -1 \\ 1 \end{bmatrix}$.

5. Which pairs of vectors from the following list are orthogonal?

$$\begin{bmatrix} 1 \\ 1 \\ 1 \end{bmatrix}, \begin{bmatrix} 0 \\ -1 \\ 2 \end{bmatrix}, \begin{bmatrix} 0 \\ 1 \\ -1 \end{bmatrix}, \begin{bmatrix} 2 \\ 1 \\ -3 \end{bmatrix}, \begin{bmatrix} 4 \\ 1 \\ 2 \end{bmatrix}, \begin{bmatrix} -1 \\ 1 \\ 1 \end{bmatrix},$$

$$\begin{bmatrix} 0 \\ 0 \\ 0 \end{bmatrix}, \begin{bmatrix} 5 \\ -1 \\ 2 \end{bmatrix}, \begin{bmatrix} -3 \\ 0 \\ 6 \end{bmatrix}, \begin{bmatrix} -3 \\ -3 \\ 6 \end{bmatrix}, \begin{bmatrix} 1 \\ 1 \\ -2 \end{bmatrix}, \begin{bmatrix} 3 \\ 3 \\ 1 \end{bmatrix}.$$

6. Let S be a subspace of $\mathbb{R}^n$. Show that the set $\{\boldsymbol{y} \in \mathbb{R}^n : \boldsymbol{y} \cdot \boldsymbol{x} = 0 \text{ for every } \boldsymbol{x} \in S\}$ is a subspace of $\mathbb{R}^n$ (the orthogonal complement of S).

7. Find a linearly independent list of three eigenvectors of the (symmetric) matrix

$$\begin{bmatrix} 2 & 0 & 1 \\ 0 & 2 & 0 \\ 1 & 0 & 2 \end{bmatrix}.$$

Verify that the list obtained is an orthogonal list.

8. Find a linearly independent list of three eigenvectors of the (non-symmetric) matrix

$$\begin{bmatrix} 1 & 2 & 3 \\ 0 & 1 & 0 \\ 2 & 1 & 2 \end{bmatrix}.$$

Is the list an orthogonal list? Can you choose other eigenvectors so as to form an orthogonal list?

9. Consider whether a $n \times n$ matrix which has an orthogonal set of n eigenvectors must necessarily be symmetric. (This will be easier to answer using ideas in Chapters 21 and 22.)

Examples

21.1 Orthogonal bases.

(i) Let $\mathbf{v}_1 = \begin{bmatrix} 1 \\ 3 \\ -1 \end{bmatrix}, \quad \mathbf{v}_2 = \begin{bmatrix} -2 \\ 1 \\ 1 \end{bmatrix}, \quad \mathbf{v}_3 = \begin{bmatrix} 4 \\ 1 \\ 7 \end{bmatrix}.$

Then $(\mathbf{v}_1, \mathbf{v}_2, \mathbf{v}_3)$ is an orthogonal basis for $\mathbb{R}^3$. To demonstrate this, the following is sufficient.

$$\begin{aligned} \mathbf{v}_1 \cdot \mathbf{v}_2 &= -2+3-1=0, \\ \mathbf{v}_2 \cdot \mathbf{v}_3 &= -8+1+7=0, \\ \mathbf{v}_3 \cdot \mathbf{v}_1 &= 4+3-7=0. \end{aligned}$$

Thus $(\mathbf{v}_1, \mathbf{v}_2, \mathbf{v}_3)$ is an orthogonal list. Hence $(\mathbf{v}_1, \mathbf{v}_2, \mathbf{v}_3)$ is LI (see Chapter 20). In $\mathbb{R}^3$, any LI list containing three vectors is a basis (see Chapter 14), so $(\mathbf{v}_1, \mathbf{v}_2, \mathbf{v}_3)$ is an orthogonal basis for $\mathbb{R}^3$.

(ii) Let $\mathbf{v}_1 = \begin{bmatrix} 1 \\ 2 \\ 1 \\ 2 \end{bmatrix}, \quad \mathbf{v}_2 = \begin{bmatrix} 1 \\ 0 \\ -1 \\ 0 \end{bmatrix},$

$\mathbf{v}_3 = \begin{bmatrix} 1 \\ -1 \\ 1 \\ 0 \end{bmatrix}, \quad \mathbf{v}_4 = \begin{bmatrix} 1 \\ 2 \\ 1 \\ -3 \end{bmatrix}.$

Then $(\mathbf{v}_1, \mathbf{v}_2, \mathbf{v}_3, \mathbf{v}_4)$ is an orthogonal basis for $\mathbb{R}^4$. Demonstration is just as above, by first verifying that all possible pairs of these vectors are orthogonal pairs, i.e. that

$$\mathbf{v}_1 \cdot \mathbf{v}_2 = \mathbf{v}_1 \cdot \mathbf{v}_3 = \mathbf{v}_1 \cdot \mathbf{v}_4 = 0,$$

and $\quad \mathbf{v}_2 \cdot \mathbf{v}_3 = \mathbf{v}_2 \cdot \mathbf{v}_4 = \mathbf{v}_3 \cdot \mathbf{v}_4 = 0,$

and then using rules from Chapters 20 and 14, as in part (i) above.

21.2 Find an orthogonal basis for the subspace of $\mathbb{R}^3$ spanned by the list $(\mathbf{v}_1, \mathbf{v}_2)$, where

$$\mathbf{v}_1 = \begin{bmatrix} 1 \\ -1 \\ 2 \end{bmatrix} \quad \text{and} \quad \mathbf{v}_2 = \begin{bmatrix} 0 \\ 3 \\ -1 \end{bmatrix}.$$

Notice of course that $(\mathbf{v}_1, \mathbf{v}_2)$ is a basis for this subspace, because it is a LI list. Is it an orthogonal basis? The answer to this is negative, because $\mathbf{v}_1 \cdot \mathbf{v}_2 = -5$, which is non-zero. The orthogonal basis which we seek will have the form

$$(\mathbf{v}_1, a\mathbf{v}_1 + b\mathbf{v}_2),$$

where a and b are chosen so that $\mathbf{v}_1 \cdot (a\mathbf{v}_1 + b\mathbf{v}_2) = 0$.

Here $\quad a\mathbf{v}_1 + b\mathbf{v}_2 = \begin{bmatrix} a \\ -a+3b \\ 2a-b \end{bmatrix},$

21

Orthogonality

In three-dimensional geometry it is customary to choose coordinate axes which are mutually perpendicular. Coordinate geometry is still possible, however, with axes which are not perpendicular, but it is not very convenient, and many of the usual processes become quite complicated. The analogy between coordinate axes in geometry and bases in algebra has already been drawn (see Chapter 17), and of course the vectors which constitute a basis for $\mathbb{R}^n$ or for a subspace of $\mathbb{R}^n$ need not form an orthogonal list. Some originally geometrical ideas relating to rectangular coordinate systems apply also in $\mathbb{R}^n$ and subspaces of $\mathbb{R}^n$, for any n. First let us consider bases.

Definition
Let S be a subspace of $\mathbb{R}^n$. A basis B for S is an *orthogonal basis* if it is an orthogonal list, i.e. if every vector in B is orthogonal to every other vector in B.

The standard basis for $\mathbb{R}^n$ is an orthogonal basis. Other examples are given in Example 21.1.

There are some general rules which can guide us here. We know (Chapter 20) that an orthogonal list is linearly independent. We also know (Chapter 14) that, in a space of dimension k, any LI list of k vectors is a basis. This rule from Chapter 14 can therefore be modified as follows.

Rule
Let S be a subspace of $\mathbb{R}^n$ with dimension k. Then any orthogonal list of k vectors from S is a basis for S.

One situation where this arises is the case of a $n \times n$ real symmetric matrix which has n distinct eigenvalues. A list of corresponding

so $$\boldsymbol{v}_1 \cdot (a\boldsymbol{v}_1 + b\boldsymbol{v}_2) = a - (-a + 3b) + 2(2a - b) = 6a - 5b.$$

Consequently, if we choose a and b so that $6a - 5b = 0$, then the two vectors will be orthogonal. So choose $a = 5$ and $b = 6$.

Then $$a\boldsymbol{v}_1 + b\boldsymbol{v}_2 = \begin{bmatrix} 5 \\ 13 \\ 4 \end{bmatrix}.$$

Last, then, because it is a LI list of two vectors in a space of dimension 2,

$$\left(\begin{bmatrix} 1 \\ -1 \\ 2 \end{bmatrix}, \begin{bmatrix} 5 \\ 13 \\ 4 \end{bmatrix} \right)$$

is a basis, and so an orthogonal basis as required.

21.3 Examples of orthonormal bases.

(i) $$\left(\begin{bmatrix} \frac{3}{5} \\ \frac{4}{5} \end{bmatrix}, \begin{bmatrix} -\frac{4}{5} \\ \frac{3}{5} \end{bmatrix} \right), \quad \text{for } \mathbb{R}^2.$$

(ii) $$\left(\begin{bmatrix} -\frac{1}{3} \\ \frac{2}{3} \\ \frac{2}{3} \end{bmatrix}, \begin{bmatrix} \frac{2}{3} \\ -\frac{1}{3} \\ \frac{2}{3} \end{bmatrix}, \begin{bmatrix} \frac{2}{3} \\ \frac{2}{3} \\ -\frac{1}{3} \end{bmatrix} \right), \quad \text{for } \mathbb{R}^3.$$

(iii) $$\left(\begin{bmatrix} 1/\sqrt{3} \\ -1/\sqrt{3} \\ 1/\sqrt{3} \end{bmatrix}, \begin{bmatrix} -1/\sqrt{2} \\ 0 \\ 1/\sqrt{2} \end{bmatrix}, \begin{bmatrix} 1/\sqrt{6} \\ 2/\sqrt{6} \\ 1/\sqrt{6} \end{bmatrix} \right), \quad \text{for } \mathbb{R}^3.$$

(iv) $$\left(\begin{bmatrix} -\frac{1}{2} \\ \frac{1}{2} \\ \frac{1}{2} \\ \frac{1}{2} \end{bmatrix}, \begin{bmatrix} \frac{1}{2} \\ -\frac{1}{2} \\ \frac{1}{2} \\ \frac{1}{2} \end{bmatrix}, \begin{bmatrix} \frac{1}{2} \\ \frac{1}{2} \\ -\frac{1}{2} \\ \frac{1}{2} \end{bmatrix}, \begin{bmatrix} \frac{1}{2} \\ \frac{1}{2} \\ \frac{1}{2} \\ -\frac{1}{2} \end{bmatrix} \right), \quad \text{for } \mathbb{R}^4.$$

(v) We know (Example 21.1(ii)) that

$$\left(\begin{bmatrix} 1 \\ 2 \\ 1 \\ 2 \end{bmatrix}, \begin{bmatrix} 1 \\ 0 \\ -1 \\ 0 \end{bmatrix}, \begin{bmatrix} 1 \\ -1 \\ 1 \\ 0 \end{bmatrix}, \begin{bmatrix} 1 \\ 2 \\ 1 \\ -3 \end{bmatrix} \right)$$

is an orthogonal basis for $\mathbb{R}^4$. The lengths of these vectors are respectively $\sqrt{10}$, $\sqrt{2}$, $\sqrt{3}$ and $\sqrt{15}$, so it is certainly not an orthonormal basis. To obtain an orthonormal basis, divide each vector by its length.

$$\left(\begin{bmatrix} 1/\sqrt{10} \\ 2/\sqrt{10} \\ 1/\sqrt{10} \\ 2/\sqrt{10} \end{bmatrix}, \begin{bmatrix} 1/\sqrt{2} \\ 0 \\ -1/\sqrt{2} \\ 0 \end{bmatrix}, \begin{bmatrix} 1/\sqrt{3} \\ -1/\sqrt{3} \\ 1/\sqrt{3} \\ 0 \end{bmatrix}, \begin{bmatrix} 1/\sqrt{15} \\ 2/\sqrt{15} \\ 1/\sqrt{15} \\ -3/\sqrt{15} \end{bmatrix} \right)$$

is an orthonormal basis for $\mathbb{R}^4$.

eigenvectors is an orthogonal list (see Chapter 20), so is a basis for $\mathbb{R}^n$, by the above rule.

Given a basis, how can we find an orthogonal basis for the same space? Example 21.2 shows one way of proceeding in the simple case of two dimensions. For the first member of our orthogonal basis choose either one of the given basis vectors. For the second, find a linear combination of the two given basis vectors which is orthogonal to the chosen one. This process will always work: there will always be an orthogonal basis which will be found this way. There is another way (the Gram–Schmidt process), which we shall describe shortly, which works better when we come to deal with more than two dimensions. But first we define a new notion.

Definition

Let S be a subspace of $\mathbb{R}^n$. A basis B for S is called an *orthonormal basis* if

(i) B is an orthogonal basis, and
(ii) every vector in B has length 1.

Again, the standard basis for $\mathbb{R}^n$ is an orthonormal basis. Other examples are given in Examples 21.3. There is an easy way to convert an orthogonal basis into an orthonormal basis for the same space: just 'divide' each vector in the basis by its length. If $(\boldsymbol{v}_1, \ldots, \boldsymbol{v}_r)$ is an orthogonal basis for a space S, then $\left(\dfrac{\boldsymbol{v}_1}{|\boldsymbol{v}_1|}, \ldots, \dfrac{\boldsymbol{v}_r}{|\boldsymbol{v}_r|}\right)$ is an orthonormal basis for S. Verification of this is left as an exercise. This process is known as *normalisation*.

21.4 Application of the Gram–Schmidt process for finding an orthonormal basis.

Let $\boldsymbol{v}_1 = \begin{bmatrix} 1 \\ 1 \\ 1 \end{bmatrix}, \quad \boldsymbol{v}_2 = \begin{bmatrix} -1 \\ 0 \\ -1 \end{bmatrix}, \quad \boldsymbol{v}_3 = \begin{bmatrix} -1 \\ 2 \\ 3 \end{bmatrix}.$

Then $(\boldsymbol{v}_1, \boldsymbol{v}_2, \boldsymbol{v}_3)$ spans $\mathbb{R}^3$ (and is LI). We find an orthonormal basis for $\mathbb{R}^3$ by applying the Gram–Schmidt process to this list.

Let $\boldsymbol{w}_1 = \dfrac{\boldsymbol{v}_1}{|\boldsymbol{v}_1|} = (1/\sqrt{3})\begin{bmatrix} 1 \\ 1 \\ 1 \end{bmatrix}.$

Let $\boldsymbol{w}_2 = \dfrac{\boldsymbol{v}_2 - (\boldsymbol{v}_2 \cdot \boldsymbol{w}_1)\boldsymbol{w}_1}{|\boldsymbol{v}_2 - (\boldsymbol{v}_2 \cdot \boldsymbol{w}_1)\boldsymbol{w}_1|}.$

Now $\boldsymbol{v}_2 \cdot \boldsymbol{w}_1 = -1/\sqrt{3} + 0 - 1/\sqrt{3} = -2/\sqrt{3},$

so $\boldsymbol{v}_2 - (\boldsymbol{v}_2 \cdot \boldsymbol{w}_1)\boldsymbol{w}_1 = \begin{bmatrix} -1 \\ 0 \\ -1 \end{bmatrix} + (2/\sqrt{3})(1/\sqrt{3})\begin{bmatrix} 1 \\ 1 \\ 1 \end{bmatrix} = \begin{bmatrix} -\frac{1}{3} \\ \frac{2}{3} \\ -\frac{1}{3} \end{bmatrix}.$

Then $|\boldsymbol{v}_2 - (\boldsymbol{v}_2 \cdot \boldsymbol{w}_1)\boldsymbol{w}_1| = \sqrt{\frac{1}{9} + \frac{4}{9} + \frac{1}{9}} = \sqrt{6}/3.$

Hence $\boldsymbol{w}_2 = (3/\sqrt{6})\begin{bmatrix} -\frac{1}{3} \\ \frac{2}{3} \\ -\frac{1}{3} \end{bmatrix} = \begin{bmatrix} -1/\sqrt{6} \\ 2/\sqrt{6} \\ -1/\sqrt{6} \end{bmatrix}.$

Next, $\boldsymbol{w}_3 = \dfrac{\boldsymbol{v}_3 - (\boldsymbol{v}_3 \cdot \boldsymbol{w}_2)\boldsymbol{w}_2 - (\boldsymbol{v}_3 \cdot \boldsymbol{w}_1)\boldsymbol{w}_1}{|\boldsymbol{v}_3 - (\boldsymbol{v}_3 \cdot \boldsymbol{w}_2)\boldsymbol{w}_2 - (\boldsymbol{v}_3 \cdot \boldsymbol{w}_1)\boldsymbol{w}_1|}.$

Now $\boldsymbol{v}_3 \cdot \boldsymbol{w}_2 = \quad 1/\sqrt{6} + 4/\sqrt{6} - 3/\sqrt{6} = 2/\sqrt{6}$

and $\boldsymbol{v}_3 \cdot \boldsymbol{w}_1 = -1/\sqrt{3} + 2/\sqrt{3} + 3/\sqrt{3} = 4/\sqrt{3},$

so $\boldsymbol{v}_3 - (\boldsymbol{v}_3 \cdot \boldsymbol{w}_2)\boldsymbol{w}_2 - (\boldsymbol{v}_3 \cdot \boldsymbol{w}_1)\boldsymbol{w}_1$

$$= \begin{bmatrix} -1 \\ 2 \\ 3 \end{bmatrix} - (2/\sqrt{6})\begin{bmatrix} -1/\sqrt{6} \\ 2/\sqrt{6} \\ -1-\sqrt{6} \end{bmatrix} - (4/\sqrt{3})\begin{bmatrix} 1/\sqrt{3} \\ 1/\sqrt{3} \\ 1/\sqrt{3} \end{bmatrix}$$

$$= \begin{bmatrix} -1 + \frac{1}{3} - \frac{4}{3} \\ 2 - \frac{2}{3} - \frac{4}{3} \\ 3 + \frac{1}{3} - \frac{4}{3} \end{bmatrix} = \begin{bmatrix} -2 \\ 0 \\ 2 \end{bmatrix}.$$

Then $|\boldsymbol{v}_3 - (\boldsymbol{v}_3 \cdot \boldsymbol{w}_2)\boldsymbol{w}_2 - (\boldsymbol{v}_3 \cdot \boldsymbol{w}_1)\boldsymbol{w}_1| = \sqrt{4 + 0 + 4} = \sqrt{8},$

and so $\boldsymbol{w}_3 = (1/\sqrt{8})\begin{bmatrix} -2 \\ 0 \\ 2 \end{bmatrix} = \begin{bmatrix} -1/\sqrt{2} \\ 0 \\ 1/\sqrt{2} \end{bmatrix}.$

Thus the list

$$\left(\begin{bmatrix} 1/\sqrt{3} \\ 1/\sqrt{3} \\ 1/\sqrt{3} \end{bmatrix}, \begin{bmatrix} -1/\sqrt{6} \\ 2/\sqrt{6} \\ -1/\sqrt{6} \end{bmatrix}, \begin{bmatrix} -1/\sqrt{2} \\ 0 \\ 1/\sqrt{2} \end{bmatrix}\right)$$

is an orthonormal basis as required.

Now we come to the process for constructing orthonormal bases. It is given below as the justification for the following rule.

Rule

Every subspace of $\mathbb{R}^n$ has an orthonormal basis.

The process here described converts a given (arbitrary) basis for a subspace S of $\mathbb{R}^n$ into an orthonormal basis. This is the Gram–Schmidt process, referred to above. By the results of Chapter 14, there must always be a basis, so let $B = (\boldsymbol{v}_1, \ldots, \boldsymbol{v}_r)$ be a basis for S. Construct a list $(\boldsymbol{w}_1, \ldots, \boldsymbol{w}_r)$ as follows. (See Example 21.4.)

Let $$\boldsymbol{w}_1 = \frac{\boldsymbol{v}_1}{|\boldsymbol{v}_1|}.$$

Then the vector $\boldsymbol{v}_2 - (\boldsymbol{v}_2 \cdot \boldsymbol{w}_1)\boldsymbol{w}_1$ is orthogonal to $\boldsymbol{w}_1$ (check it).

So let $$\boldsymbol{w}_2 = \frac{\boldsymbol{v}_2 - (\boldsymbol{v}_2 \cdot \boldsymbol{w}_1)\boldsymbol{w}_1}{|\boldsymbol{v}_2 - (\boldsymbol{v}_2 \cdot \boldsymbol{w}_1)\boldsymbol{w}_1|}.$$

The vector $\boldsymbol{v}_3 - (\boldsymbol{v}_3 \cdot \boldsymbol{w}_2)\boldsymbol{w}_2 - (\boldsymbol{v}_3 \cdot \boldsymbol{w}_1)\boldsymbol{w}_1$ is orthogonal to both $\boldsymbol{w}_1$ and to $\boldsymbol{w}_2$ (check these).

So let $$\boldsymbol{w}_3 = \frac{\boldsymbol{v}_3 - (\boldsymbol{v}_3 \cdot \boldsymbol{w}_2)\boldsymbol{w}_2 - (\boldsymbol{v}_3 \cdot \boldsymbol{w}_1)\boldsymbol{w}_1}{|\boldsymbol{v}_3 - (\boldsymbol{v}_3 \cdot \boldsymbol{w}_2)\boldsymbol{w}_2 - (\boldsymbol{v}_3 \cdot \boldsymbol{w}_1)\boldsymbol{w}_1|}.$$

Similarly form $\boldsymbol{w}_4$ by dividing the vector

$$\boldsymbol{v}_4 - (\boldsymbol{v}_4 \cdot \boldsymbol{w}_3)\boldsymbol{w}_3 - (\boldsymbol{v}_4 \cdot \boldsymbol{w}_2)\boldsymbol{w}_2 - (\boldsymbol{v}_4 \cdot \boldsymbol{w}_1)\boldsymbol{w}_1$$

by its length, and so on, until the list $(\boldsymbol{w}_1, \ldots, \boldsymbol{w}_r)$ has been constructed.

Our next task is to integrate the idea of an orthogonal matrix into this context. Recall that a square matrix A is orthogonal if $A^{\mathrm{T}}A = AA^{\mathrm{T}} = I$, and that if A is an orthogonal matrix then A is invertible, with $A^{-1} = A^{\mathrm{T}}$. It may help to return to the notion from Chapter 16 of a matrix representing a linear transformation, and to the particular case of three dimensions. If A is a 3×3 matrix then A represents the function (linear transformation) which takes each 3-vector $\boldsymbol{x}$ to the 3-vector $A\boldsymbol{x}$. The images of the standard basis vectors $\boldsymbol{e}_1, \boldsymbol{e}_2, \boldsymbol{e}_3$ are the vectors $A\boldsymbol{e}_1$, $A\boldsymbol{e}_2$ and $A\boldsymbol{e}_3$, which as we have noted before are just the columns of A. If A is an invertible matrix then these three vectors will form a basis for $\mathbb{R}^3$. What has to be true about A if the images of $\boldsymbol{e}_1$, $\boldsymbol{e}_2$ and $\boldsymbol{e}_3$ are to form an orthonormal basis for $\mathbb{R}^3$? Precisely this: A must be an orthogonal matrix. In Example 21.5 we show that a 3×3 matrix A is orthogonal if and only if its columns form an orthonormal basis for $\mathbb{R}^3$. The methods used there extend to the case of $\mathbb{R}^n$ in general.

21.5 Proof that a 3×3 matrix is orthogonal if and only if its columns form an orthonormal basis for $\mathbb{R}^3$.

Let $$A = \begin{bmatrix} a_1 & b_1 & c_1 \\ a_2 & b_2 & c_2 \\ a_3 & b_3 & c_3 \end{bmatrix}.$$

Denote the columns of A by $\boldsymbol{a}$, $\boldsymbol{b}$ and $\boldsymbol{c}$ respectively.

Then $$A^{\mathrm{T}}A = \begin{bmatrix} a_1 & a_2 & a_3 \\ b_1 & b_2 & b_3 \\ c_1 & c_2 & c_3 \end{bmatrix}\begin{bmatrix} a_1 & b_1 & c_1 \\ a_2 & b_2 & c_2 \\ a_3 & b_3 & c_3 \end{bmatrix}.$$

The (1, 1)-entry in $A^{\mathrm{T}}A$ is $a_1^2 + a_2^2 + a_3^2$, i.e. $|\boldsymbol{a}|^2$.
The (2, 2)-entry in $A^{\mathrm{T}}A$ is $b_1^2 + b_2^2 + b_3^2$, i.e. $|\boldsymbol{b}|^2$.
The (3, 3)-entry in $A^{\mathrm{T}}A$ is $c_1^2 + c_2^2 + c_3^2$, i.e. $|\boldsymbol{c}|^2$.

The (1, 2)-entry in $A^{\mathrm{T}}A$ is $a_1b_1 + a_2b_2 + a_3b_3$, i.e. $\boldsymbol{a}\cdot\boldsymbol{b}$. Similarly the other off-diagonal elements of $A^{\mathrm{T}}A$ are equal to $\boldsymbol{b}\cdot\boldsymbol{c}$ or to $\boldsymbol{c}\cdot\boldsymbol{a}$ or to $\boldsymbol{a}\cdot\boldsymbol{b}$. In fact

$$A^{\mathrm{T}}A = \begin{bmatrix} |\boldsymbol{a}|^2 & \boldsymbol{a}\cdot\boldsymbol{b} & \boldsymbol{a}\cdot\boldsymbol{c} \\ \boldsymbol{b}\cdot\boldsymbol{a} & |\boldsymbol{b}|^2 & \boldsymbol{b}\cdot\boldsymbol{c} \\ \boldsymbol{c}\cdot\boldsymbol{a} & \boldsymbol{c}\cdot\boldsymbol{b} & |\boldsymbol{c}|^2 \end{bmatrix}.$$

Suppose that A is orthogonal. Then $A^{\mathrm{T}}A = I$, so

$$|\boldsymbol{a}|^2 = |\boldsymbol{b}|^2 = |\boldsymbol{c}|^2 = 1,$$

and $$\boldsymbol{a}\cdot\boldsymbol{b} = \boldsymbol{b}\cdot\boldsymbol{c} = \boldsymbol{c}\cdot\boldsymbol{a} = 0,$$

and consequently $(\boldsymbol{a}, \boldsymbol{b}, \boldsymbol{c})$ is an orthonormal basis for $\mathbb{R}^3$.
Conversely, suppose that $(\boldsymbol{a}, \boldsymbol{b}, \boldsymbol{c})$ is an orthonormal basis for $\mathbb{R}^3$. Then, by the above, we must have $A^{\mathrm{T}}A = I$. This is sufficient for A to be orthogonal (if $A^{\mathrm{T}}A = I$ then $A^{-1} = A^{\mathrm{T}}$, so $AA^{\mathrm{T}} = I$ also: see Chapter 5).

21.6 Orthogonal transformations.

(i) In $\mathbb{R}^2$, a rotation about the origin is an orthogonal transformation. (See Chapter 16.)
A rotation through angle α is represented by the matrix

$$A = \begin{bmatrix} \cos\alpha & -\sin\alpha \\ \sin\alpha & \cos\alpha \end{bmatrix}.$$

$$\begin{aligned} A^{\mathrm{T}}A &= \begin{bmatrix} \cos\alpha & \sin\alpha \\ -\sin\alpha & \cos\alpha \end{bmatrix}\begin{bmatrix} \cos\alpha & -\sin\alpha \\ \sin\alpha & \cos\alpha \end{bmatrix} \\ &= \begin{bmatrix} \cos^2\alpha + \sin^2\alpha & 0 \\ 0 & \cos^2\alpha + \sin^2\alpha \end{bmatrix} \\ &= I, \end{aligned}$$

so this matrix A is an orthogonal matrix. Indeed $\left(\begin{bmatrix} \cos\alpha \\ \sin\alpha \end{bmatrix}, \begin{bmatrix} -\sin\alpha \\ \cos\alpha \end{bmatrix}\right)$ is an orthonormal basis for $\mathbb{R}^2$, whatever value α takes.
(Note: a rotation clearly preserves lengths and angles.)

Rule

A $n \times n$ matrix A is orthogonal if and only if its columns form an orthonormal basis for $\mathbb{R}^n$.

An orthogonal matrix represents a linear transformation which takes the standard basis to an orthonormal basis. What this means geometrically in three dimensions is that an orthogonal matrix will transform unit vectors in the directions of the coordinate axes into unit vectors which are themselves mutually perpendicular, and so may be taken as the directions of new coordinate axes. See Exercise 8 at the end of this chapter. A linear transformation which is represented by an orthogonal matrix is called an *orthogonal transformation.* An orthogonal transformation preserves lengths and angles, in a way made precise in the following rule.

Rule

Let A be a $n \times n$ orthogonal matrix, and let $\mathbf{x}$, $\mathbf{y}$ be non-zero vectors in $\mathbb{R}^n$. Then

(i) $|\mathbf{x}| = |A\mathbf{x}|$.

(ii) The angle between $A\mathbf{x}$ and $A\mathbf{y}$ is equal to the angle between $\mathbf{x}$ and $\mathbf{y}$.

Demonstrations of these properties are useful exercises.

(i)
$$\begin{aligned} |A\mathbf{x}|^2 &= (A\mathbf{x})^{\mathrm{T}}A\mathbf{x} \\ &= \mathbf{x}^{\mathrm{T}}A^{\mathrm{T}}A\mathbf{x} \\ &= \mathbf{x}^{\mathrm{T}}\mathbf{x} = |\mathbf{x}|^2. \end{aligned}$$

(ii) Let the angle between $A\mathbf{x}$ and $A\mathbf{y}$ be θ. Then

$$\begin{aligned} \cos\theta &= \frac{A\mathbf{x}\cdot A\mathbf{y}}{|A\mathbf{x}||A\mathbf{y}|} = \frac{(A\mathbf{x})^{\mathrm{T}}(A\mathbf{y})}{|\mathbf{x}||\mathbf{y}|} \\ &= \frac{\mathbf{x}^{\mathrm{T}}A^{\mathrm{T}}A\mathbf{y}}{|\mathbf{x}||\mathbf{y}|} \\ &= \frac{\mathbf{x}^{\mathrm{T}}\mathbf{y}}{|\mathbf{x}||\mathbf{y}|} = \frac{\mathbf{x}\cdot\mathbf{y}}{|\mathbf{x}||\mathbf{y}|}, \end{aligned}$$

which is (by definition) the cosine of the angle between $\mathbf{x}$ and $\mathbf{y}$.

Examples 21.6 provide illustration of orthogonal matrices and transformations.

(ii) Let $A = \begin{bmatrix} \frac{1}{3} & -\frac{2}{3} & -\frac{2}{3} \\ -\frac{2}{3} & \frac{1}{3} & -\frac{2}{3} \\ -\frac{2}{3} & -\frac{2}{3} & \frac{1}{3} \end{bmatrix}$

This matrix represents the linear transformation which takes each point (x, y, z) to its reflection in the plane with equation $x + y + z = 0$. It is easy to verify that A is orthogonal, and that

$$\left(\begin{bmatrix} \frac{1}{3} \\ -\frac{2}{3} \\ -\frac{2}{3} \end{bmatrix}, \begin{bmatrix} -\frac{2}{3} \\ \frac{1}{3} \\ -\frac{2}{3} \end{bmatrix}, \begin{bmatrix} -\frac{2}{3} \\ -\frac{2}{3} \\ \frac{1}{3} \end{bmatrix} \right)$$

is an orthonormal basis for $\mathbb{R}^3$. These are the images of the standard basis vectors. Notice that this is an example of an orthogonal transformation which is not a rotation.

It is a worthwhile exercise to find the eigenvalues and eigenvectors of this matrix A and to put the results in a geometrical context.

Summary
It is shown how to find an orthogonal basis for a given subspace of $\mathbb{R}^n$. The idea of an orthonormal basis is introduced and the Gram–Schmidt process for converting a given basis into an orthonormal basis is described and used. The relationship between orthogonal matrices and orthonormal bases for $\mathbb{R}^n$ is shown. Finally, there is a brief description of the properties of orthogonal transformations.

Exercises

1. (i) Write down three different orthogonal bases for $\mathbb{R}^2$.
   (ii) Write down three different orthogonal bases for $\mathbb{R}^3$.
   (iii) Write down an orthogonal basis for $\mathbb{R}^4$ (other than the standard basis).
   (iv) Is any of the bases you have chosen above an orthonormal basis? By the normalisation process convert any which are not orthonormal bases into orthonormal bases.
2. Let $X = (\boldsymbol{v}_1, \ldots, \boldsymbol{v}_r)$ be an orthonormal basis for a subspace S of $\mathbb{R}^n$ (for some n). We know that any vector $\boldsymbol{x}$ in S may be represented uniquely as
$$\boldsymbol{x} = a_1\boldsymbol{v}_1 + \cdots + a_r\boldsymbol{v}_r.$$
Prove that $a_i = \boldsymbol{x}\cdot\boldsymbol{v}_i$, for $1 \leqslant i \leqslant r$.
3. The following are pairs of unit vectors which are orthogonal. In each case find a third unit vector so that with the given two vectors an orthonormal basis for $\mathbb{R}^3$ is formed.

(i) $\begin{bmatrix} 0 \\ 1\sqrt{2} \\ 1\sqrt{2} \end{bmatrix}, \begin{bmatrix} 1/3 \\ 2/3 \\ -2/3 \end{bmatrix}.$ (ii) $\begin{bmatrix} 1 \\ 0 \\ 0 \end{bmatrix}, \begin{bmatrix} 0 \\ 1/\sqrt{2} \\ -1/\sqrt{2} \end{bmatrix}.$

(iii) $\begin{bmatrix} 1/2 \\ 1/2 \\ 1/\sqrt{2} \end{bmatrix}, \begin{bmatrix} -1/2 \\ -1/2 \\ 1/\sqrt{2} \end{bmatrix}.$ (iv) $\begin{bmatrix} 1/\sqrt{6} \\ 2/\sqrt{6} \\ -1/\sqrt{6} \end{bmatrix}, \begin{bmatrix} -2/\sqrt{5} \\ 1/\sqrt{5} \\ 0 \end{bmatrix}.$

4. In each case below, apply the Gram–Schmidt process to the given basis of $\mathbb{R}^3$ to obtain an orthonormal basis.

(i) $\begin{bmatrix} 1 \\ 1 \\ 1 \end{bmatrix}, \begin{bmatrix} 0 \\ 1 \\ 1 \end{bmatrix}, \begin{bmatrix} 1 \\ 1 \\ 0 \end{bmatrix}.$ (ii) $\begin{bmatrix} 1 \\ 1 \\ 0 \end{bmatrix}, \begin{bmatrix} 1 \\ 0 \\ 1 \end{bmatrix}, \begin{bmatrix} 0 \\ 1 \\ 1 \end{bmatrix}.$

(iii) $\begin{bmatrix} -2 \\ 0 \\ 1 \end{bmatrix}, \begin{bmatrix} 1 \\ 1 \\ 1 \end{bmatrix}, \begin{bmatrix} 3 \\ -1 \\ -1 \end{bmatrix}.$ (iv) $\begin{bmatrix} 1 \\ 3 \\ 1 \end{bmatrix}, \begin{bmatrix} 4 \\ 1 \\ 4 \end{bmatrix}, \begin{bmatrix} 2 \\ 3 \\ 0 \end{bmatrix}.$

5. Find an orthonormal basis for $\mathbb{R}^3$ which contains the vector $\begin{bmatrix} 1/3 \\ 2/3 \\ 2/3 \end{bmatrix}$. Do the same for $\begin{bmatrix} 0 \\ 3/5 \\ 4/5 \end{bmatrix}$.
6. Find an orthonormal basis for the subspace of $\mathbb{R}^4$ spanned by the list
$$\left(\begin{bmatrix} 2 \\ 0 \\ 0 \\ 0 \end{bmatrix}, \begin{bmatrix} 1 \\ 3 \\ 3 \\ 0 \end{bmatrix}, \begin{bmatrix} 0 \\ 4 \\ 6 \\ 1 \end{bmatrix} \right).$$
7. Repeat Exercise 6 with the list
$$\left(\begin{bmatrix} 1 \\ -1 \\ 1 \\ -1 \end{bmatrix}, \begin{bmatrix} 1 \\ 2 \\ 2 \\ 0 \end{bmatrix}, \begin{bmatrix} 1 \\ 0 \\ 1 \\ 0 \end{bmatrix} \right).$$
8. Let $A = \begin{bmatrix} 2/3 & 2/3 & -1/3 \\ 2/3 & -1/3 & 2/3 \\ -1/3 & 2/3 & 2/3 \end{bmatrix}$.

   Show that A is an orthogonal matrix. Let f be the linear transformation from $\mathbb{R}^3$ to $\mathbb{R}^3$ given by $f(\boldsymbol{x}) = A\boldsymbol{x}$. Write down the images under f of the standard basis vectors. Evaluate $A\boldsymbol{x}$ and $A\boldsymbol{y}$, where
$$\boldsymbol{x} = \begin{bmatrix} 1 \\ 1 \\ 1 \end{bmatrix} \quad \text{and} \quad \boldsymbol{y} = \begin{bmatrix} -1 \\ 1 \\ 1 \end{bmatrix}.$$
   Verify that $|\boldsymbol{x}| = |A\boldsymbol{x}|$ and $|\boldsymbol{y}| = |A\boldsymbol{y}|$. Verify also that the cosine of the angle between $A\boldsymbol{x}$ and $A\boldsymbol{y}$ is equal to the cosine of the angle between $\boldsymbol{x}$ and $\boldsymbol{y}$.
9. (See Example 21.6(ii).) Find the eigenvalues and corresponding eigenvectors of the matrix
$$\begin{bmatrix} 1/3 & -2/3 & -2/3 \\ -2/3 & 1/3 & -2/3 \\ -2/3 & -2/3 & 1/3 \end{bmatrix}$$
   (which is an orthogonal matrix). Interpret your answers geometrically.

Examples

22.1 Find an orthogonal matrix P and a diagonal matrix D such that $P^{\mathrm{T}}AP = D$, where

$$A = \begin{bmatrix} 1 & -3 & 1 \\ -3 & 1 & 1 \\ 1 & 1 & -3 \end{bmatrix}.$$

First find eigenvalues.

$$\det(A - kI) = \begin{vmatrix} 1-k & -3 & 1 \\ -3 & 1-k & 1 \\ 1 & 1 & -3-k \end{vmatrix}$$

$$\begin{aligned} &= (1-k)[(1-k)(-3-k)-1] + 3[-3(-3-k)-1] \\ &\quad + [-3-(1-k)] \\ &= (1-k)(-4+2k+k^2) + (8+3k) + (-4+k) \\ &= 16 + 16k - k^2 - k^3 \\ &= (1+k)(16-k^2) \quad \text{(by observing that } k=-1 \text{ gives value 0)} \\ &= (1+k)(4+k)(4-k). \end{aligned}$$

Hence the eigenvalues are $-1, -4, 4$.

Next find corresponding eigenvectors.

$k = -1$: Solve $\begin{bmatrix} 2 & -3 & 1 \\ -3 & 2 & 1 \\ 1 & 1 & -2 \end{bmatrix}\begin{bmatrix} x \\ y \\ z \end{bmatrix} = \begin{bmatrix} 0 \\ 0 \\ 0 \end{bmatrix}$.

obtaining $x = t$, $y = t$ and $z = t$ $(t \in \mathbb{R})$.

$k = -4$: Solve $\begin{bmatrix} -3 & -3 & 1 \\ -3 & -3 & 1 \\ 1 & 1 & -7 \end{bmatrix}\begin{bmatrix} x \\ y \\ z \end{bmatrix} = \begin{bmatrix} 0 \\ 0 \\ 0 \end{bmatrix}$,

obtaining $x = -t$, $y = t$ and $z = 0$ $(t \in \mathbb{R})$.

$k = 4$: Solve $\begin{bmatrix} 5 & -3 & 1 \\ -3 & 5 & 1 \\ 1 & 1 & 1 \end{bmatrix}\begin{bmatrix} x \\ y \\ z \end{bmatrix} = \begin{bmatrix} 0 \\ 0 \\ 0 \end{bmatrix}$,

obtaining $x = -\frac{1}{2}t$, $y = -\frac{1}{2}t$ and $z = t$ $(t \in \mathbb{R})$.

Hence corresponding eigenvectors are respectively

$$\begin{bmatrix} t \\ t \\ t \end{bmatrix} \quad (t \in \mathbb{R},\ t \neq 0),$$

$$\begin{bmatrix} -t \\ t \\ 0 \end{bmatrix} \quad (t \in \mathbb{R},\ t \neq 0),$$

$$\begin{bmatrix} -\frac{1}{2}t \\ -\frac{1}{2}t \\ t \end{bmatrix} \quad (t \in \mathbb{R},\ t \neq 0).$$

Last, form the matrix P by taking as columns a list of eigenvectors which forms an orthonormal basis for $\mathbb{R}^3$. Pick any eigenvector for each eigenvalue, say

22

Diagonalisation 2

In Chapter 19 we saw how to find, given a $n \times n$ matrix A, an invertible matrix P and a diagonal matrix D such that $D = P^{-1}AP$. We saw that this is possible when there exists a LI list of n eigenvectors of A, and that the columns of the matrix P are the vectors in such a list. In fact the matrix P may be taken to be an orthogonal matrix. This is expressed precisely in the next rule, for which justification is given below, and illustration is given in Example 22.1.

Rule

Let A be a $n \times n$ real symmetric matrix which has n distinct eigenvalues. Then there is an orthogonal matrix P and a diagonal matrix D such that $D = P^{\mathrm{T}}AP$. (Recall that for an orthogonal matrix P, $P^{-1} = P^{\mathrm{T}}$.)

The matrix P has eigenvectors of A as columns. In order for P to be orthogonal, its columns must form an orthogonal list, and each column must have unit length. In Chapter 20 we saw that eigenvectors corresponding to distinct eigenvalues are orthogonal, so the columns of P must form an orthogonal list. Also, we know from Chapter 20 (Exercise 3) that when choosing eigenvectors we are free to choose them to have unit length. The matrix D remains as before: the diagonal entries are the eigenvalues of A. Remember that it is because the matrix A is symmetric that we can be sure that the eigenvectors are orthogonal. From Chapter 21 we know that the columns of a $n \times n$ orthogonal matrix form an orthonormal basis for $\mathbb{R}^n$, so this gives another aspect to the argument above – the matrix A has a list of n eigenvectors which form an orthonormal basis for $\mathbb{R}^n$, and these eigenvectors form the orthogonal matrix P.

$$\begin{bmatrix}1\\1\\1\end{bmatrix},\ \begin{bmatrix}-1\\1\\0\end{bmatrix},\ \begin{bmatrix}-1\\-1\\2\end{bmatrix}.$$

These form an orthogonal (and therefore LI) list (see Chapter 20). Next obtain eigenvectors as we require by dividing each of these by its length, thus:

$$\begin{bmatrix}1/\sqrt{3}\\1/\sqrt{3}\\1/\sqrt{3}\end{bmatrix},\ \begin{bmatrix}-1/\sqrt{2}\\1/\sqrt{2}\\0\end{bmatrix},\ \begin{bmatrix}-1/\sqrt{6}\\-1/\sqrt{6}\\2/\sqrt{6}\end{bmatrix}.$$

Finally, let

$$P=\begin{bmatrix}1/\sqrt{3} & -1/\sqrt{2} & -1/\sqrt{6}\\1/\sqrt{3} & 1/\sqrt{2} & -1/\sqrt{6}\\1/\sqrt{3} & 0 & 2/\sqrt{6}\end{bmatrix}.$$

Then (this requires no verification)

$$P^{T}AP=\begin{bmatrix}-1 & 0 & 0\\0 & -4 & 0\\0 & 0 & 4\end{bmatrix}.$$

This last matrix is the required matrix D, the order in which the eigenvalues occur in D being determined by the order of the corresponding eigenvectors as columns of P. Notice that the matrices

$$P=\begin{bmatrix}-1/\sqrt{6} & -1/\sqrt{2} & 1/\sqrt{3}\\-1/\sqrt{6} & 1/\sqrt{2} & 1/\sqrt{3}\\2/\sqrt{6} & 0 & 1/\sqrt{3}\end{bmatrix},\qquad D=\begin{bmatrix}4 & 0 & 0\\0 & -4 & 0\\0 & 0 & -1\end{bmatrix}$$

also satisfy $P^{T}AP=D$, and similarly with other ways of ordering the columns of the two matrices.

22.2 Diagonalisation when the characteristic equation has a repeated root. Find an orthogonal matrix P and a diagonal matrix D such that $P^{T}AP=D$, where

$$A=\begin{bmatrix}4 & 1 & -1\\1 & 4 & -1\\-1 & -1 & 4\end{bmatrix}.$$

The characteristic equation turns out to be

$$54-45k+12k^2-k^3=0,$$

i.e. $\quad (3-k)(18-9k+k^2)=0,$

i.e. $\quad (3-k)(3-k)(6-k)=0.$

Consequently there are only two eigenvalues, 3 and 6. The eigenvalue 3 is a 'repeated' eigenvalue. Find eigenvectors as usual.

$k=6$: Solve $\begin{bmatrix}-2 & 1 & -1\\1 & -2 & -1\\-1 & -1 & -2\end{bmatrix}\begin{bmatrix}x\\y\\z\end{bmatrix}=\begin{bmatrix}0\\0\\0\end{bmatrix}.$

The above rule in fact holds for all real symmetric matrices, even in cases where the requirement about distinct eigenvalues is not met. A proof of this in general is beyond the scope of this book, but we show how the matrices P and D may be obtained in such cases. In Example 22.2, A is a 3×3 real symmetric matrix which has two distinct eigenvalues, one of them a repeated root of the characteristic equation of A. In order to construct P, we need a LI list of three eigenvectors of A, which can be converted into an orthonormal basis for $\mathbb{R}^3$ as required for the columns of P. The repeated eigenvalue leads to a solution of the equation $(A - kI)\mathbf{x} = \mathbf{0}$ in which two parameters appear. By assigning values to these parameters we obtain two eigenvectors, not multiples of each other, both corresponding to the same eigenvalue. Moreover, both of these eigenvectors are orthogonal to any eigenvector corresponding to the other eigenvalue (see Chapter 20).

More briefly, what we do is to find, for a repeated eigenvalue k, a basis for the eigenspace corresponding to k. It turns out (although we shall not prove it) that, when A is symmetric, this basis will contain as many vectors as the multiplicity of k. The multiplicity of an eigenvalue k_0 is the power to which $(k_0 - k)$ occurs in the characteristic equation once its left-hand side has been fully factorised. For example, if the characteristic equation of a given 6×6 matrix is

$$(3 - k)^2(1 + k)^3(2 - k) = 0,$$

then 3 is an eigenvalue with multiplicity 2, and -1 is an eigenvalue with multiplicity 3 (and 2 is a 'normal' eigenvalue). When we have found a basis for the eigenspace corresponding to k, we use the techniques of Chapter 21 to find an orthonormal basis for the eigenspace. This may include use of the Gram–Schmidt process, but usually only if the eigenspace has dimension 3 or more. This procedure is easier to carry out than to describe, so the reader should gain facility by working the exercises.

Obtain $x=-t$, $y=-t$ and $z=t$ ($t\in\mathbb{R}$), so all vectors

$$\begin{bmatrix}-t\\-t\\t\end{bmatrix}\qquad (t\in\mathbb{R},\ t\neq 0)$$

are eigenvectors.

$k=3$: Solve $\begin{bmatrix}1&1&-1\\1&1&-1\\-1&-1&1\end{bmatrix}\begin{bmatrix}x\\y\\z\end{bmatrix}=\begin{bmatrix}0\\0\\0\end{bmatrix}$.

Obtain $x=u-t$, $y=u$ and $z=t$ ($t,u\in\mathbb{R}$). All vectors

$$\begin{bmatrix}u-t\\u\\t\end{bmatrix}\qquad (t,u\in\mathbb{R},\ \text{not both zero})$$

are eigenvectors corresponding to the eigenvalue 3. (The eigenspace has dimension 2.)

To form the matrix P we must pick one eigenvector corresponding to $k=6$ and two eigenvectors corresponding to $k=3$, so as to obtain an orthonormal list which spans $\mathbb{R}^3$. First pick (say) $\begin{bmatrix}-1\\-1\\1\end{bmatrix}$, which gives $\begin{bmatrix}-1/\sqrt{3}\\-1/\sqrt{3}\\1/\sqrt{3}\end{bmatrix}$ when divided by its length.

Next, for $k=3$, we need to pick two eigenvectors which are orthogonal to each other. Pick any one, say $\begin{bmatrix}1\\1\\0\end{bmatrix}$, and find values of t and u for another which is orthogonal to this one. Thus we shall require

$$u-t+u=0,\qquad \text{i.e. } t=2u.$$

So take, say, $t=2$ and $u=1$, giving the eigenvector $\begin{bmatrix}-1\\1\\2\end{bmatrix}$. Now divide each by its length, giving

$$\begin{bmatrix}1/\sqrt{2}\\1/\sqrt{2}\\0\end{bmatrix},\ \begin{bmatrix}-1/\sqrt{6}\\1/\sqrt{6}\\2/\sqrt{6}\end{bmatrix}.$$

Let $$P=\begin{bmatrix}-1/\sqrt{3}&1/\sqrt{2}&-1/\sqrt{6}\\-1/\sqrt{3}&1/\sqrt{2}&1/\sqrt{6}\\1/\sqrt{3}&0&2/\sqrt{6}\end{bmatrix}.$$

Then $$P^{\mathrm{T}}AP=\begin{bmatrix}6&0&0\\0&3&0\\0&0&3\end{bmatrix}=D.$$

22.3 Examples of finding an orthonormal basis for an eigenspace.

(i) Let $A=\begin{bmatrix} 2 & -1 & 2 \\ -1 & 2 & -2 \\ 2 & -2 & 5 \end{bmatrix}$.

The characteristic equation is $(1-k)^2(7-k)=0$. So 1 is an eigenvalue with multiplicity 2. The eigenspace corresponding to $k=1$ is the set of solutions to the equation

$$\begin{bmatrix} 1 & -1 & 2 \\ -1 & 1 & -2 \\ 2 & -2 & 4 \end{bmatrix}\begin{bmatrix} x \\ y \\ z \end{bmatrix}=\begin{bmatrix} 0 \\ 0 \\ 0 \end{bmatrix}.$$

This set is

$$\left\{\begin{bmatrix} u-2t \\ u \\ t \end{bmatrix} : t, u \in \mathbb{R}\right\}.$$

To find an orthogonal basis for this eigenspace, first pick any eigenvector, say (taking $u=1$ and $t=1$),

$$\begin{bmatrix} -1 \\ 1 \\ 1 \end{bmatrix}.$$

Next find values of t and u such that $\begin{bmatrix} u-2t \\ u \\ t \end{bmatrix}$ is orthogonal to this first vector.

So we must satisfy

$$-(u-2t)+u+t=0, \qquad \text{i.e. } 3t=0.$$

Take, therefore, $u=1$ (say) and $t=0$, obtaining the eigenvector $\begin{bmatrix} 1 \\ 1 \\ 0 \end{bmatrix}$, which is orthogonal to $\begin{bmatrix} -1 \\ 1 \\ 1 \end{bmatrix}$. Now divide each of these vectors by its length, to obtain

$$\left(\begin{bmatrix} 1/\sqrt{2} \\ 1/\sqrt{2} \\ 0 \end{bmatrix}, \begin{bmatrix} -1/\sqrt{3} \\ 1/\sqrt{3} \\ 1/\sqrt{3} \end{bmatrix}\right),$$

an orthonormal basis for the eigenspace corresponding to the eigenvalue 1.

(ii) $A=\begin{bmatrix} 3 & -1 & 2 & 1 \\ 2 & 0 & 4 & 2 \\ 1 & -1 & 4 & 1 \\ -1 & 1 & -2 & 1 \end{bmatrix}$.

Given that 2 is an eigenvalue of A with multiplicity 3, find an orthonormal basis for the eigenspace of A corresponding to 2.

Solving $\begin{bmatrix} 1 & -1 & 2 & 1 \\ 2 & -2 & 4 & 2 \\ 1 & -1 & 2 & 1 \\ -1 & 1 & -2 & -1 \end{bmatrix}\begin{bmatrix} x_1 \\ x_2 \\ x_3 \\ x_4 \end{bmatrix}=\begin{bmatrix} 0 \\ 0 \\ 0 \\ 0 \end{bmatrix}$

Examples 22.3 give further illustration. Notice that in the matrix D the repeated eigenvalue occurs more than once, in the positions determined by the order of the columns of P.

It is important to realise that these methods do not work for all matrices. We have a rule above which says that they will work at least for all symmetric matrices. For matrices which are not symmetric, however, we cannot be certain of the outcome. The matrix

$$A = \begin{bmatrix} 1 & 0 & 1 \\ 0 & 1 & 1 \\ -1 & 1 & 2 \end{bmatrix}$$

has characteristic equation

$$(1-k)^2(2-k)=0,$$

but, when we seek eigenvectors corresponding to the repeated eigenvalue $k=1$, we find an eigenspace of dimension only 1. The details of this appear in Example 19.4, where it is shown that A is not diagonalisable. Matrices which are not symmetric may be diagonalisable, or they may not.

gives the eigenspace

$$\left\{\begin{bmatrix} x_1 \\ x_2 \\ x_3 \\ x_4 \end{bmatrix} \in \mathbb{R}^4 : x_1 - x_2 + 2x_3 + x_4 = 0\right\}.$$

Pick a LI list of three vectors in this subspace, to constitute a basis, say

$$\boldsymbol{v}_1 = \begin{bmatrix} 1 \\ 1 \\ 0 \\ 0 \end{bmatrix}, \quad \boldsymbol{v}_2 = \begin{bmatrix} 1 \\ -1 \\ 1 \\ 0 \end{bmatrix}, \quad \boldsymbol{v}_3 = \begin{bmatrix} 0 \\ 1 \\ 0 \\ 1 \end{bmatrix}.$$

Now apply the Gram–Schmidt process, to obtain an orthonormal basis $(\boldsymbol{w}_1, \boldsymbol{w}_2, \boldsymbol{w}_3)$.

$$\boldsymbol{w}_1 = \begin{bmatrix} 1/\sqrt{2} \\ 1/\sqrt{2} \\ 0 \\ 0 \end{bmatrix}.$$

$$\boldsymbol{w}_2 = \begin{bmatrix} 1/\sqrt{3} \\ -1/\sqrt{3} \\ 1/\sqrt{3} \\ 0 \end{bmatrix}.$$

(Notice that $\boldsymbol{v}_1$ and $\boldsymbol{v}_2$ are already orthogonal, so this stage is particularly simple here.)

$$\boldsymbol{w}_3 = \begin{bmatrix} -1/\sqrt{42} \\ 1/\sqrt{42} \\ 2/\sqrt{42} \\ 6/\sqrt{42} \end{bmatrix}.$$

22.4 Diagonalisation of a linear transformation.

Let f be the linear transformation from $\mathbb{R}^3$ to $\mathbb{R}^3$ represented (with respect to the standard basis) by the matrix

$$A = \begin{bmatrix} 1 & 1 & 0 \\ 1 & 0 & 1 \\ 0 & 1 & 1 \end{bmatrix}. \text{ (See Example 19.2(ii).)}$$

Let $$P = \begin{bmatrix} -1 & 1 & 1 \\ 0 & 1 & -2 \\ 1 & 1 & 1 \end{bmatrix},$$

whose columns are a LI list of eigenvectors of A, and consider the change of coordinates given by

$$\boldsymbol{x} = P\boldsymbol{u}, \qquad \boldsymbol{u} = P^{-1}\boldsymbol{x}.$$

Let $\boldsymbol{u} = \begin{bmatrix} u_1 \\ u_2 \\ u_3 \end{bmatrix}$ be the component vector of some element $\boldsymbol{x}$ of $\mathbb{R}^3$ with respect to the basis B consisting of the columns of P. Find the component vector with respect to B of the element $f(\boldsymbol{x})$.

What is the purpose of diagonalisation? We shall see uses of it in geometry and in differential equations in the next two chapters. But there is perhaps a more fundamental reason for its usefulness, concerning linear transformations.

A linear transformation f may be represented by a matrix A, so that $f(\boldsymbol{x}) = A\boldsymbol{x}$. Suppose that we wished to refer vectors in $\mathbb{R}^n$ to a basis B other than the standard basis (see Chapter 17). A vector $\boldsymbol{x}$ in $\mathbb{R}^n$ would then have a column vector $\boldsymbol{u}$ of components with respect to B, with

$$\boldsymbol{x} = P\boldsymbol{u} \qquad \text{and} \qquad \boldsymbol{u} = P^{-1}\boldsymbol{x},$$

where P is the matrix whose columns are the vectors in the basis B. Can the component vector $\boldsymbol{u}$ (of $\boldsymbol{x}$) and the component vector of $f(\boldsymbol{x})$ with respect to B be conveniently related by a matrix equation? The answer is 'yes', and we show how. See Example 22.4 for a particular case. Start with a vector $\boldsymbol{u}$ of components with respect to B (of some vector $\boldsymbol{x} \in \mathbb{R}^n$). Then $\boldsymbol{x} = P\boldsymbol{u}$. The linear transformation f, applied to $\boldsymbol{x}$, yields $A\boldsymbol{x}$, i.e. $AP\boldsymbol{u}$. Now obtain the vector of components of this with respect to B, i.e. $P^{-1}(AP\boldsymbol{u})$. With respect to B, then, f has the effect

$$\boldsymbol{u} \to P^{-1}AP\boldsymbol{u}.$$

We know that for certain matrices A it is possible to choose a matrix P so that $P^{-1}AP$ is a diagonal matrix. In such cases, $P^{-1}AP\boldsymbol{u}$ is a very simple expression indeed. If the ith entry of $\boldsymbol{u}$ is u_i (for $1 \leqslant i \leqslant n$) and if $P^{-1}AP$ is a diagonal matrix whose (i, i)-entry is k_i (for $1 \leqslant i \leqslant n$), then

$$P^{-1}AP\boldsymbol{u} = \begin{bmatrix} k_1u_1 \\ k_2u_2 \\ \vdots \\ k_nu_n \end{bmatrix}.$$

First, $\boldsymbol{x}$ is represented by $P\boldsymbol{u}$ with respect to the standard basis, and

$$P\boldsymbol{u} = \begin{bmatrix} -u_1 + u_2 + u_3 \\ u_2 - 2u_3 \\ u_1 + u_2 + u_3 \end{bmatrix}.$$

The image of this under f is $AP\boldsymbol{u}$,

i.e. $$\begin{bmatrix} 1 & 1 & 0 \\ 1 & 0 & 1 \\ 0 & 1 & 1 \end{bmatrix}\begin{bmatrix} -u_1 + u_2 + u_3 \\ u_2 - 2u_3 \\ u_1 + u_2 + u_3 \end{bmatrix}$$

i.e. $$\begin{bmatrix} -u_1 + 2u_2 - u_3 \\ 2u_2 + 2u_3 \\ u_1 + 2u_2 - u_3 \end{bmatrix},$$

with respect to the standard basis.

Last, with respect to B, this vector is represented by the component vector $P^{-1}(AP\boldsymbol{u})$,

i.e. $$\begin{bmatrix} -1 & 1 & 1 \\ 0 & 1 & -2 \\ 1 & 1 & 1 \end{bmatrix}^{-1}\begin{bmatrix} -u_1 + 2u_2 - u_3 \\ 2u_2 + 2u_3 \\ u_1 + 2u_2 - u_3 \end{bmatrix}$$

i.e. $$\tfrac{1}{6}\begin{bmatrix} -3 & 0 & 3 \\ 2 & 2 & 2 \\ 1 & -2 & 1 \end{bmatrix}\begin{bmatrix} -u_1 + 2u_2 - u_3 \\ 2u_2 + 2u_3 \\ u_1 + 2u_2 - u_3 \end{bmatrix}$$

i.e. $$\tfrac{1}{6}\begin{bmatrix} 6u_1 \\ 12u_2 \\ -6u_3 \end{bmatrix}, \quad \text{which is equal to} \begin{bmatrix} u_1 \\ 2u_2 \\ -u_3 \end{bmatrix}.$$

This last vector is equal to

$$\begin{bmatrix} 1 & 0 & 0 \\ 0 & 2 & 0 \\ 0 & 0 & -1 \end{bmatrix}\begin{bmatrix} u_1 \\ u_2 \\ u_3 \end{bmatrix},$$

in which the multiplying matrix is a diagonal matrix, namely the matrix obtained in Example 19.2(i) by the standard diagonalisation process.

22.5 Diagonalisation of a linear transformation.
For $\boldsymbol{x} \in \mathbb{R}^3$, let $f(\boldsymbol{x}) = A\boldsymbol{x}$, where

$$A = \begin{bmatrix} 1 & -3 & 1 \\ -3 & 1 & 1 \\ 1 & 1 & -3 \end{bmatrix}.$$

Find a basis B for $\mathbb{R}^3$ with respect to which f is represented by a diagonal matrix D, i.e. so that, with respect to B, the image of the vector with components $\begin{bmatrix} u_1 \\ u_2 \\ u_3 \end{bmatrix}$ has components $\begin{bmatrix} k_1u_1 \\ k_2u_2 \\ k_3u_3 \end{bmatrix}$, for some numbers k_1, k_2 and k_3.

The required labour has been carried out in Example 22.1. An answer is

This process is known as diagonalisation of a linear transformation. It provides a simple expression for the function values, by referring everything to an appropriate basis. Example 22.5 provides illustration. Be warned, however. The diagonalisation process works only for certain matrices (as seen in Chapter 19), so it may be expected to work only for certain linear transformations.

Summary

An extension of the earlier diagonalisation process is described whereby, for certain matrices A, an orthogonal matrix P may be found such that $P^{\mathrm{T}}AP$ is a diagonal matrix. In particular, this may be done when A is a real symmetric matrix. The situation in which there is a repeated eigenvalue is dealt with. There is a brief discussion of the idea of diagonalisation of a linear transformation, via an appropriate change of basis.

Exercises

1. For each of the following matrices, find a list of eigenvectors which constitutes an orthonormal basis for $\mathbb{R}^3$.

   (i) $\begin{bmatrix} -1 & 2 & 2 \\ 2 & 1 & 0 \\ 2 & 0 & 1 \end{bmatrix}$. (ii) $\begin{bmatrix} 2 & 0 & 1 \\ 0 & 2 & 0 \\ 1 & 0 & 2 \end{bmatrix}$.

   (iii) $\begin{bmatrix} 1 & 3 & 4 \\ 3 & 1 & 0 \\ 4 & 0 & 1 \end{bmatrix}$. (iv) $\begin{bmatrix} 7 & -2 & 0 \\ -2 & 6 & -2 \\ 0 & -2 & 5 \end{bmatrix}$.

2. For each of the matrices A given in Exercise 1, write down an orthogonal matrix P and a diagonal matrix D such that $P^{\mathrm{T}}AP = D$.
3. Each of the following matrices has a repeated eigenvalue. Find in each case a list of eigenvectors which constitutes an orthonormal basis for $\mathbb{R}^3$.

   (i) $\begin{bmatrix} 2 & -1 & 2 \\ -1 & 2 & -2 \\ 2 & -2 & 5 \end{bmatrix}$. (ii) $\begin{bmatrix} -2 & 2 & 2 \\ 2 & 1 & 4 \\ 2 & 4 & 1 \end{bmatrix}$. (iii) $\begin{bmatrix} 0 & 2 & 3 \\ 2 & 3 & 6 \\ 3 & 6 & 8 \end{bmatrix}$.

4. For each of the matrices A given in Exercise 3, write down an orthogonal matrix P and a diagonal matrix D such that $P^{\mathrm{T}}AP = D$.
5. Find an orthogonal (4×4) matrix P and a diagonal matrix D such that $P^{\mathrm{T}}AP = D$, where A is the matrix

$$\begin{bmatrix} 1 & 3 & 3 & -3 \\ 3 & 1 & -3 & 3 \\ 3 & -3 & 1 & 3 \\ -3 & 3 & 3 & 1 \end{bmatrix}.$$

$$B=\left(\begin{bmatrix}1/\sqrt{3}\\1/\sqrt{3}\\1/\sqrt{3}\end{bmatrix},\begin{bmatrix}-1/\sqrt{2}\\1/\sqrt{2}\\0\end{bmatrix},\begin{bmatrix}-1/\sqrt{6}\\-1/\sqrt{6}\\2/\sqrt{6}\end{bmatrix}\right).$$

These vectors are the columns of the matrix P found in Example 22.1 such that

$$P^{\mathrm{T}}AP=D=\begin{bmatrix}-1&0&0\\0&-4&0\\0&0&4\end{bmatrix}.$$

In this case (because A is symmetric) a basis B can be found which is orthonormal. The numbers k_1, k_2 and k_3 required are the eigenvalues -1, -4 and 4.

6. Repeat Exercise 5 with the matrix

$$\begin{bmatrix} 1 & 4 & 0 & 0 \\ 4 & 1 & 0 & 0 \\ 0 & 0 & 3 & 4 \\ 0 & 0 & 4 & -3 \end{bmatrix}.$$

7. Let f be the linear transformation from $\mathbb{R}^3$ to $\mathbb{R}^3$ given by $f(\boldsymbol{x}) = A\boldsymbol{x}$, where

$$A = \begin{bmatrix} 1 & 1 & 2 \\ 0 & 1 & 0 \\ 0 & 1 & 3 \end{bmatrix}.$$

Find a basis X for $\mathbb{R}^3$ such that, when vectors in $\mathbb{R}^3$ are represented as column vectors of components with respect to X, f is given by $f(\boldsymbol{u}) = D\boldsymbol{u}$, where D is a diagonal matrix.

8. Repeat Exercise 7 with the linear transformation g given by $g(\boldsymbol{x}) = B\boldsymbol{x}$, where

$$B = \begin{bmatrix} 2 & -1 & 2 \\ -1 & 2 & -2 \\ 2 & -2 & 5 \end{bmatrix}.$$

Notice that the basis obtained here is an orthogonal list, whereas that obtained in Exercise 7 was not.

Examples

23.1 Simplify the equation

$$x^2 + 4xy + y^2 + 3 = 0$$

by means of the change of variables

$$x = \frac{1}{\sqrt{2}}u + \frac{1}{\sqrt{2}}v, \qquad y = -\frac{1}{\sqrt{2}}u + \frac{1}{\sqrt{2}}v.$$

Substituting, we obtain

$$\tfrac{1}{2}u^2 + uv + \tfrac{1}{2}v^2 - 2u + 2v + \tfrac{1}{2}u^2 - uv + \tfrac{1}{2}v^2 + 3 = 0$$

i.e. $$-u^2 + 3v^2 + 3 = 0$$

i.e. $$\frac{u^2}{3} - v^2 = 1.$$

This is now recognisable as in the standard form for an equation of a hyperbola. In effect the new equation is a consequence of the change of basis

$$\begin{bmatrix} x \\ y \end{bmatrix} = \begin{bmatrix} 1/\sqrt{2} & 1/\sqrt{2} \\ -1/\sqrt{2} & 1/\sqrt{2} \end{bmatrix} \begin{bmatrix} u \\ v \end{bmatrix}.$$

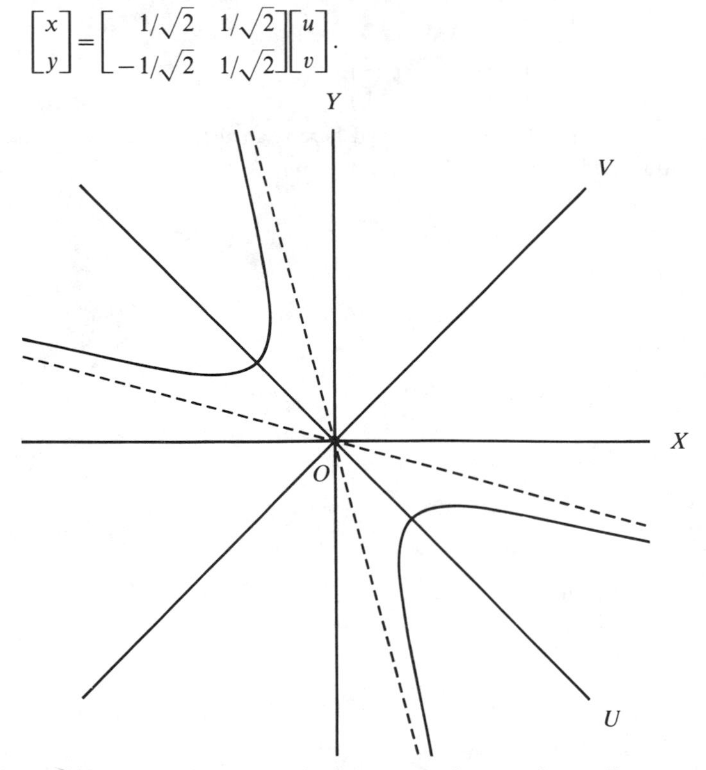

This change of basis is, in geometrical terms, a change of coordinate axes. The new axes are OU and OV, along the directions of the eigenvectors of the matrix $\begin{bmatrix} 1 & 2 \\ 2 & 1 \end{bmatrix}$. This matrix is obtained from the quadratic form $x^2 + 4xy + y^2$ as described in the text.

23

Geometry

In the coordinate geometry of two dimensions, a second degree equation in x and y generally represents a conic. We shall not be concerned with the geometrical background, but we shall need to be familiar with some standard forms of equation:

$x^2 + y^2 = r^2$ circle, $\dfrac{x^2}{a^2} + \dfrac{y^2}{b^2} = 1$ ellipse,

$\dfrac{x^2}{a^2} - \dfrac{y^2}{b^2} = 1$ hyperbola, $\dfrac{y^2}{b^2} - \dfrac{x^2}{a^2} = 1$ hyperbola.

Each of these figures has centre at the origin. Of course there is another sort of conic, the parabola, but the methods developed here do not apply in that case. Ellipses and hyperbolas have a symmetry which makes our algebraic methods appropriate. This symmetry is reflected in the form of the equations above, in the sense that x and y occur only in terms involving x^2 and y^2. But ellipses and hyperbolas with centre at the origin can have equations which have a rather different form. For example,

$$x^2 + y^2 + 4xy + 3 = 0$$

is an equation for a hyperbola with centre at the origin, and

$$5x^2 + 2y^2 + 4xy - 6 = 0$$

is an equation for an ellipse with centre at the origin.

Our familiar diagonalisation process provides a routine by which we can simplify such equations and decide which sort of conic they represent. And it can be used in three dimensions for an analogous procedure involving quadric surfaces and their equations. The process is equivalent to changing to new coordinate axes to simplify the equation. But the diagonalisation process is what tells us which new coordinate axes to choose. Example 23.1 shows how an equation can be simplified into one of the four standard forms above by an appropriate change of coordinates. Before we can see how the new axes are chosen, we must discuss a new notion.

23.2 Examples of quadratic forms.

$$2x^2 - y^2 + 4xy$$
$$3x^2 - 4xy$$
$$4x^2 - 3y^2$$
$$x^2 - 4y^2 + z^2 + 16yz - 6zx + 16xy$$
$$3y^2 - 4xz + 6xy$$
$$x^2 + y^2 + z^2 + xz$$
$$15y^2 + z^2 - 8xy.$$

23.3 Matrix expression for a quadratic form.

Let $A = \begin{bmatrix} 1 & 2 \\ 3 & 4 \end{bmatrix}$ and $\boldsymbol{x} = \begin{bmatrix} x \\ y \end{bmatrix}$.

Then
$$\begin{aligned} \boldsymbol{x}^{\mathrm{T}} A \boldsymbol{x} &= [x \quad y] \begin{bmatrix} 1 & 2 \\ 3 & 4 \end{bmatrix} \begin{bmatrix} x \\ y \end{bmatrix} \\ &= [x \quad y] \begin{bmatrix} x + 2y \\ 3x + 4y \end{bmatrix} \\ &= [x(x + 2y) + y(3x + 4y)]. \end{aligned}$$

The single entry in this 1×1 matrix is equal to

$$x^2 + 2xy + 3xy + 4y^2$$

i.e. $x^2 + 5xy + 4y^2$.

Definition

A *quadratic form* in two variables x and y is an expression of the form

$$ax^2 + by^2 + cxy,$$

where a, b and c are real numbers. A quadratic form in n variables $x_1, \ldots, x_n$ is a sum of terms each of which is of the second degree, i.e. is of the form ax_i^2 or hx_ix_j.

We shall deal with only two or three variables. Some examples are given in Example 23.2.

Now we make the link with matrices and diagonalisation. See Example 23.3 for illustration of the following. Let A be the 2×2 matrix

$$\begin{bmatrix} a_{11} & a_{12} \\ a_{21} & a_{22} \end{bmatrix}, \qquad \text{and let} \qquad \boldsymbol{x} = \begin{bmatrix} x \\ y \end{bmatrix}.$$

Then $\boldsymbol{x}^T A \boldsymbol{x}$ is a 1×1 matrix, which we can evaluate as follows.

$$\begin{aligned} \boldsymbol{x}^T A \boldsymbol{x} &= [x \quad y] \begin{bmatrix} a_{11} & a_{12} \\ a_{21} & a_{22} \end{bmatrix} \begin{bmatrix} x \\ y \end{bmatrix} \\ &= [x \quad y] \begin{bmatrix} a_{11}x + a_{12}y \\ a_{21}x + a_{22}y \end{bmatrix} \\ &= a_{11}x^2 + a_{21}xy + a_{12}xy + a_{22}y^2 \\ &= a_{11}x^2 + (a_{12} + a_{21})xy + a_{22}y^2. \end{aligned}$$

Here we do not distinguish between a 1×1 matrix and the number which is its sole entry. The important thing is that $\boldsymbol{x}^T A \boldsymbol{x}$ is a quadratic form in x and y.

We can do this also with any square matrix A and any appropriately sized vector $\boldsymbol{x}$. If A is a 3×3 matrix with (i, j)-entry a_{ij} and $\boldsymbol{x}$ is a 3-vector with entries x, y and z, then (details omitted)

$$\begin{aligned} \boldsymbol{x}^T A \boldsymbol{x} = a_{11}x^2 + a_{22}y^2 + a_{33}z^2 &+ (a_{23} + a_{32})yz \\ &+ (a_{31} + a_{13})zx + (a_{12} + a_{21})xy. \end{aligned}$$

This process can be reversed. Given a quadratic form, we can write down a matrix A such that $\boldsymbol{x}^T A \boldsymbol{x}$ is equal to the given quadratic form. For example

$$ax^2 + by^2 + cxy = [x \quad y] \begin{bmatrix} a & \frac{1}{2}c \\ \frac{1}{2}c & b \end{bmatrix} \begin{bmatrix} x \\ y \end{bmatrix}.$$

But note also that

$$ax^2 + by^2 + cxy = [x \quad y] \begin{bmatrix} a & c \\ 0 & b \end{bmatrix} \begin{bmatrix} x \\ y \end{bmatrix},$$

so the matrix A is not uniquely determined. The coefficient c has to be 'shared out' between the (1, 2)-entry and the (2, 1)-entry.

23.4 Matrices representing quadratic forms.

(i) Find matrices A such that the quadratic form

$$2x^2 - 6xy - y^2$$

is given by the product $\mathbf{x}^{\mathrm{T}} A \mathbf{x}$.
Take $A = [a_{ij}]_{2 \times 2}$, say, where

$$a_{11} = 2, \qquad a_{22} = -1 \qquad \text{and} \qquad a_{12} + a_{21} = -6.$$

The following (among infinitely many possible matrices) will do:

$$\begin{bmatrix} 2 & 0 \\ -6 & -1 \end{bmatrix}, \begin{bmatrix} 2 & -2 \\ -4 & -1 \end{bmatrix}, \begin{bmatrix} 2 & 2 \\ -8 & -1 \end{bmatrix}, \begin{bmatrix} 2 & -3 \\ -3 & -1 \end{bmatrix}.$$

The last is significant, being the only symmetric matrix A with the required property.

(ii) Find matrices A such that the quadratic form

$$x^2 - 3y^2 + 7z^2 - 3yz + 2zx + 8xy$$

is given by the product $\mathbf{x}^{\mathrm{T}} A \mathbf{x}$.
Take $A = [a_{ij}]_{3 \times 3}$, say, where

$$a_{11} = 1, \qquad a_{22} = -3, \qquad a_{33} = 7, \qquad a_{12} + a_{21} = 8,$$
$$a_{23} + a_{32} = -3, \qquad a_{31} + a_{13} = 2.$$

The following are among the possible answers:

$$\begin{bmatrix} 1 & 1 & 2 \\ 7 & -3 & -1 \\ 0 & -2 & 7 \end{bmatrix}, \begin{bmatrix} 1 & -2 & 1 \\ 10 & -3 & -4 \\ 1 & 1 & 7 \end{bmatrix}, \begin{bmatrix} 1 & 4 & 1 \\ 4 & -3 & -\frac{3}{2} \\ 1 & -\frac{3}{2} & 7 \end{bmatrix}.$$

The last one of these is the symmetric matrix which has the required property.

(iii) Find a symmetric matrix A such that the quadratic form

$$ax^2 + by^2 + cz^2 + dyz + ezx + fxy$$

is given by the product $\mathbf{x}^{\mathrm{T}} A \mathbf{x}$.

$$\text{Answer: } A = \begin{bmatrix} a & \frac{1}{2}f & \frac{1}{2}e \\ \frac{1}{2}f & b & \frac{1}{2}d \\ \frac{1}{2}e & \frac{1}{2}d & c \end{bmatrix}.$$

It is easily verified that this matrix has the required property.

An example of finding a matrix for a quadratic form in three variables is given in Example 23.4.

It will be particularly convenient, in view of the rule in Chapter 22 about when a matrix is diagonalisable, to choose our matrix to be symmetric. This can always be done, as is shown for two variables above, and for three variables in Example 23.4.

Observe that quadratic forms in which there are no cross-terms, i.e. in which only squares of the variables occur, give rise to diagonal matrices, and vice-versa. For example,

$$3x^2 - 2y^2 = [x \quad y]\begin{bmatrix} 3 & 0 \\ 0 & -2 \end{bmatrix}\begin{bmatrix} x \\ y \end{bmatrix},$$

$$x^2 - \tfrac{1}{2}y^2 + 4z^2 = [x \quad y \quad z]\begin{bmatrix} 1 & 0 & 0 \\ 0 & -\tfrac{1}{2} & 0 \\ 0 & 0 & 4 \end{bmatrix}\begin{bmatrix} x \\ y \\ z \end{bmatrix}.$$

Also notice that the left-hand sides of our standard forms of equation for conics are such quadratic forms. What we seek to do, therefore, is to convert a quadratic form $\boldsymbol{x}^{\mathrm{T}}A\boldsymbol{x}$, by an appropriate change of basis (i.e. change of axes), to a quadratic form $\boldsymbol{u}^{\mathrm{T}}D\boldsymbol{u}$, where D is a diagonal matrix. This is now easy for us, at least for symmetric matrices A.

23.5 Diagonalisation of quadratic forms.

(i) Find a matrix P such that the change of variables given by

$$\begin{bmatrix} x \\ y \end{bmatrix} = P \begin{bmatrix} u \\ v \end{bmatrix}$$

will simplify the quadratic form

$$6x^2 - 4xy + 3y^2$$

into a quadratic form which has no cross-term.
First find a symmetric matrix A such that $\mathbf{x}^{\mathrm{T}} A \mathbf{x}$ yields the given quadratic form. As in earlier examples,

$$A = \begin{bmatrix} 6 & -2 \\ -2 & 3 \end{bmatrix}.$$

By the standard methods, find eigenvalues and eigenvectors of A, and an orthogonal matrix P such that $P^{\mathrm{T}}AP$ is a diagonal matrix. We obtain

$$P^{\mathrm{T}}AP = \begin{bmatrix} 7 & 0 \\ 0 & 2 \end{bmatrix}, \qquad \text{with } P = \begin{bmatrix} -2/\sqrt{5} & 1/\sqrt{5} \\ 1/\sqrt{5} & 2/\sqrt{5} \end{bmatrix}.$$

This matrix P is the one with the required property (no need to verify this), and the simplified quadratic form is

$$7u^2 + 2v^2.$$

(ii) Repeat part (i), but with the quadratic form

$$x^2 + 4xy + y^2.$$

Here $A = \begin{bmatrix} 1 & 2 \\ 2 & 1 \end{bmatrix}$. Find the eigenvalues and eigenvectors, and a matrix P such that $P^{\mathrm{T}}AP$ is a diagonal matrix. By the usual methods we obtain

$$P^{\mathrm{T}}AP = \begin{bmatrix} -1 & 0 \\ 0 & 3 \end{bmatrix}, \qquad \text{with } P = \begin{bmatrix} 1/\sqrt{2} & 1/\sqrt{2} \\ -1/\sqrt{2} & 1/\sqrt{2} \end{bmatrix}.$$

The eigenvalues are -1 and 3, and the vectors

$$\begin{bmatrix} 1/\sqrt{2} \\ -1/\sqrt{2} \end{bmatrix} \quad \text{and} \quad \begin{bmatrix} 1/\sqrt{2} \\ 1/\sqrt{2} \end{bmatrix}$$

are corresponding eigenvectors, respectively.
Notice that the usual process could lead to a slightly different matrix P. For example, P might have been chosen as

$$\begin{bmatrix} -1/\sqrt{2} & 1/\sqrt{2} \\ 1/\sqrt{2} & 1/\sqrt{2} \end{bmatrix}.$$

This makes no difference. Also, P might have been chosen as

$$\begin{bmatrix} 1/\sqrt{2} & 1/\sqrt{2} \\ 1/\sqrt{2} & -1/\sqrt{2} \end{bmatrix}$$

with the columns in the other order. This would have the effect of making $P^{\mathrm{T}}AP = \begin{bmatrix} 3 & 0 \\ 0 & -1 \end{bmatrix}$, and of interchanging the new coordinates u and v.

We have a method for finding an orthogonal matrix P such that $P^T AP$ is a diagonal matrix. Therefore if we specify a change of basis by

$$\boldsymbol{x} = P\boldsymbol{u} \qquad (\text{so that } \boldsymbol{x}^T = \boldsymbol{u}^T P^T),$$

then $\quad \boldsymbol{x}^T A\boldsymbol{x} = (\boldsymbol{u}^T P^T)A(P\boldsymbol{u}) = \boldsymbol{u}^T(P^T AP)\boldsymbol{u}.$

Let us write this out as a rule. See Example 23.5.

Rule

Let $\boldsymbol{x}^T A\boldsymbol{x}$ be a quadratic form, where A is a symmetric $n \times n$ matrix and $\boldsymbol{x}$ is a n-vector with entries $x_1, \ldots, x_n$. Then the change of basis $\boldsymbol{x} = P\boldsymbol{u}$, where P is an orthogonal matrix whose columns are eigenvectors of A leads to the simplification

$$\boldsymbol{x}^T A\boldsymbol{x} = \boldsymbol{u}^T D\boldsymbol{u} = k_1 u_1^2 + \cdots + k_n u_n^2,$$

where $k_1, \ldots, k_n$ are the eigenvalues of A (possibly including repetitions) and $u_1, \ldots, u_n$ are the new coordinates. The new basis consists of the columns of the matrix P.

In our two-dimensional situation, what this amounts to is choosing new axes in the directions of the eigenvectors. It is the eigenvectors which determine the change of coordinates which will simplify the quadratic form. And notice that the eigenvalues appear as the coefficients in the simplified form.

Example 23.5(ii) is a re-working of Example 23.1 using these ideas. It turns out that the new coordinate axes (in the directions of the eigenvectors) are the axes of symmetry of the conic, sometimes called the principal axes.

In three dimensions, an equation of the form

$$ax^2 + by^2 + cz^2 + 2fyz + 2gzx + 2hxy + d = 0$$

generally represents a surface called a quadric surface. There are standard types, represented by standard forms of equation, for example

$$\frac{x^2}{a^2} + \frac{y^2}{b^2} + \frac{z^2}{c^2} = 1 \qquad \text{ellipsoid},$$

$$\frac{x^2}{a^2} + \frac{y^2}{b^2} - \frac{z^2}{c^2} = 1 \qquad \text{hyperboloid of one sheet},$$

$$\frac{x^2}{a^2} - \frac{y^2}{b^2} - \frac{z^2}{c^2} = 1 \qquad \text{hyperboloid of two sheets}.$$

There are obviously other possible combinations of positive and negative signs. (A sphere is a special case of an ellipsoid, when a, b and c are all equal.)

23.6 Diagonalisation of a quadratic form in three variables.
Find a matrix P such that the change of variables given by

$$\begin{bmatrix} x \\ y \\ z \end{bmatrix} = P \begin{bmatrix} u \\ v \\ w \end{bmatrix}$$

will simplify the quadratic form

$$x^2 + y^2 - 3z^2 + 2yz + 2zx - 6xy$$

into a quadratic form which has no cross-terms.
This quadratic form may be represented as $\mathbf{x}^{\mathrm{T}} A \mathbf{x}$, with

$$A = \begin{bmatrix} 1 & -3 & 1 \\ -3 & 1 & 1 \\ 1 & 1 & -3 \end{bmatrix}.$$

We have already done the work required, in Example 22.1. The matrix P which we found in Example 22.1 is the one needed here. Without specific verification we know that the change of variables

$$\begin{bmatrix} x \\ y \\ z \end{bmatrix} = \begin{bmatrix} 1/\sqrt{3} & -1/\sqrt{2} & -1/\sqrt{6} \\ 1/\sqrt{3} & 1/\sqrt{2} & -1/\sqrt{6} \\ 1/\sqrt{3} & 0 & 2/\sqrt{6} \end{bmatrix} \begin{bmatrix} u \\ v \\ w \end{bmatrix}$$

will convert the given quadratic form into

$$-u^2 - 4v^2 + 4w^2.$$

The coefficients in this quadratic form are the eigenvalues of the matrix A.

$$P^{\mathrm{T}} A P = D = \begin{bmatrix} -1 & 0 & 0 \\ 0 & -4 & 0 \\ 0 & 0 & 4 \end{bmatrix},$$

and $\quad \mathbf{u}^{\mathrm{T}} D \mathbf{u} = [-u^2 - 4v^2 + 4w^2].$

Without going into the geometry, we can apply our methods to a given equation such as the general one above, to simplify it into one of the standard forms, and thereby to recognise the type of quadric surface and to find some of its simple properties. This is done in Example 23.6. The eigenvectors again turn out to determine axes of symmetry (i.e. the principal axes) and the eigenvalues become the coefficients in the simplified form. One general rule, therefore, is: we obtain an ellipsoid when and only when all three eigenvalues are positive (presuming that the equation is given in the form $\mathbf{x}^{\mathrm{T}}A\mathbf{x}=c$, with $c>0$).

Summary

Standard forms of equation are given for circles, ellipses and hyperbolas. In general, a locus of one of these kinds has an equation which involves a quadratic form. It is shown how to transform a quadratic form into a sum of square terms, using the diagonalisation process from the preceding chapter, and hence to convert a given equation into one of the standard forms. This procedure is extended to the three-dimensional situation, where the methods are applied to equations for quadric surfaces.

Exercises

1. For each of the quadratic forms listed below, find a symmetric matrix A such that the quadratic form may be expressed as $\mathbf{x}^{\mathrm{T}}A\mathbf{x}$, where $\mathbf{x}$ is either $\begin{bmatrix} x \\ y \end{bmatrix}$ or $\begin{bmatrix} x \\ y \\ z \end{bmatrix}$, depending on the situation.

   (i) $2x^2 - y^2 + 4xy$.

   (ii) $3x^2 - 4xy$.

   (iii) $4x^2 - 3y^2$.

   (iv) $x^2 - 4y^2 + z^2 + 16yz - 6zx + 16xy$.

   (v) $3y^2 - 4xz + 6xy$.

   (vi) $x^2 + y^2 + z^2 + xz$.

   (vii) $15y^2 + z^2 - 8xy$.

   (These are the quadratic forms listed in Example 23.2.)

2. For each of the quadratic forms given in Exercise 1, find an orthogonal matrix P which determines a change of coordinates $\mathbf{x} = P\mathbf{u}$ which transforms the given quadratic form into a sum of squares.

3. Each of the following equations represents a conic (an ellipse or a hyperbola) with centre at the origin. Find the directions of the principal axes in each case, write down an equation for the conic with respect to the principal axes, and say which sort of conic it is.

(i) $x^2 + y^2 - 4xy = 3$.

(ii) $3x^2 + 3y^2 + 2xy = 1$.

(iii) $1 + 4xy - x^2 - 3y^2 = 0$.

(iv) $3x^2 + 4xy + 4 = 0$.

4. Each of the following equations represents a quadric surface. Find the directions of the principal axes in each case, write down an equation for the surface with respect to the principal axes, and say what sort of surface it is.

(i) $-x^2 + y^2 + z^2 + 4xy + 4xz = 3$.

(ii) $x^2 + y^2 + z^2 + xz = 1$.

(iii) $x^2 + y^2 - 3z^2 + 2yz + 2zx - 6xy = 4$.

(iv) $2x^2 + 2y^2 + 5z^2 - 4yz + 4zx - 2xy = 7$.

Is any of these surfaces a surface of revolution?

Examples

24.1 Illustration of simultaneous differential equations.

(i)
$$\frac{dx}{dt}=y$$
$$\frac{dy}{dt}=x. \qquad \text{These equations are satisfied by}$$
$$x=e^{t}, \qquad y=e^{t},$$
and by $x=e^{-t}, \qquad y=-e^{-t}.$

(ii)
$$\frac{dx}{dt}= x+2y$$
$$\frac{dy}{dt}=2x+y. \qquad \text{These equations are satisfied by}$$
$$x=e^{3t}+e^{-t}, \qquad y=e^{3t}-e^{-t},$$
and by $x=2e^{3t}, \qquad y=2e^{3t}.$

(iii)
$$\frac{dx}{dt}=y+z$$
$$\frac{dy}{dt}=x+z$$
$$\frac{dz}{dt}=x+y. \qquad \text{These equations are satisfied by}$$
$$x=2e^{-t}, \qquad y=-e^{-t}, \qquad z=-e^{-t}.$$

Indeed they are satisfied by any three functions x, y and z of the form

$$x=2ae^{-t}+be^{2t},$$
$$y=-ae^{-t}+be^{2t},$$
$$z=-ae^{-t}+be^{2t},$$

where a and b are arbitrary real numbers.

24.2 Solve the simultaneous differential equations

$$\frac{dx}{dt}=6x-2y$$
$$\frac{dy}{dt}=2x+3y.$$

Try $\begin{bmatrix} x \\ y \end{bmatrix}=\begin{bmatrix} q_1 \\ q_2 \end{bmatrix}e^{kt},$

so that $\dfrac{dx}{dt}=q_1ke^{kt}$ and $\dfrac{dy}{dt}=q_2ke^{kt}.$

Substituting into the original equations, we obtain

$$\begin{bmatrix} q_1 \\ q_2 \end{bmatrix}ke^{kt}=\begin{bmatrix} 6x-2y \\ 2x+3y \end{bmatrix},$$

24

Differential equations

Functions x and y of an independent variable t may have their derivatives given as linear functions of x and y themselves:

$$\frac{dx}{dt} = ax + by,$$

$$\frac{dy}{dt} = cx + dy.$$

These may be written as a matrix equation

$$\frac{d\mathbf{x}}{dt} = A\mathbf{x},$$

where $A = \begin{bmatrix} a & b \\ c & d \end{bmatrix}$ and $\mathbf{x} = \begin{bmatrix} x \\ y \end{bmatrix}$.

Example 24.1 gives an illustration of such equations and the sort of functions x and y which satisfy them.

There is a formal similarity between the matrix form of differential equation above and the simple differential equation

$$\frac{dx}{dt} = ax.$$

Its solution is

$$x = \alpha e^{at} \qquad (\alpha \text{ an arbitrary constant}).$$

The similarity does not end here. We may obtain the general solution to the matrix equation by trying out

$$\mathbf{x} = \mathbf{q}e^{kt},$$

i.e. $x = q_1 e^{kt}$ and $y = q_2 e^{kt}$,

so that $\dfrac{dx}{dt} = q_1 k e^{kt}$ and $\dfrac{dy}{dt} = q_2 k e^{kt}$.

i.e. $$k\begin{bmatrix} q_1 \\ q_2 \end{bmatrix} e^{kt} = \begin{bmatrix} 6 & -2 \\ 2 & 3 \end{bmatrix} \begin{bmatrix} x \\ y \end{bmatrix} = \begin{bmatrix} 6 & -2 \\ 2 & 3 \end{bmatrix} \begin{bmatrix} q_1 \\ q_2 \end{bmatrix} e^{kt}.$$

So $$k\begin{bmatrix} q_1 \\ q_2 \end{bmatrix} = \begin{bmatrix} 6 & -2 \\ 2 & 3 \end{bmatrix} \begin{bmatrix} q_1 \\ q_2 \end{bmatrix}.$$

Thus k must be an eigenvalue of the matrix $\begin{bmatrix} 6 & -2 \\ 2 & 3 \end{bmatrix}$, and $\begin{bmatrix} q_1 \\ q_2 \end{bmatrix}$ must be a corresponding eigenvector.

The standard process shows that 7 and 2 are the eigenvalues of this matrix, with corresponding eigenvectors

$$\begin{bmatrix} -2u \\ u \end{bmatrix} \quad (u \in \mathbb{R},\ u \neq 0) \qquad \text{and} \qquad \begin{bmatrix} \frac{1}{2}u \\ u \end{bmatrix} \quad (u \in \mathbb{R},\ u \neq 0).$$

Consequently,

$$\begin{bmatrix} x \\ y \end{bmatrix} = \begin{bmatrix} -2a \\ a \end{bmatrix} e^{7t} \qquad \text{and} \qquad \begin{bmatrix} x \\ y \end{bmatrix} = \begin{bmatrix} \frac{1}{2}b \\ b \end{bmatrix} e^{2t}$$

are solutions to the original equations, for any choice of the numbers a and b.

24.3 Find solutions to the differential equations

$$\frac{dx}{dt} = x - 2y$$

$$\frac{dy}{dt} = x + 4y.$$

We can abbreviate the process of Example 24.2. Find eigenvalues and eigenvectors of the matrix $\begin{bmatrix} 1 & -2 \\ 1 & 4 \end{bmatrix}$. The characteristic equation is

$$\begin{vmatrix} 1-k & -2 \\ 1 & 4-k \end{vmatrix} = 0$$

i.e. $$4 - 5k + k^2 + 2 = 0$$

i.e. $$6 - 5k + k^2 = 0$$

i.e. $$(3-k)(2-k) = 0.$$

So the eigenvalues are 3 and 2. Now find the eigenvectors.

$k = 2$: Solve $\begin{bmatrix} -1 & -2 \\ 1 & 2 \end{bmatrix} \begin{bmatrix} x \\ y \end{bmatrix} = \begin{bmatrix} 0 \\ 0 \end{bmatrix}$, giving $\begin{bmatrix} -2u \\ u \end{bmatrix}$ $(u \in \mathbb{R},\ u \neq 0)$.

$k = 3$: Solve $\begin{bmatrix} -2 & -2 \\ 1 & 1 \end{bmatrix} \begin{bmatrix} x \\ y \end{bmatrix} = \begin{bmatrix} 0 \\ 0 \end{bmatrix}$, giving $\begin{bmatrix} -u \\ u \end{bmatrix}$ $(u \in \mathbb{R},\ u \neq 0)$.

Consequently

$$\begin{bmatrix} x \\ y \end{bmatrix} = \begin{bmatrix} -2a \\ a \end{bmatrix} e^{2t} \qquad \text{and} \qquad \begin{bmatrix} x \\ y \end{bmatrix} = \begin{bmatrix} -b \\ b \end{bmatrix} e^{3t}$$

are solutions, for any real numbers a and b. These may be written

$$x = -2ae^{2t} \quad \text{and} \quad x = -be^{3t},$$
$$y = ae^{2t} \quad \text{and} \quad y = be^{3t}.$$

Substituting these, we obtain

$$\boldsymbol{q}ke^{kt} = A\boldsymbol{q}e^{kt},$$

giving $\quad \boldsymbol{q}k = A\boldsymbol{q},$

i.e. $\quad A\boldsymbol{q} = k\boldsymbol{q}.$

Thus $\boldsymbol{x} = \boldsymbol{q}e^{kt}$ is a solution to our matrix differential equation just when k is an eigenvalue of the matrix A and $\boldsymbol{q}$ is a corresponding eigenvector (or of course when $\boldsymbol{q} = \boldsymbol{0}$, which yields the trivial solution $\boldsymbol{x} = \boldsymbol{0}$). See Example 24.2.

Example 24.3 shows this technique in practice. Find the eigenvalues k_1 and k_2, and corresponding eigenvectors $\boldsymbol{q}_1$ and $\boldsymbol{q}_2$. Then

$$\boldsymbol{x} = \boldsymbol{q}_1 e^{k_1 t} \quad \text{and} \quad \boldsymbol{x} = \boldsymbol{q}_2 e^{k_2 t}$$

are solutions to the differential equations. We know from our previous work that the eigenvectors $\boldsymbol{q}_1$ and $\boldsymbol{q}_2$ have an element of arbitrariness about them. For any eigenvalue k, if $\boldsymbol{q}$ is an eigenvector corresponding to k, then any multiple of $\boldsymbol{q}$ is also an eigenvector corresponding to k. Thus arbitrary constants arise, as they customarily do in the process of solving differential equations, in the solutions found by this process.

24.4 Illustrations of the general solution to a set of differential equations.

(i) The general solution in Example 24.2 is

$$\begin{bmatrix} x \\ y \end{bmatrix} = a\begin{bmatrix} -2 \\ 1 \end{bmatrix} e^{7t} + b\begin{bmatrix} \frac{1}{2} \\ 1 \end{bmatrix} e^{2t}$$

or

$$x = -2ae^{7t} + \tfrac{1}{2}be^{2t},$$
$$y = \quad ae^{7t} + \ be^{2t}.$$

(ii) The general solution in Example 24.3 is

$$\begin{bmatrix} x \\ y \end{bmatrix} = a\begin{bmatrix} -2 \\ 1 \end{bmatrix} e^{2t} + b\begin{bmatrix} -1 \\ 1 \end{bmatrix} e^{3t},$$

or

$$x = -2ae^{2t} - be^{3t},$$
$$y = \quad ae^{2t} + be^{3t}.$$

24.5 Find the general solution to the set of simultaneous differential equations

$$\frac{dx}{dt} = 2x + 2y + 3z$$
$$\frac{dy}{dt} = \ x + 2y + \ z$$
$$\frac{dz}{dt} = 2x - 2y + \ z.$$

The answer is

$$\begin{bmatrix} x \\ y \\ z \end{bmatrix} = a\boldsymbol{q}_1 e^{k_1 t} + b\boldsymbol{q}_2 e^{k_2 t} + c\boldsymbol{q}_3 e^{k_3 t},$$

where a, b and c are arbitrary constants, k_1, k_2 and k_3 are the three eigenvalues and $\boldsymbol{q}_1$, $\boldsymbol{q}_2$ and $\boldsymbol{q}_3$ are corresponding eigenvectors of the matrix

$$A = \begin{bmatrix} 2 & 2 & 3 \\ 1 & 2 & 1 \\ 2 & -2 & 1 \end{bmatrix}.$$

These are found by the usual process. The work has been done in Example 19.1. Take

$$k_1 = -1, \qquad k_2 = 2, \qquad k_3 = 4,$$

and

$$\boldsymbol{q}_1 = \begin{bmatrix} 2 \\ 0 \\ -2 \end{bmatrix}, \quad \boldsymbol{q}_2 = \begin{bmatrix} 4 \\ 6 \\ -4 \end{bmatrix}, \quad \boldsymbol{q}_3 = \begin{bmatrix} 8 \\ 5 \\ 2 \end{bmatrix}.$$

The required general solution may then be written

$$\begin{bmatrix} x \\ y \\ z \end{bmatrix} = a\begin{bmatrix} 2 \\ 0 \\ -2 \end{bmatrix} e^{-t} + b\begin{bmatrix} 4 \\ 6 \\ -4 \end{bmatrix} e^{2t} + c\begin{bmatrix} 8 \\ 5 \\ 2 \end{bmatrix} e^{4t}.$$

It is not hard to show (an exercise for the reader) that if we have any two solutions $\boldsymbol{x}_1$ and $\boldsymbol{x}_2$ to the equation $\frac{d\boldsymbol{x}}{dt} = A\boldsymbol{x}$, then their sum $\boldsymbol{x}_1 + \boldsymbol{x}_2$ is also a solution. In fact it can be shown (beyond the scope of this book) that the most general form of solution is given by a sum of solutions as found above. This may be stated as a rule for writing down the general solution.

Rule

Let A be a 2×2 real matrix with distinct real eigenvalues k_1 and k_2. Let $\boldsymbol{q}_1$ and $\boldsymbol{q}_2$ be any chosen eigenvectors corresponding to k_1 and k_2 respectively. Then the general solution to the differential equation

$$\frac{d\boldsymbol{x}}{dt} = A\boldsymbol{x}$$

is $$\boldsymbol{x} = a\boldsymbol{q}_1 e^{k_1 t} + b\boldsymbol{q}_2 e^{k_2 t},$$

where a and b are arbitrary real constants.

See Examples 24.4.

These ideas apply also where there are more than two functions and equations. For example, we may have $\boldsymbol{x} = \begin{bmatrix} x \\ y \\ z \end{bmatrix}$ and A may be a 3×3 matrix. Then the equation

$$\frac{d\boldsymbol{x}}{dt} = A\boldsymbol{x}$$

represents a set of three linear differential equations. If k_1, k_2 and k_3 are distinct eigenvalues of A, with corresponding eigenvectors $\boldsymbol{q}_1$, $\boldsymbol{q}_2$ and $\boldsymbol{q}_3$, then the general solution is

$$\boldsymbol{x} = a\boldsymbol{q}_1 e^{k_1 t} + b\boldsymbol{q}_2 e^{k_2 t} + c\boldsymbol{q}_3 e^{k_3 t},$$

where a, b and c are arbitrary real constants. This process generalises to work for larger sets of equations, provided that the number of equations is the same as the number of unknown functions, so that the matrix of coefficients is square.

A calculation in the 3×3 case is given in Example 24.5.

24.6 Solve
$$\frac{dx}{dt} = 2x - y + 2z$$
$$\frac{dy}{dt} = -x + 2y - 2z$$
$$\frac{dz}{dt} = 2x - 2y + 5z.$$

The matrix of coefficients in this case is
$$\begin{bmatrix} 2 & -1 & 2 \\ -1 & 2 & -2 \\ 2 & -2 & 5 \end{bmatrix}.$$

In Example 22.3(i) we saw that the eigenvalues of this matrix are 1 (with multiplicity 2) and 7, and that corresponding eigenvectors are
$$\begin{bmatrix} u-2v \\ u \\ v \end{bmatrix} \qquad (u, v \in \mathbb{R}, \text{ not both } 0)$$
and
$$\begin{bmatrix} u \\ -u \\ 2u \end{bmatrix} \qquad (u \in \mathbb{R},\ u \neq 0).$$

These lead to solutions to the differential equations
$$\begin{bmatrix} x \\ y \\ z \end{bmatrix} = \begin{bmatrix} a-2b \\ a \\ b \end{bmatrix} e^{t} \qquad \text{and} \qquad \begin{bmatrix} x \\ y \\ z \end{bmatrix} = \begin{bmatrix} c \\ -c \\ 2c \end{bmatrix} e^{7t}.$$

The general solution (the sum of these) may be written
$$\begin{bmatrix} x \\ y \\ z \end{bmatrix} = a\begin{bmatrix} 1 \\ 1 \\ 0 \end{bmatrix} e^{t} + b\begin{bmatrix} -2 \\ 0 \\ 1 \end{bmatrix} e^{t} + c\begin{bmatrix} 1 \\ -1 \\ 2 \end{bmatrix} e^{7t}.$$

There are complications in this: the same sort of complications which were mentioned in relation to the diagonalisation process itself in Chapter 19. There may not be enough eigenvalues, or there may be repeated eigenvalues, or there may be complex eigenvalues. We can cope, however, with repeated eigenvalues, provided that there are enough linearly independent eigenvectors to make the matrix diagonalisable. Example 24.6 provides illustration. Each $e^{k_i t}$ occurs in the general solution multiplied by an eigenvector of the most general form possible, i.e. a linear combination of some basis for the eigenspace corresponding to the k_i.

24.7 A case where the eigenvalues and eigenvectors are complex numbers. See Example 18.6. Find the general solution to the system of differential equations

$$\frac{dx}{dt} = x + y$$

$$\frac{dy}{dt} = -x + y.$$

The eigenvalues of the matrix of coefficients are $1+i$ and $1-i$, with corresponding eigenvectors (say) $\begin{bmatrix} -i \\ 1 \end{bmatrix}$ and $\begin{bmatrix} i \\ 1 \end{bmatrix}$ respectively. The general solution is then

$$\begin{aligned}
\begin{bmatrix} x \\ y \end{bmatrix} &= 2\,\mathrm{Re}\left((a+ib)\begin{bmatrix} -i \\ 1 \end{bmatrix} e^{(1+i)t} \right) \\
&= 2\,\mathrm{Re}\left((a+ib)\begin{bmatrix} -i \\ 1 \end{bmatrix} e^{t} e^{it} \right) \\
&= 2e^{t}\,\mathrm{Re}\left((a+ib)\begin{bmatrix} -i \\ 1 \end{bmatrix} (\cos t + i \sin t) \right) \\
&= 2e^{t}\,\mathrm{Re}\left((a+ib)\begin{bmatrix} -i\cos t + \sin t \\ \cos t + i \sin t \end{bmatrix} \right) \\
&= 2e^{t}\,\mathrm{Re}\left(\begin{bmatrix} -ai\cos t + a\sin t + b\cos t + bi\sin t \\ a\cos t + ai\sin t + bi\cos t - b\sin t \end{bmatrix} \right) \\
&= 2e^{t}\begin{bmatrix} a\sin t + b\cos t \\ a\cos t - b\sin t \end{bmatrix},
\end{aligned}$$

i.e.

$$x = 2e^{t}(a \sin t + b \cos t)$$

$$y = 2e^{t}(a \cos t - b \sin t).$$

We can even cope with complex eigenvalues. We illustrate this with a 2×2 case. See Example 24.7. Notice that in that example, the eigenvalues are complex conjugates of each other. This reflects the fact that complex solutions to polynomial equations always come in conjugate pairs. The eigenvalues of a matrix are the roots of the characteristic equation, so complex eigenvalues will always come in conjugate pairs. The method given here will apply to such a pair of eigenvalues irrespective of the size of the matrix.

Consider $\dfrac{d\mathbf{x}}{dt} = A\mathbf{x}$, where A is a 2×2 matrix, and suppose that the eigenvalues of A are complex, say

$$k_1 = u + iv \qquad \text{and} \qquad k_2 = u - iv \qquad (u, v \in \mathbb{R}).$$

We saw in Chapter 18 that we can find complex eigenvectors corresponding to these eigenvalues by following the usual process. Indeed they will be complex conjugates of each other. Let $\mathbf{q}_1$ correspond with k_1. Then $\bar{\mathbf{q}}_1$ is an eigenvector corresponding to k_2.

$$\mathbf{x} = \mathbf{q}_1 e^{k_1 t} \qquad \text{and} \qquad \mathbf{x} = \bar{\mathbf{q}}_1 e^{k_2 t}$$

are (at least in formal terms) solutions. Indeed the general complex solution is

$$\mathbf{x} = G\mathbf{q}_1 e^{(u+iv)t} + H\bar{\mathbf{q}}_1 e^{(u-iv)t},$$

where, this time, we can think of G and H as arbitrary complex numbers. What we seek, of course, are real solutions. If we choose H to be the complex conjugate of G, then

$$G\mathbf{q}_1 e^{(u+iv)t} \qquad \text{and} \qquad H\bar{\mathbf{q}}_1 e^{(u-iv)t}$$

will be complex conjugates of each other, and their sum will be real:

$$\mathbf{x} = 2\,\mathrm{Re}(G\mathbf{q}_1 e^{(u+iv)t}).$$

It is not very helpful to try to simplify this formula in its general form. In any particular situation, the appropriate values can be inserted for $\mathbf{q}_1$, u and v, and G can be taken to be $a + ib$ (where a and b are real numbers), and then to carry out a simplification process starting from here, to find the real part, which turns out to be expressible in terms of trigonometric functions. This is done in Example 24.7.

24.8 Find the general solution to the system of differential equations

$$\frac{dx}{dt} = 2x - 2y + z$$

$$\frac{dy}{dt} = 2x + y - 2z$$

$$\frac{dz}{dt} = x + 2y + 2z.$$

By the usual process we find the eigenvalues and corresponding eigenvectors of the matrix of coefficients.

$k = 3$: Eigenvectors are multiples of $\begin{bmatrix} 1 \\ 0 \\ 1 \end{bmatrix}$.

$k = 1 + 2\sqrt{2}\,i$: Eigenvectors are multiples of $\begin{bmatrix} -1 \\ \sqrt{2}\,i \\ 1 \end{bmatrix}$.

$k = 1 - 2\sqrt{2}\,i$: Eigenvectors are multiples of $\begin{bmatrix} -1 \\ -\sqrt{2}\,i \\ 1 \end{bmatrix}$.

From the two complex eigenvalues we can obtain a real solution as in Example 24.7.

$$\begin{bmatrix} x \\ y \\ z \end{bmatrix} = 2\,\mathrm{Re}\left((a+ib) \begin{bmatrix} -1 \\ \sqrt{2}\,i \\ 1 \end{bmatrix} e^{(1+2\sqrt{2}\,i)t} \right)$$

$$= 2e^t\,\mathrm{Re}\left((a+ib) \begin{bmatrix} -1 \\ \sqrt{2}\,i \\ 1 \end{bmatrix} (\cos(2\sqrt{2}\,t) + i\sin(2\sqrt{2}\,t)) \right)$$

$$= 2e^t \begin{bmatrix} -a\cos(2\sqrt{2}\,t) + b\sin(2\sqrt{2}\,t) \\ -a\sqrt{2}\sin(2\sqrt{2}\,t) - b\sqrt{2}\cos(2\sqrt{2}\,t) \\ a\cos(2\sqrt{2}\,t) - b\sin(2\sqrt{2}\,t) \end{bmatrix}.$$

The complete general solution is then

$$\begin{bmatrix} x \\ y \\ z \end{bmatrix} = 2e^t \begin{bmatrix} -a\cos(2\sqrt{2}\,t) + b\sin(2\sqrt{2}\,t) \\ -a\sqrt{2}\sin(2\sqrt{2}\,t) - b\sqrt{2}\cos(2\sqrt{2}\,t) \\ a\cos(2\sqrt{2}\,t) - b\sin(2\sqrt{2}\,t) \end{bmatrix} + c \begin{bmatrix} 1 \\ 0 \\ 1 \end{bmatrix} e^{3t},$$

where a, b and c are arbitrary real constants. This may be written in a different form, which might be easier to comprehend:

$$x = -2ae^t\cos(2\sqrt{2}\,t) + 2be^t\sin(2\sqrt{2}\,t) + ce^{3t},$$

$$y = -2a\sqrt{2}\,e^t\sin(2\sqrt{2}\,t) - 2b\sqrt{2}\,e^t\cos(2\sqrt{2}\,t),$$

$$z = 2ae^t\cos(2\sqrt{2}\,t) - 2be^t\sin(2\sqrt{2}\,t) + ce^{3t}.$$

Example 24.8 shows how this method is carried out in a 3×3 case where there are two complex eigenvalues and one real eigenvalue.

Summary

A routine procedure is described and illustrated for finding the general solution to a system of simultaneous linear first-order differential equations. It uses the ideas of eigenvalue and eigenvector. The case where eigenvalues are complex numbers is discussed.

Exercises

1. Find the general solution to each of the following sets of simultaneous differential equations.

(i)
$$\frac{dx}{dt} = x - y, \qquad \frac{dy}{dt} = 2x + 4y$$

(ii)
$$\frac{dx}{dt} = x + 4y, \qquad \frac{dy}{dt} = x - 2y$$

(iii)
$$\frac{dx}{dt} = x + y, \qquad \frac{dy}{dt} = x + y$$

(iv)
$$\frac{dx}{dt} = -x, \qquad \frac{dy}{dt} = -y$$

(v)
$$\frac{dx}{dt} = x - y - z, \qquad \frac{dy}{dt} = 3y + 2z, \qquad \frac{dz}{dt} = -y$$

(vi)
$$\frac{dx}{dt} = 2x - 2y + 3z, \qquad \frac{dy}{dt} = 3y \quad 2z, \qquad \frac{dz}{dt} = -y + 2z$$

(vii)
$$\frac{dx}{dt} = x + y - 2z, \qquad \frac{dy}{dt} = 4x + 4z, \qquad \frac{dz}{dt} = x - y + 4z$$

(viii)
$$\frac{dx}{dt} = 2x - y + 2z, \qquad \frac{dy}{dt} = -x + 2y - 2z, \qquad \frac{dz}{dt} = 2x - 2y + 5z$$

(ix)
$$\frac{dx}{dt} = \sqrt{3}x + z, \qquad \frac{dy}{dt} = y, \qquad \frac{dz}{dt} = -x + \sqrt{3}z$$

(x)
$$\frac{dx}{dt} = -y + z, \qquad \frac{dy}{dt} = y, \qquad \frac{dz}{dt} = -x + 2z$$

ANSWERS TO EXERCISES

Chapter 1

(i) $x=1$, $y=-1$. (ii) $x=2$, $y=-3$.
(iii) $x_1=3$, $x_2=-2$, $x_3=1$. (iv) $x_1=2$, $x_2=1$, $x_3=-1$.
(v) $x_1=2$, $x_2=1$, $x_3=2$. (vi) $x_1=1$, $x_2=0$, $x_3=1$.
(vii) $x_1=1$, $x_2=-1$, $x_3=3$. (viii) $x_1=0$, $x_2=0$, $x_3=0$.
(ix) $x_1=3$, $x_2=1$, $x_3=0$. (x) $x_1=-1$, $x_2=0$, $x_3=0$.
(xi) $x_1=2$, $x_2=1$, $x_3=0$, $x_4=-1$.
(xii) $x_1=2$, $x_2=2$, $x_3=1$, $x_4=-3$.

Chapter 2

2. (i) $x=2+3t$, $y=t$ $(t\in\mathbb{R})$.
(ii) $x=-\frac{1}{2}-\frac{3}{2}t$, $y=t$ $(t\in\mathbb{R})$.
(iii) $x_1=4-3t$, $x_2=1+2t$, $x_3=t$ $(t\in\mathbb{R})$.
(iv) $x_1=1-2t$, $x_2=-1+t$, $x_3=t$ $(t\in\mathbb{R})$.
(v) $x_1=1+t$, $x_2=t$, $x_3=1$ $(t\in\mathbb{R})$.
(vi) $x_1=-2-3t$, $x_2=2+t$, $x_3=t$ $(t\in\mathbb{R})$.

3. (i) $x=2$, $y=1$. (ii) $x=3$, $y=2$.
(iii) $x_1=-1$, $x_2=1$, $x_3=2$.
(iv) $x_1=2$, $x_2=-1$, $x_3=0$.
(v) $x_1=-2$, $x_2=1$, $x_3=1$.
(vi) $x_1=4$, $x_2=3$, $x_3=2$, $x_4=1$.

4. (i) Unique solution. (ii) Infinitely many solutions.
(iii) Inconsistent. (iv) Inconsistent.
(v) Infinitely many solutions.

5. (i) $c=1$: infinitely many solutions. $c\neq 1$: inconsistent.
(ii) Consistent only for $k=6$.

6. (i) $c=-2$: no solution. $c=2$: infinitely many solutions. Otherwise: unique solution.
(ii) $c=\pm\sqrt{3}$: no solution. Otherwise: unique solution.

(iii) $c=-3$: no solution. $c=3$: infinitely many solutions. Otherwise: unique solution.

(iv) $c=0$, $c=\sqrt{6}$, $c=-\sqrt{6}$: infinitely many solutions. Otherwise: unique solution.

Chapter 3

2. (i) $\begin{bmatrix}-1\\-2\\2\end{bmatrix}$. (ii) $\begin{bmatrix}4\\-1\\3\end{bmatrix}$. (iii) $\begin{bmatrix}-3\\-3\end{bmatrix}$. (iv) $\begin{bmatrix}4\\-1\\3\\3\\0\end{bmatrix}$.

(v) $\begin{bmatrix}15\\15\\15\end{bmatrix}$. (vi) $\begin{bmatrix}-2\\1\\3\end{bmatrix}$.

3. $AB=\begin{bmatrix}19 & -1 & 6 & 13\\16 & -8 & -8 & 6\end{bmatrix}$, $AD=\begin{bmatrix}21 & -7\\13 & -16\end{bmatrix}$,

$BC=\begin{bmatrix}0 & 3 & 7\\8 & -4 & 6\\9 & 3 & 3\end{bmatrix}$, $CB=\begin{bmatrix}8 & -1 & -1 & 6\\4 & -8 & -4 & -6\\9 & 0 & -3 & 9\\4 & 0 & 4 & 2\end{bmatrix}$,

$CD=\begin{bmatrix}8 & -5\\1 & -8\\9 & -6\\5 & 0\end{bmatrix}$.

DA also exists. $A(BC)$ of course should be the same as $(AB)C$.

4. (i) $\begin{bmatrix}11 & 10\\2 & -2\\19 & 16\end{bmatrix}$. (ii) $\begin{bmatrix}3 & 4 & 2\\-2 & -2 & 10\\7 & 4 & 3\\8 & 8 & 4\end{bmatrix}$.

(iii) $\begin{bmatrix}3 & 4\\1 & 2\end{bmatrix}$. (iv) $\begin{bmatrix}2\\-6\end{bmatrix}$.

5. $\begin{bmatrix}8 & 8 & 13\\8 & 5 & 8\\13 & 8 & 8\end{bmatrix}$.

6. A must be $p\times q$ and B must be $q\times p$, for some numbers p and q.

Chapter 4

1. $A^2=\begin{bmatrix}1 & 2 & 3\\0 & 1 & 2\\0 & 0 & 1\end{bmatrix}$, $A^3=\begin{bmatrix}1 & 3 & 6\\0 & 1 & 3\\0 & 0 & 1\end{bmatrix}$,

$$A^4=\begin{bmatrix}1&4&10\\0&1&4\\0&0&1\end{bmatrix},\quad B^2=\begin{bmatrix}1&0&0\\2&1&0\\3&2&1\end{bmatrix},$$

$$B^3=\begin{bmatrix}1&0&0\\3&1&0\\6&3&1\end{bmatrix},\quad B^4=\begin{bmatrix}1&0&0\\4&1&0\\10&4&1\end{bmatrix}.$$

3. (i) $\begin{bmatrix}0&-1&5\\0&4&-3\\0&0&-1\end{bmatrix}$. (ii) $\begin{bmatrix}x_1+2x_2+x_3\\x_2-2x_3\\x_3\end{bmatrix}$.

(iii) $\begin{bmatrix}-1&0&0\\-3&-2&0\\9&5&3\end{bmatrix}$. (iv) $[6\quad 5\quad 3]$.

4. $x_1=2, x_2=6, x_3=1$.

6. Symmetric:

$$\begin{bmatrix}1&2\\2&3\end{bmatrix},\ \begin{bmatrix}1&0\\0&-1\end{bmatrix},\ \begin{bmatrix}2&3&1\\3&0&-1\\1&-1&2\end{bmatrix},$$

$$\begin{bmatrix}1&0&1\\0&1&0\\1&0&1\end{bmatrix},\ \begin{bmatrix}1&1&0\\1&0&-1\\0&-1&-1\end{bmatrix}.$$

Skew-symmetric:

$$\begin{bmatrix}0&2\\-2&0\end{bmatrix},\ \begin{bmatrix}0&1&-2\\-1&0&3\\2&-3&0\end{bmatrix}.$$

9. (i) Interchange rows 2 and 3.
(ii) Add twice row 3 to row 1.
(iii) Subtract three times row 1 from row 3.
(iv) Multiply row 2 by -2.

10. $$T=\begin{bmatrix}1&0&0\\0&1&0\\0&0&\frac{1}{6}\end{bmatrix}\begin{bmatrix}1&0&0\\0&1&0\\0&1&1\end{bmatrix}\begin{bmatrix}1&0&0\\0&1&0\\-2&0&1\end{bmatrix}\begin{bmatrix}0&1&0\\1&0&0\\0&0&1\end{bmatrix},$$

and

$$TA=\begin{bmatrix}1&2&-1\\0&1&3\\0&0&1\end{bmatrix}.$$

11. $$T=\begin{bmatrix}1&0&0\\0&1&0\\0&0&-\frac{1}{6}\end{bmatrix}\begin{bmatrix}1&0&0\\0&1&0\\0&-3&1\end{bmatrix}\begin{bmatrix}1&0&0\\0&\frac{1}{2}&0\\0&0&1\end{bmatrix}$$

$$\times\begin{bmatrix}1&0&0\\0&1&0\\-2&0&1\end{bmatrix}\begin{bmatrix}1&0&0\\1&1&0\\0&0&1\end{bmatrix}.$$

and

$$TA=\begin{bmatrix}1&-1&2&1\\0&1&1&1\\0&0&1&1\end{bmatrix}.$$

Chapter 5

1. Invertible: (i), (ii), (iv), (v), (vi), (viii), (xi).

2. $\begin{bmatrix} 1/a & 0 & 0 \\ 0 & 1/b & 0 \\ 0 & 0 & 1/c \end{bmatrix}$.

Chapter 6

1. $x=-1$, $y=2$.

2. (i) Yes. (ii) No. (iii) Yes. (iv) No.

3. In each case we give the coefficients in a linear combination which is equal to $\mathbf{0}$ (the vectors taken in the order given).
(i) 2, −1. (ii) 6, −7, 2. (iii) 4, −1, 3.
(iv) −11, 7, 16. (v) 5, −4, 3. (vi) 1, −13, 5.

4. LD: (ii), (iii), (vi).
LI: (i), (iv), (v), (vii).

5. 1, 2, 2, 1, 2, 2, 3, 3, 2, 1, 1, 2, 1, 2, 3, 4, 4, 2, 3.

6. Rank of $\mathbf{x}\mathbf{y}^{\mathrm{T}}$ is 1. This holds for any p.

Chapter 7

1. −7, −2, 3, 0, −28, −7, 7, −3.

2. 2, −4, −9, 0, −10, 10, 0, 0, −3.

3. −16, −35.

4. No, only for $p \times p$ skew-symmetric matrices with *odd* p.

8. $\det A = 2$. $\operatorname{adj} A = \begin{bmatrix} -1 & 1 & 1 \\ 1 & -1 & 1 \\ 1 & 1 & -1 \end{bmatrix}$.

Chapter 8

(i) Ranks 2, 2; unique solution.
(ii) Ranks 1, 2; inconsistent.
(iii) Ranks 1, 1; infinitely many solutions.
(iv) Ranks 2, 2; unique solution.
(v) Ranks 3, 3; unique solution.
(vi) Ranks 3, 3; unique solution.
(vii) Ranks 2, 3; inconsistent.
(viii) Ranks 2, 2; infinitely many solutions.
(ix) Ranks 2, 2; infinitely many solutions.
(x) Ranks 2, 2; infinitely many solutions.
(xi) Ranks 2, 2; infinitely many solutions.
(xii) Ranks 3, 3; unique solution.

Chapter 9

1. C: $\boldsymbol{b}-\boldsymbol{a}$, D: $-\boldsymbol{a}$, E: $-\boldsymbol{b}$, F: $\boldsymbol{a}-\boldsymbol{b}$.

5. (i) $\begin{bmatrix} 2 \\ -1 \\ 3 \end{bmatrix}$. (ii) $\begin{bmatrix} -2 \\ 1 \\ -3 \end{bmatrix}$. (iii) $\begin{bmatrix} -2 \\ -2 \\ -2 \end{bmatrix}$. (iv) $\begin{bmatrix} 0 \\ -2 \\ 1 \end{bmatrix}$.

(v) $\begin{bmatrix} 1 \\ 0 \\ -1 \end{bmatrix}$.

6. (i) $\begin{bmatrix} 0 \\ 1 \\ 4 \end{bmatrix}$. (ii) $\begin{bmatrix} 8 \\ 5 \\ 3 \end{bmatrix}$. (iii) $\begin{bmatrix} 6 \\ 6 \\ 6 \end{bmatrix}$. (iv) $\begin{bmatrix} 12 \\ -10 \\ 16 \end{bmatrix}$.

(v) $\begin{bmatrix} 10 \\ 5 \\ -5 \end{bmatrix}$.

9. (i) $\begin{bmatrix} \frac{1}{\sqrt{2}} \\ 0 \\ -\frac{1}{\sqrt{2}} \end{bmatrix}$. (ii) $\begin{bmatrix} \frac{2}{3} \\ \frac{2}{3} \\ \frac{1}{3} \end{bmatrix}$. (iii) $\begin{bmatrix} \frac{1}{\sqrt{6}} \\ -\frac{2}{\sqrt{6}} \\ -\frac{1}{\sqrt{6}} \end{bmatrix}$. (iv) $\begin{bmatrix} \frac{1}{\sqrt{3}} \\ \frac{1}{\sqrt{3}} \\ \frac{1}{\sqrt{3}} \end{bmatrix}$.

Chapter 10

1. (i) $x=t,\ y=1-t,\ z=3-2t \quad (t\in\mathbb{R})$.
(ii) $x=1,\ y=1+t,\ z=-2+2t \quad (t\in\mathbb{R})$.
(iii) $x=-1,\ y=2,\ z=4-11t \quad (t\in\mathbb{R})$.
(iv) $x=1+t,\ y=1+t,\ z=1+t \quad (t\in\mathbb{R})$.
(v) $x=3t,\ y=-t,\ z=2t \quad (t\in\mathbb{R})$.

2. (i) $\begin{bmatrix} -1 \\ 2 \\ -5 \end{bmatrix}$. (ii) $\begin{bmatrix} 2 \\ -1 \\ 1 \end{bmatrix}$. (iii) $\begin{bmatrix} -3 \\ 0 \\ -1 \end{bmatrix}$.

3. (i) Intersect at (0, 1, 1).
(ii) Intersect at (0, 3, 1).
(iii) Do not intersect.
(iv) These two sets of equations represent the same line.

4. (i) $\frac{2}{\sqrt{6}}$. (ii) $-\frac{1}{6}$. (iii) $\frac{2}{\sqrt{42}}$. (iv) $-\frac{2}{\sqrt{6}}$.

5. $\cos\hat{A}=-\frac{1}{\sqrt{2}}$, $\cos\hat{B}=\frac{5}{\sqrt{34}}$, $\cos\hat{C}=\frac{4}{\sqrt{17}}$.
The largest angle is $\hat{A}$, which is $3\pi/4$ radians.

6. For example,

$$\begin{bmatrix} 1 \\ 2 \\ 3 \end{bmatrix}.$$

7. 7 units.

8. (i) $\sqrt{2}$. (ii) $\sqrt{24}$. (iii) 0 (the point lies on the line).

9. (i) $x-y+2z-3=0$. (ii) $4x+5y+6z+1=0$.
(iii) $y+3z-18=0$. (iv) $x+y+z-3=0$.

10. (i) 0 (the planes are perpendicular).
(ii) $\dfrac{4}{\sqrt{180}}$. (iii) $\dfrac{1}{2}$. (iv) $\dfrac{2}{\sqrt{14}}$.

11. (i) Straight line. (ii) Single point.
(iii) Empty. (iv) Single point.

12. (i) 0. (ii) $\dfrac{6}{\sqrt{14}}$. (iii) $\sqrt{3}$. (iv) $\dfrac{6}{\sqrt{42}}$. (v) $\dfrac{1}{\sqrt{42}}$.

Chapter 11

1. (i) $\begin{bmatrix} -3 \\ 1 \\ 4 \end{bmatrix}$. (ii) $\begin{bmatrix} 3 \\ -7 \\ 6 \end{bmatrix}$. (iii) $\begin{bmatrix} -4 \\ -6 \\ -3 \end{bmatrix}$.

(iv) $\begin{bmatrix} 3 \\ -1 \\ -4 \end{bmatrix}$. (v) $\begin{bmatrix} 12 \\ -4 \\ -16 \end{bmatrix}$. (vi) $\begin{bmatrix} 3 \\ 0 \\ -3 \end{bmatrix}$.

2. (i) $\frac{1}{2}\sqrt{26}$. (ii) $\frac{1}{2}\sqrt{94}$. (iii) $\frac{1}{2}\sqrt{61}$.
(iv) $\frac{1}{2}\sqrt{26}$. (v) $\frac{1}{2}\sqrt{416}$. (vi) $\frac{1}{2}\sqrt{18}$.

3. (i) $\frac{1}{2}\sqrt{35}$. (ii) $\frac{1}{2}\sqrt{66}$.

4. (i) $\begin{bmatrix} -2 \\ 7 \\ 3 \end{bmatrix}$. (ii) $\begin{bmatrix} 1 \\ -3 \\ -10 \end{bmatrix}$.

5. (i) $3y-z-2=0$. (ii) $x+y+z-2=0$.
(iii) $x-y+z=0$.

6. 3 units^3.

7. 4 units^3.

8. (i), (ii) and (iv) are coplanar. The others are not.

Chapter 12

1. (i) 2, 1, 1. (ii) No solution. (iii) $t, 1-2t, t$. (iv) $-1, 2, 1, -1$.
3. (i) 3, yes. (ii) 3, yes. (iii) 2, no. (iv) 3, yes.
4. (i) LD. (ii) LD. (iii) LI. (iv) LI. (v) LD. (vi) LD.

Chapter 13

1. (i) Yes. (ii) Yes. (iii) No. (iv) No. (v) Yes. (vi) Yes. (vii) No.
2. (i), (iii).
3. (i), (ii), (iii).
7. (ii), (iv).
8. (i) Plane, $\begin{bmatrix} 1 & -1 & -1 \\ 0 & 0 & 0 \\ 0 & 0 & 0 \end{bmatrix}$. (ii) Plane, $\begin{bmatrix} 1 & 0 & 0 \\ 0 & 0 & 0 \\ 0 & 0 & 0 \end{bmatrix}$.

(iii) Plane, $\begin{bmatrix} 2 & -1 & 3 \\ 0 & 0 & 0 \\ 0 & 0 & 0 \end{bmatrix}$. (iv) Line, $\begin{bmatrix} 2 & -1 & -1 \\ 1 & -1 & -2 \\ 0 & 0 & 0 \end{bmatrix}$.

Chapter 14

2. $\begin{bmatrix} 0 \\ 1 \\ 0 \end{bmatrix}, \begin{bmatrix} 3 \\ 2 \\ 1 \end{bmatrix}, \begin{bmatrix} 5 \\ 3 \\ -1 \end{bmatrix}$.
For the other two the coefficients are 1, 1 and $\frac{3}{2}, -\frac{1}{2}$.

3. (i) $\left(\begin{bmatrix} 2 \\ 1 \end{bmatrix}, \begin{bmatrix} 1 \\ 1 \end{bmatrix}\right)$. (ii) $\left(\begin{bmatrix} 1 \\ 1 \end{bmatrix}\right)$. (iii) $\left(\begin{bmatrix} 2 \\ 1 \\ 5 \end{bmatrix}, \begin{bmatrix} -1 \\ 2 \\ 0 \end{bmatrix}\right)$.

(iv) $\left(\begin{bmatrix} 3 \\ 1 \\ 0 \end{bmatrix}, \begin{bmatrix} 1 \\ -3 \\ 2 \end{bmatrix}\right)$. (v) $\left(\begin{bmatrix} 1 \\ 1 \\ 0 \end{bmatrix}, \begin{bmatrix} 1 \\ 0 \\ 1 \end{bmatrix}, \begin{bmatrix} 0 \\ 1 \\ 1 \end{bmatrix}\right)$.

5. (i) $\left(\begin{bmatrix} -2 \\ 3 \end{bmatrix}, \begin{bmatrix} 1 \\ 2 \end{bmatrix}\right)$. (ii) $\left(\begin{bmatrix} 1 \\ 1 \end{bmatrix}, \begin{bmatrix} 1 \\ -1 \end{bmatrix}\right)$. (iii) $\left(\begin{bmatrix} 2 \\ -2 \\ 2 \end{bmatrix}, \begin{bmatrix} -3 \\ 3 \\ 3 \end{bmatrix}, \begin{bmatrix} 1 \\ -1 \\ 2 \end{bmatrix}\right)$.

(iv) $\left(\begin{bmatrix} 0 \\ 1 \\ -1 \end{bmatrix}, \begin{bmatrix} -1 \\ 1 \\ 0 \end{bmatrix}, \begin{bmatrix} 1 \\ 2 \\ -1 \end{bmatrix}\right)$. (v) $\left(\begin{bmatrix} 1 \\ -1 \\ 2 \\ 1 \end{bmatrix}, \begin{bmatrix} 0 \\ 1 \\ 1 \\ 3 \end{bmatrix}, \begin{bmatrix} -1 \\ 0 \\ 1 \\ 1 \end{bmatrix}\right)$.

6. Coefficients: (i) $-\frac{1}{2}, \frac{1}{2}, \frac{1}{2}$. (ii) $\frac{1}{4}, -\frac{1}{2}, \frac{3}{4}$. (iii) $\frac{2}{9}, \frac{2}{9}, -\frac{1}{9}$. (iv) $\frac{1}{2}, -\frac{1}{2}, 0$.

8. (i) $\left(\begin{bmatrix} 1 \\ 1 \\ 0 \end{bmatrix}, \begin{bmatrix} 2 \\ 1 \\ 1 \end{bmatrix}\right)$. (ii) $\left(\begin{bmatrix} 0 \\ 1 \\ 0 \end{bmatrix}, \begin{bmatrix} 0 \\ 0 \\ 1 \end{bmatrix}\right)$. (iii) $\left(\begin{bmatrix} 1 \\ 2 \\ 0 \end{bmatrix}, \begin{bmatrix} 0 \\ 3 \\ 1 \end{bmatrix}\right)$. (iv) $\left(\begin{bmatrix} 1 \\ 3 \\ -1 \end{bmatrix}\right)$.

Chapter 15

3. (i) LI. (ii) LD. (iii) LD. (iv) LI.
4. (i) 2. (ii) 3. (iii) 2. (iv) 3. (v) 2. (vi) 3.

Chapter 16

1. (iii) $A = \begin{bmatrix} 1 & 0 \\ 1 & 1 \\ -1 & 2 \end{bmatrix}$.

2. $\begin{bmatrix} 1 & 1 & 1 & 0 \\ 0 & 1 & 1 & 1 \\ -1 & 0 & 1 & 1 \end{bmatrix}$.

3. f is not one-to-one. im f is not equal to $\mathbb{R}^3$.
4. (i) Yes. (ii) No. (iii) Yes. (iv) No. (v) No. (vi) Yes. (vii) No. (viii) No.

Chapter 17

1. (i) 3, -1. (ii) -2, 3. (iii) -2, 3, 1. (iv) 2, -1, 1.

2. $\begin{bmatrix} 1 & 2 \\ 2 & 1 \end{bmatrix}$.

3. $\begin{bmatrix} 1 & 2 & 1 \\ 2 & 5 & 1 \\ 1 & 0 & 4 \end{bmatrix}$.

4. $\begin{bmatrix} 1 & 1 & 0 \\ 0 & 2 & 1 \\ 0 & 1 & 0 \end{bmatrix}$.

5. $\begin{bmatrix} 1 & 2 & 3 \\ 0 & 1 & 2 \\ 1 & 0 & 3 \end{bmatrix}$.

6. $\begin{bmatrix} 1 \\ -1 \\ 2 \end{bmatrix}, \begin{bmatrix} 3 \\ -2 \\ 4 \end{bmatrix}, \begin{bmatrix} -1 \\ 2 \\ -3 \end{bmatrix}$.

Chapter 18

1. (i) 3, $\begin{bmatrix} 2t \\ t \end{bmatrix}$; -1, $\begin{bmatrix} -2t \\ t \end{bmatrix}$. (ii) 0, $\begin{bmatrix} t \\ -t \end{bmatrix}$; 2, $\begin{bmatrix} t \\ t \end{bmatrix}$.

(iii) 3, $\begin{bmatrix} t \\ -2t \end{bmatrix}$; 2, $\begin{bmatrix} -t \\ t \end{bmatrix}$. (iv) 2, $\begin{bmatrix} 4t \\ t \end{bmatrix}$; -3, $\begin{bmatrix} -t \\ t \end{bmatrix}$.

(v) $1, \begin{bmatrix} t \\ 0 \\ 0 \end{bmatrix}$; $-1, \begin{bmatrix} t \\ -2t \\ 4t \end{bmatrix}$; $4, \begin{bmatrix} -t \\ 2t \\ t \end{bmatrix}$. (vi) $1, \begin{bmatrix} 2t \\ 2t \\ 3t \end{bmatrix}$; $-1, \begin{bmatrix} 0 \\ t \\ t \end{bmatrix}$; $0, \begin{bmatrix} t \\ 2t \\ -2t \end{bmatrix}$.

(vii) $3, \begin{bmatrix} t \\ 0 \\ 0 \end{bmatrix}$; $-3, \begin{bmatrix} t \\ -6t \\ 0 \end{bmatrix}$; $1, \begin{bmatrix} t \\ -2t \\ 8t \end{bmatrix}$. (viii) $3, \begin{bmatrix} 0 \\ 0 \\ t \end{bmatrix}$; $1, \begin{bmatrix} 0 \\ t \\ -t \end{bmatrix}$; $-1, \begin{bmatrix} 4t \\ -4t \\ t \end{bmatrix}$.

2. (i) $1, \begin{bmatrix} 0 \\ t \end{bmatrix}$. (ii) $-1, \begin{bmatrix} -2t-3u \\ t \\ u \end{bmatrix}$. (iii) $1, \begin{bmatrix} t \\ t \end{bmatrix}$.

(iv) $-3, \begin{bmatrix} -2t-2u \\ t \\ u \end{bmatrix}$. (v) $2, \begin{bmatrix} 2t+u-v \\ t \\ u \\ v \end{bmatrix}$.

3. (i) $1, \begin{bmatrix} -t \\ 2t \\ t \end{bmatrix}$. (ii) $3, \begin{bmatrix} t \\ 0 \\ t \end{bmatrix}$. (iii) $-1, \begin{bmatrix} t \\ u \end{bmatrix}$. (iv) None.

4. (i) $i, \begin{bmatrix} -it \\ t \end{bmatrix}$; $-i, \begin{bmatrix} it \\ t \end{bmatrix}$. (ii) $1+2i\sqrt{2}, \begin{bmatrix} -t \\ i\sqrt{2}\,t \\ t \end{bmatrix}$; $1-2i\sqrt{2}, \begin{bmatrix} -t \\ -i\sqrt{2}\,t \\ t \end{bmatrix}$.

(iv) $i\sqrt{6}, \begin{bmatrix} 3t \\ i\sqrt{6}\,t \end{bmatrix}$; $-i\sqrt{6}, \begin{bmatrix} 3t \\ -i\sqrt{6}\,t \end{bmatrix}$.

5. $(a+d)^2 \geqslant 4(bc-ad)^2$.

Chapter 19

1. (i) $\begin{bmatrix} 2 & -2 \\ 1 & 1 \end{bmatrix}$. (ii) $\begin{bmatrix} 1 & 1 \\ -1 & 1 \end{bmatrix}$. (iii) $\begin{bmatrix} 1 & -1 \\ -2 & 1 \end{bmatrix}$. (iv) $\begin{bmatrix} 4 & -1 \\ 1 & 1 \end{bmatrix}$.

(v) $\begin{bmatrix} 1 & 1 & -1 \\ 0 & -2 & 2 \\ 0 & 4 & 1 \end{bmatrix}$. (vi) $\begin{bmatrix} 2 & 0 & 1 \\ 2 & 1 & 2 \\ 3 & 1 & -2 \end{bmatrix}$. (vii) $\begin{bmatrix} 1 & 1 & 1 \\ 0 & -6 & -2 \\ 0 & 0 & 8 \end{bmatrix}$.

(viii) $\begin{bmatrix} 0 & 0 & 4 \\ 0 & 1 & -4 \\ 1 & -1 & 1 \end{bmatrix}$.

2. (iii) $\begin{bmatrix} -2 & -3 \\ 1 & 1 \end{bmatrix}$. (vii) $\begin{bmatrix} 1 & 1 & 2 \\ 1 & 1 & 5 \\ -1 & 1 & -4 \end{bmatrix}$. (viii) $\begin{bmatrix} 0 & 1 & 2 \\ 1 & 1 & -1 \\ -1 & 1 & -1 \end{bmatrix}$.

(ix) $\begin{bmatrix} 2 & 0 & 0 \\ -2 & 1 & 0 \\ 1 & -2 & 1 \end{bmatrix}$. (x) $\begin{bmatrix} -1 & 0 & 1 \\ 3 & 2 & 0 \\ 1 & 1 & -1 \end{bmatrix}$.

Chapter 20

1. (i) 5. (ii) -1. (iii) 0. (iv) 0. (v) 0. (vi) 1. (vii) 14. (viii) -7. (ix) -1. (x) 3.

2. $\sqrt{2}$, $\sqrt{10}$, 5, $\sqrt{3}$, $\sqrt{14}$, 5, $\sqrt{6}$, 9, 2, $\sqrt{10}$, $\sqrt{15}$, $\sqrt{43}$.

3. $\begin{bmatrix} 1/\sqrt{3} \\ 1/\sqrt{3} \\ 1/\sqrt{3} \end{bmatrix}, \begin{bmatrix} 1/\sqrt{2} \\ 0 \\ -1/\sqrt{2} \end{bmatrix}, \begin{bmatrix} 1/\sqrt{6} \\ -2/\sqrt{6} \\ 1/\sqrt{6} \end{bmatrix}.$

4. (i) $\frac{3}{5}$. (ii) $6/\sqrt{13}$. (iii) $1/\sqrt{12}$. (iv) $-1/\sqrt{14}$. (v) $1/\sqrt{2}$. (vi) $-5/\sqrt{99}$. (vii) 0. (viii) $\frac{1}{5}$.

7. $\begin{bmatrix} 1 \\ 0 \\ -1 \end{bmatrix}, \begin{bmatrix} 0 \\ 1 \\ 0 \end{bmatrix}, \begin{bmatrix} 1 \\ 0 \\ 1 \end{bmatrix}.$

8. $\begin{bmatrix} 0 \\ 1 \\ 0 \end{bmatrix}, \begin{bmatrix} 3 \\ 0 \\ -2 \end{bmatrix}, \begin{bmatrix} 1 \\ 0 \\ 1 \end{bmatrix}$, no, no.

Chapter 21

3. (i) $\begin{bmatrix} 4/\sqrt{18} \\ -1/\sqrt{18} \\ 1/\sqrt{18} \end{bmatrix}$. (ii) $\begin{bmatrix} 0 \\ 1/\sqrt{2} \\ 1/\sqrt{2} \end{bmatrix}$. (iii) $\begin{bmatrix} 1/\sqrt{2} \\ -1/\sqrt{2} \\ 0 \end{bmatrix}$. (iv) $\begin{bmatrix} -1/\sqrt{14} \\ 2/\sqrt{14} \\ 3/\sqrt{14} \end{bmatrix}$.

4. (i) $\begin{bmatrix} 1/\sqrt{3} \\ 1/\sqrt{3} \\ 1/\sqrt{3} \end{bmatrix}, \begin{bmatrix} -2/\sqrt{6} \\ 1/\sqrt{6} \\ 1/\sqrt{6} \end{bmatrix}, \begin{bmatrix} 0 \\ 1/\sqrt{2} \\ -1/\sqrt{2} \end{bmatrix}$. (ii) $\begin{bmatrix} 1/\sqrt{2} \\ 1/\sqrt{2} \\ 0 \end{bmatrix}, \begin{bmatrix} 1/\sqrt{6} \\ -1/\sqrt{6} \\ 2/\sqrt{6} \end{bmatrix}, \begin{bmatrix} -1/\sqrt{3} \\ 1/\sqrt{3} \\ 1/\sqrt{3} \end{bmatrix}$.

(iii) $\begin{bmatrix} -2/\sqrt{5} \\ 0 \\ 1/\sqrt{5} \end{bmatrix}, \begin{bmatrix} -1/\sqrt{6} \\ 1/\sqrt{6} \\ 2/\sqrt{6} \end{bmatrix}, \begin{bmatrix} 1/\sqrt{14} \\ -3/\sqrt{14} \\ 2/\sqrt{14} \end{bmatrix}$.

(iv) $\begin{bmatrix} 1/\sqrt{11} \\ 3/\sqrt{11} \\ 1/\sqrt{11} \end{bmatrix}, \begin{bmatrix} 3/\sqrt{22} \\ -2/\sqrt{22} \\ 3/\sqrt{22} \end{bmatrix}, \begin{bmatrix} 1/\sqrt{2} \\ 0 \\ -1/\sqrt{2} \end{bmatrix}$.

5. $\begin{bmatrix} \frac{1}{3} \\ \frac{2}{3} \\ \frac{2}{3} \end{bmatrix}, \begin{bmatrix} \frac{2}{3} \\ \frac{1}{3} \\ -\frac{2}{3} \end{bmatrix}, \begin{bmatrix} \frac{2}{3} \\ -\frac{2}{3} \\ \frac{1}{3} \end{bmatrix}; \begin{bmatrix} 0 \\ \frac{3}{5} \\ \frac{4}{5} \end{bmatrix}, \begin{bmatrix} 0 \\ -\frac{4}{5} \\ \frac{3}{5} \end{bmatrix}, \begin{bmatrix} 1 \\ 0 \\ 0 \end{bmatrix}.$

6. $\begin{bmatrix} 1 \\ 0 \\ 0 \\ 0 \end{bmatrix}, \begin{bmatrix} 0 \\ 1/\sqrt{2} \\ 1/\sqrt{2} \\ 0 \end{bmatrix}, \begin{bmatrix} 0 \\ 1/\sqrt{3} \\ -1/\sqrt{3} \\ 1/\sqrt{3} \end{bmatrix}.$

7. Hint: observe that $\left(\begin{bmatrix}1\\0\\1\\0\end{bmatrix}, \begin{bmatrix}0\\1\\0\\1\end{bmatrix}, \begin{bmatrix}1\\2\\2\\1\end{bmatrix}\right)$ spans the same space.

9. $1, \begin{bmatrix}-t-u\\t\\u\end{bmatrix}$ (eigenspace is plane of reflection).

$-1, \begin{bmatrix}t\\t\\t\end{bmatrix}$ (eigenspace is line normal to the plane).

Chapter 22

1. (i) $\begin{bmatrix}0\\1/\sqrt{2}\\-1/\sqrt{2}\end{bmatrix}, \begin{bmatrix}1/\sqrt{3}\\1/\sqrt{3}\\1/\sqrt{3}\end{bmatrix}, \begin{bmatrix}2/\sqrt{6}\\-1/\sqrt{6}\\-1/\sqrt{6}\end{bmatrix}$. (ii) $\begin{bmatrix}1/\sqrt{2}\\0\\-1/\sqrt{2}\end{bmatrix}, \begin{bmatrix}0\\1\\0\end{bmatrix}, \begin{bmatrix}1/\sqrt{2}\\0\\1/\sqrt{2}\end{bmatrix}$.

(iii) $\begin{bmatrix}0\\\frac{4}{5}\\-\frac{3}{5}\end{bmatrix}, \begin{bmatrix}-1/\sqrt{2}\\3/\sqrt{50}\\4/\sqrt{50}\end{bmatrix}, \begin{bmatrix}1/\sqrt{2}\\3/\sqrt{50}\\4/\sqrt{50}\end{bmatrix}$. (iv) $\begin{bmatrix}\frac{1}{3}\\\frac{2}{3}\\\frac{2}{3}\end{bmatrix}, \begin{bmatrix}\frac{2}{3}\\\frac{1}{3}\\-\frac{2}{3}\end{bmatrix}, \begin{bmatrix}\frac{2}{3}\\-\frac{2}{3}\\\frac{1}{3}\end{bmatrix}$.

3. (i) $\begin{bmatrix}-1/\sqrt{3}\\1/\sqrt{3}\\1/\sqrt{3}\end{bmatrix}, \begin{bmatrix}1/\sqrt{2}\\1/\sqrt{2}\\0\end{bmatrix}, \begin{bmatrix}1/\sqrt{6}\\-1/\sqrt{6}\\2/\sqrt{6}\end{bmatrix}$. (ii) $\begin{bmatrix}0\\1/\sqrt{2}\\-1/\sqrt{2}\end{bmatrix}, \begin{bmatrix}-4/\sqrt{18}\\1/\sqrt{18}\\1/\sqrt{18}\end{bmatrix}, \begin{bmatrix}\frac{1}{3}\\\frac{2}{3}\\\frac{2}{3}\end{bmatrix}$.

(iii) $\begin{bmatrix}-2/\sqrt{5}\\1/\sqrt{5}\\0\end{bmatrix}, \begin{bmatrix}3/\sqrt{70}\\6/\sqrt{70}\\-5/\sqrt{70}\end{bmatrix}, \begin{bmatrix}1/\sqrt{14}\\2/\sqrt{14}\\3/\sqrt{14}\end{bmatrix}$.

5. $\begin{bmatrix}1/\sqrt{2} & 0 & \frac{1}{2} & -\frac{1}{2}\\0 & 1/\sqrt{2} & \frac{1}{2} & \frac{1}{2}\\0 & -1/\sqrt{2} & \frac{1}{2} & \frac{1}{2}\\-1/\sqrt{2} & 0 & \frac{1}{2} & -\frac{1}{2}\end{bmatrix}, \begin{bmatrix}3 & 0 & 0 & 0\\0 & 3 & 0 & 0\\0 & 0 & 3 & 0\\0 & 0 & 0 & -8\end{bmatrix}$.

6. $\begin{bmatrix}1/\sqrt{2} & 0 & 0 & 1/\sqrt{2}\\1/\sqrt{2} & 0 & 0 & -1/\sqrt{2}\\0 & 2/\sqrt{5} & 1/\sqrt{5} & 0\\0 & 1/\sqrt{5} & -2/\sqrt{5} & 0\end{bmatrix}, \begin{bmatrix}5 & 0 & 0 & 0\\0 & 5 & 0 & 0\\0 & 0 & -5 & 0\\0 & 0 & 0 & -3\end{bmatrix}$.

7. $\begin{bmatrix}1 & 0 & 1\\0 & 2 & 0\\0 & -1 & 1\end{bmatrix}$.

8. $\begin{bmatrix}1 & -1 & 1\\1 & 1 & -1\\0 & 1 & 2\end{bmatrix}$.

Chapter 23

1. (i) $\begin{bmatrix} 2 & 2 \\ 2 & -1 \end{bmatrix}$. (ii) $\begin{bmatrix} 3 & -2 \\ -2 & 0 \end{bmatrix}$. (iii) $\begin{bmatrix} 4 & 0 \\ 0 & -3 \end{bmatrix}$. (iv) $\begin{bmatrix} 1 & -8 & -3 \\ -8 & -4 & 8 \\ -3 & 8 & 1 \end{bmatrix}$.

(v) $\begin{bmatrix} 1 & 0 & \frac{1}{2} \\ 0 & 1 & 0 \\ \frac{1}{2} & 0 & 1 \end{bmatrix}$. (vi) $\begin{bmatrix} 1 & 0 & 0 \\ 0 & 1 & 0 \\ 0 & 0 & 1 \end{bmatrix}$. (vii) $\begin{bmatrix} 0 & -4 & 0 \\ -4 & 15 & 0 \\ 0 & 0 & 1 \end{bmatrix}$.

2. (i) $\begin{bmatrix} 2/\sqrt{5} & -1/\sqrt{5} \\ 1/\sqrt{5} & 2/\sqrt{5} \end{bmatrix}$. (ii) $\begin{bmatrix} 2/\sqrt{5} & 1/\sqrt{5} \\ -1/\sqrt{5} & 2/\sqrt{5} \end{bmatrix}$. (iii) $\begin{bmatrix} 1 & 0 \\ 0 & 1 \end{bmatrix}$.

(iv) $\begin{bmatrix} -1/\sqrt{3} & 1/\sqrt{6} & 1/\sqrt{2} \\ 1/\sqrt{3} & 2/\sqrt{6} & 0 \\ 1/\sqrt{3} & -1/\sqrt{6} & 1/\sqrt{2} \end{bmatrix}$. (v) $\begin{bmatrix} 1/\sqrt{2} & 0 & 1/\sqrt{2} \\ 0 & 1 & 0 \\ -1/\sqrt{2} & 0 & 1/\sqrt{2} \end{bmatrix}$.

(vi) $\begin{bmatrix} 1 & 0 & 0 \\ 0 & 1 & 0 \\ 0 & 0 & 1 \end{bmatrix}$. (vii) $\begin{bmatrix} 0 & -1/\sqrt{17} & 4/\sqrt{17} \\ 0 & 4/\sqrt{17} & 1/\sqrt{17} \\ 1 & 0 & 0 \end{bmatrix}$.

3. (i) Directions of $\begin{bmatrix} 1 \\ -1 \end{bmatrix}, \begin{bmatrix} 1 \\ 1 \end{bmatrix}$; $u^2 - \dfrac{v^2}{3} = 1$; hyperbola.

(ii) $\begin{bmatrix} 1 \\ -1 \end{bmatrix}, \begin{bmatrix} 1 \\ 1 \end{bmatrix}$; $2u^2 + 4v^2 = 1$; ellipse.

(iii) $\begin{bmatrix} -2 \\ 1+\sqrt{5} \end{bmatrix}, \begin{bmatrix} 1+\sqrt{5} \\ 2 \end{bmatrix}$; $(2+\sqrt{5})u^2 - (\sqrt{5}-2)v^2 = 1$; hyperbola.

(iv) $\begin{bmatrix} 2 \\ 1 \end{bmatrix}, \begin{bmatrix} -1 \\ 2 \end{bmatrix}$; $u^2 - \dfrac{v^2}{4} = 1$; hyperbola.

4. (i) $\begin{bmatrix} 0 \\ 1 \\ -1 \end{bmatrix}, \begin{bmatrix} 1 \\ 1 \\ 1 \end{bmatrix}, \begin{bmatrix} 2 \\ -1 \\ -1 \end{bmatrix}$; $\dfrac{u^2}{3} + v^2 - w^2 = 1$; hyperboloid of one sheet.

(ii) $\begin{bmatrix} 0 \\ 1 \\ 0 \end{bmatrix}, \begin{bmatrix} 1 \\ 0 \\ 1 \end{bmatrix}, \begin{bmatrix} -1 \\ 0 \\ -1 \end{bmatrix}$; $u^2 + \dfrac{3v^2}{2} + \dfrac{w^2}{2} = 1$; ellipsoid.

(iii) $\begin{bmatrix} 1 \\ -1 \\ 0 \end{bmatrix}, \begin{bmatrix} 1 \\ 1 \\ -2 \end{bmatrix}, \begin{bmatrix} 1 \\ 1 \\ 1 \end{bmatrix}$; $u^2 - v^2 - \dfrac{w^2}{4} = 1$; hyperboloid of two sheets.

(iv) $\begin{bmatrix} 1 \\ 1 \\ 0 \end{bmatrix}, \begin{bmatrix} -1 \\ 1 \\ 1 \end{bmatrix}, \begin{bmatrix} 1 \\ -1 \\ 2 \end{bmatrix}$ (say); $\dfrac{u^2}{7} + \dfrac{v^2}{7} + w^2 = 1$; ellipsoid.

(Only (iv) is a surface of revolution.)

Chapter 24

1. (i) $\boldsymbol{x} = a\begin{bmatrix} 1 \\ -1 \end{bmatrix} e^{2t} + b\begin{bmatrix} -1 \\ 2 \end{bmatrix} e^{3t}$.

(ii) $\boldsymbol{x} = a\begin{bmatrix} 4 \\ 1 \end{bmatrix} e^{2t} + b\begin{bmatrix} 1 \\ -1 \end{bmatrix} e^{-3t}$.

(iii) $\boldsymbol{x} = a\begin{bmatrix} 1 \\ -1 \end{bmatrix} + b\begin{bmatrix} 1 \\ 1 \end{bmatrix} e^{2t}$.

(iv) $\boldsymbol{x} = \begin{bmatrix} a \\ b \end{bmatrix} e^{-t}$.

(v) $\boldsymbol{x} = \begin{bmatrix} a \\ b \\ -b \end{bmatrix} e^{t} + c\begin{bmatrix} 1 \\ -2 \\ 1 \end{bmatrix} e^{2t}$.

(vi) $\boldsymbol{x} = a\begin{bmatrix} 1 \\ 1 \\ 1 \end{bmatrix} e^{t} + b\begin{bmatrix} 1 \\ 0 \\ 0 \end{bmatrix} e^{2t} + c\begin{bmatrix} 7 \\ -4 \\ 2 \end{bmatrix} e^{4t}$.

(vii) $\boldsymbol{x} = a\begin{bmatrix} -1 \\ 3 \\ 1 \end{bmatrix} + b\begin{bmatrix} 0 \\ 2 \\ 1 \end{bmatrix} e^{2t} + c\begin{bmatrix} 1 \\ 0 \\ -1 \end{bmatrix} e^{3t}$.

(viii) $\boldsymbol{x} = \begin{bmatrix} a-2b \\ a \\ b \end{bmatrix} e^{t} + c\begin{bmatrix} 1 \\ -1 \\ 2 \end{bmatrix} e^{7t}$.

(ix) $\boldsymbol{x} = 2e^{\sqrt{3}\,t}\begin{bmatrix} b\cos t + a\sin t \\ 0 \\ a\cos t - b\sin t \end{bmatrix} + c\begin{bmatrix} 0 \\ 1 \\ 0 \end{bmatrix} e^{t}$.

(x) $\boldsymbol{x} = 2e^{t}\begin{bmatrix} (a+b)\cos t + (a-b)\sin t \\ 0 \\ a\cos t - b\sin t \end{bmatrix} + c\begin{bmatrix} 1 \\ 1 \\ 1 \end{bmatrix} e^{t}$.

SAMPLE TEST PAPERS FOR PART 1

Paper 1

1

(i) Let X be a 3×4 matrix. What size must the matrix Y be if the product XYX is to exist? For such a matrix Y, what is the size of the matrix XYX?

Calculate AB or BA (or both, if both exist), where

$$A = \begin{bmatrix} -1 & 2 \\ 0 & 1 \\ 3 & -1 \end{bmatrix} \quad \text{and} \quad B = \begin{bmatrix} 0 & 2 & 3 \\ 1 & -1 & 0 \end{bmatrix}.$$

(ii) Find the values of t for which the following equations have (a) a unique solution, and (b) infinitely many solutions.

$$\begin{aligned} tx + 4y &= 0 \\ (t-1)x + ty &= 0. \end{aligned}$$

(iii) Let X and Y be $p \times p$ symmetric matrices. Is the matrix $XY - YX$ symmetric? Is it skew-symmetric? If P and Q are skew-symmetric matrices, what can be said about the symmetry or skew-symmetry of the matrix $PQ - QP$?

2

Show that the list

$$\left(\begin{bmatrix} 1 \\ 2 \\ 1 \end{bmatrix}, \begin{bmatrix} 1 \\ 4 \\ -1 \end{bmatrix}, \begin{bmatrix} 1 \\ -1 \\ a^2 + 3a \end{bmatrix} \right)$$

is linearly independent if and only if $a = 1$ or $a = -4$. For each of these values of a, find a non-trivial linear combination of these vectors which is equal to the zero vector.

What is the rank of the matrix

$$A = \begin{bmatrix} 1 & 1 & 1 \\ 2 & 4 & -1 \\ 1 & -1 & a^2 + 3a \end{bmatrix},$$

when $a = 1$ or $a = -4$?

Find the inverse of A when $a=0$, and hence or otherwise solve the equation

$$A\mathbf{x}=\begin{bmatrix}-1\\-2\\1\end{bmatrix}$$

in the case when $a=0$.

3

(i) Show that the determinant of a 3×3 skew-symmetric matrix is equal to zero. Do all skew-symmetric matrices have determinant equal to zero? Justify your answer.
(ii) Explain what is meant by an *elementary matrix*. Give examples of the three different kinds of elementary matrix, and explain their connection with the Gaussian elimination process.
(iii) Let A, B and C be the points $(1,0,0)$, $(0,2,0)$ and $(0,0,2)$ respectively. Using the cross product of vectors, or otherwise, find the surface area of the tetrahedron $OABC$.

4

Let $A(2,1,-4)$, $B(0,-1,-6)$, $C(3,0,-1)$ and $D(-3,-4,-3)$ be four points in space. Find parametric equations for the straight lines AB and CD. Hence show that these two lines do not intersect. Let P and Q be points on AB and CD respectively such that PQ is perpendicular to both AB and CD. Calculate the length of PQ.

Find an equation for the locus of all midpoints of line segments joining a point on AB to a point on CD. Deduce that the locus is a plane.

Paper 2

1

(i) Let

$$A=\begin{bmatrix}2 & 1\\ 0 & 1\end{bmatrix},\quad B=\begin{bmatrix}1 & -1 & 0\\ 2 & 2 & 1\end{bmatrix}\quad\text{and}\quad C=\begin{bmatrix}1 & 1\\ 2 & 0\\ -1 & -1\\ 2 & 3\end{bmatrix}.$$

Calculate all possible products of two of these matrices. Is it possible to multiply them all together? If so, in what order? Calculate any such product of all three.

(ii) Let

$$X=\begin{bmatrix}1 & 2 & -1\\ 2 & 3 & 0\\ -1 & 0 & -2\end{bmatrix}.$$

Find whether X is invertible. If it is, find its inverse.

(iii) Define a *skew-symmetric* matrix. Explain why the entries on the main diagonal of a skew-symmetric matrix must all be zero. Let H be the matrix

$$\begin{bmatrix}0 & 1\\ -1 & 0\end{bmatrix}.$$

Show that $H^2+I=0$, that H is invertible, and that $H^{-1}=H^{\mathrm{T}}$.

2

(i) Let A be a $p\times q$ matrix and let $\boldsymbol{b}$ be a p-vector. In the matrix equation $A\boldsymbol{x}=\boldsymbol{b}$, what condition must be satisfied by the rank of A and the rank of the augmented matrix $[A \mid \boldsymbol{b}]$ if the equation is to have no solutions? Prove that the following set of equations has no solutions.

$$\begin{aligned} x+2y+3z&=1\\ x+\ y+\ z&=2\\ 5x+7y+9z&=6. \end{aligned}$$

(ii) Find whether the list

$$\left(\begin{bmatrix}2\\ 1\\ -3\end{bmatrix},\ \begin{bmatrix}1\\ -2\\ -1\end{bmatrix},\ \begin{bmatrix}-1\\ -3\\ 2\end{bmatrix}\right)$$

is linearly dependent or linearly independent.

(iii) Find all values of c for which the equations

$$\begin{aligned} (c+1)x+\qquad 2y&=0\\ 3x+(c-1)y&=0 \end{aligned}$$

have a solution other than $x=y=0$.

3

(i) Evaluate the determinant

$$\begin{vmatrix}2 & 1 & 3 & -1\\ -2 & 0 & 1 & 1\\ 1 & 0 & 2 & 2\\ 3 & 0 & -1 & 1\end{vmatrix}$$

(ii) Explain what is meant by the adjoint (adj A) of a square matrix A. Show that, for any 3×3 matrix A,

$$A(\text{adj } A)=(\det A)I,$$

where I is the 3×3 identity matrix.

(iii) Find an equation for the plane containing the three points $A(2,1,1)$, $B(-1,5,9)$ and $C(4,5,-1)$.

4

Define the *dot product* $\boldsymbol{a}\cdot\boldsymbol{b}$ of two non-zero vectors $\boldsymbol{a}$ and $\boldsymbol{b}$.

(i) Let $OABC$ be a tetrahedron, O being the origin. Suppose that OC is perpendicular to AB and that OB is perpendicular to AB. Prove that OA is perpendicular to BC.

(ii) Let P and Q be the points $(1,0,-1)$ and $(0,1,1)$ respectively. Find all unit vectors $\boldsymbol{u}$ which are perpendicular to $\overrightarrow{OP}$ and which make angle $\pi/3$ (60°) with $\overrightarrow{OQ}$.

Paper 3

1

(i) Let

$$A=\begin{bmatrix}1 & 0\\ 2 & -1\\ 0 & 2\\ 1 & 3\end{bmatrix},\quad B=\begin{bmatrix}1 & 1 & 0 & -1\\ -1 & 0 & 2 & 0\\ 0 & 0 & 1 & 1\end{bmatrix}\quad \text{and}\quad C=\begin{bmatrix}1 & 0 & 1\\ -1 & 1 & 0\end{bmatrix}.$$

Evaluate the product CB. Evaluate every other product of A or C with B. There are exactly three orders in which it is possible to multiply A, B and C all together. Write these down but do not evaluate the products. State the sizes of the three product matrices.

(ii) Define the *rank* of a matrix. Calculate the rank of the matrix

$$\begin{bmatrix}1 & -3 & 4\\ 2 & -1 & 7\\ 2 & 4 & 6\end{bmatrix}.$$

Use your answer in determining (without actually solving them) whether the following equations are consistent.

$$\begin{aligned} x-3y+4z&=0\\ 2x-\ y+7z&=4\\ 2x+4y+6z&=8. \end{aligned}$$

(iii) Show that the product of two upper triangular 3×3 matrices is upper triangular.

2

Solve the system of equations

$$\begin{aligned} x+2y-\ z&=\ \ 4\\ 2x-\ y+\ z&=-3 \qquad (*)\\ -x+\ y+4z&=-7. \end{aligned}$$

Show that the inverse of an invertible symmetric matrix is symmetric, and verify this by finding the inverse of

$$P=\begin{bmatrix}1 & 1 & 0\\ 1 & 2 & 1\\ 0 & 1 & 0\end{bmatrix}.$$

Let A be a 3×3 matrix and let $\boldsymbol{b}$ be a 3-vector. Show that if $\boldsymbol{c}$ is a solution to the equation $A\boldsymbol{x}=\boldsymbol{b}$ then $P^{-1}\boldsymbol{c}$ is a solution to the equation $(AP)\boldsymbol{x}=\boldsymbol{b}$. Use your earlier results to find a solution to

$$(AP)\boldsymbol{x}=\begin{bmatrix}4\\ -3\\ -7\end{bmatrix},$$

where A is the matrix of coefficients on the left-hand sides of equations (*) above.

3

(i) Let

$$A=\begin{bmatrix} 1 & t & 0 \\ 1+t & 1 & 5 \\ 0 & -t & t \end{bmatrix}, \quad \text{where } t\in\mathbb{R}.$$

Evaluate det A and hence find all values of t for which A is singular.

(ii) Let X be a 3×1 matrix and let Y be a 1×3 matrix. Show that XY is a *singular* 3×3 matrix.

(iii) Let A, B, C and P be points with coordinates $(2, 1, 1)$, $(-4, -2, 1)$, $(1, 2, 3)$ and $(-1, -1, 2)$ respectively. Find which of the angles $B\hat{P}C$, $C\hat{P}A$ and $A\hat{P}B$ is the smallest.

4

Give the definition of the *cross product* $\boldsymbol{a}\times\boldsymbol{b}$ of two non-zero vectors $\boldsymbol{a}$ and $\boldsymbol{b}$.

Find an equation for the plane π through the points $A(1, 1, 4)$, $B(3, -2, 4)$ and $C(3, -1, 1)$.

What is the perpendicular distance of the point $X(1, -3, 5)$ from the plane π? Find the volume of the parallelepiped which has X as one vertex and A, B and C as the vertices adjacent to X. Find the coordinates of the vertex of this parallelepiped which is farthest from X.

SAMPLE TEST PAPERS FOR PART 2

Paper 4

1

(i) Show that the set

$$\left\{\begin{bmatrix} x \\ y \\ z \end{bmatrix} \in \mathbb{R}^3 : x = 0\right\}$$

is a subspace of $\mathbb{R}^3$. Find a basis for this subspace, justifying your answer. Show that the set

$$\left\{\begin{bmatrix} x \\ y \\ z \end{bmatrix} \in \mathbb{R}^3 : x = 1\right\}$$

is not a subspace of $\mathbb{R}^3$.

(ii) Show that the list

$$\left(\begin{bmatrix} 1 \\ 1 \\ -1 \end{bmatrix}, \begin{bmatrix} 2 \\ 1 \\ 0 \end{bmatrix}, \begin{bmatrix} 1 \\ 3 \\ -6 \end{bmatrix}\right)$$

is a basis for $\mathbb{R}^3$. Express the vector $\begin{bmatrix} 1 \\ 0 \\ 0 \end{bmatrix}$ as a linear combination of these basis vectors.

2

(i) Suppose that we are told that a certain linear transformation f from $\mathbb{R}^3$ to $\mathbb{R}^2$ has the properties:

$$f\left(\begin{bmatrix} 1 \\ 2 \\ -1 \end{bmatrix}\right) = \begin{bmatrix} 1 \\ -1 \end{bmatrix}, \quad f\left(\begin{bmatrix} 2 \\ 2 \\ 0 \end{bmatrix}\right) = \begin{bmatrix} 2 \\ 1 \end{bmatrix}.$$

Is it possible to deduce the value of $f\left(\begin{bmatrix} 1 \\ 4 \\ -3 \end{bmatrix}\right)$?

If so, find it. If not, explain why not.

(ii) Find the kernel and the image of the linear transformation from $\mathbb{R}^4$ to $\mathbb{R}^3$ represented by the matrix

$$\begin{bmatrix} 1 & 2 & 0 & 3 \\ 1 & 1 & -1 & 2 \\ -1 & 2 & 4 & 1 \end{bmatrix}.$$

What is the rank of this matrix?

3

(i) Give the definition of *orthogonal matrix*. Show that the determinant of an orthogonal matrix is equal to either 1 or -1. Show also that the product of two orthogonal matrices is an orthogonal matrix.

(ii) Find two vectors (not multiples of one another) which are orthogonal to $\begin{bmatrix} 1 \\ 2 \\ 2 \end{bmatrix}$.

Hence find an orthonormal basis for $\mathbb{R}^3$ which contains the vector $\begin{bmatrix} \frac{1}{3} \\ \frac{2}{3} \\ \frac{2}{3} \end{bmatrix}$.

4

(i) Find a diagonal matrix D and an invertible matrix P such that $P^{-1}AP = D$, where

$$A = \begin{bmatrix} 0 & -2 & 2 \\ 2 & -1 & 0 \\ 2 & -2 & 1 \end{bmatrix}.$$

(ii) What can be said in general about the eigenvalues of

(a) a diagonal matrix,

(b) an upper triangular matrix?

Justify your answers.

Paper 5

1

(i) Find a basis for the subspace S of $\mathbb{R}^4$ spanned by the list of vectors

$$\left(\begin{bmatrix}1\\-1\\2\\2\end{bmatrix},\begin{bmatrix}0\\1\\0\\1\end{bmatrix},\begin{bmatrix}-1\\2\\-1\\0\end{bmatrix},\begin{bmatrix}1\\1\\3\\5\end{bmatrix}\right).$$

A 4×4 matrix A is such that $A\boldsymbol{x}=\mathbf{0}$ for every vector $\boldsymbol{x}$ belonging to this subspace S. What must be the rank of A? Find such a matrix A.

(ii) Show that the set

$$\left\{\begin{bmatrix}x\\y\\z\end{bmatrix}\in\mathbb{R}^3 : x+y=z\right\}$$

is a subspace of $\mathbb{R}^3$.

2

Let $X=\left(\begin{bmatrix}2\\1\\1\end{bmatrix},\begin{bmatrix}1\\0\\3\end{bmatrix},\begin{bmatrix}1\\-1\\2\end{bmatrix}\right)$. This list is a basis for $\mathbb{R}^3$ (do not verify this). Find a matrix P which transforms a column vector of components with respect to the basis X into a column vector of components with respect to the standard basis for $\mathbb{R}^3$. Now let

$$Y=\left(\begin{bmatrix}1\\-1\\2\end{bmatrix},\begin{bmatrix}-1\\2\\-1\end{bmatrix},\begin{bmatrix}1\\-3\\1\end{bmatrix}\right).$$

This list is also a basis for $\mathbb{R}^3$ (do not verify this). Find (and evaluate fully) a matrix which transforms components with respect to the basis X into components with respect to Y.

3

Find an orthogonal matrix P such that the matrix $P^{\mathrm{T}}AP$ is a diagonal matrix, where

$$A=\begin{bmatrix}3&-2&0\\-2&5&2\\0&2&3\end{bmatrix}.$$

Describe a change of coordinate axes which will transform the equation

$$3x^2+5y^2+3z^2+4yz-4xy=21$$

into a standard form. Find such a standard form, and deduce the nature of the surface which has this equation.

4

(i) Using the Gram–Schmidt process, or otherwise, find an orthonormal basis for the subspace of $\mathbb{R}^4$ spanned by the list

$$\left(\begin{bmatrix} 1 \\ 0 \\ 1 \\ 0 \end{bmatrix}, \begin{bmatrix} 1 \\ 1 \\ 1 \\ 0 \end{bmatrix}, \begin{bmatrix} 0 \\ 1 \\ 1 \\ 1 \end{bmatrix} \right).$$

(ii) Find the general solution to the simultaneous differential equations

$$\frac{dx}{dt} = 3x - y$$

$$\frac{dy}{dt} = 2x + y.$$

Paper 6

1

(i) Let S be the subspace of $\mathbb{R}^3$ spanned by the list

$$\left(\begin{bmatrix} -2 \\ 1 \\ 2 \end{bmatrix}, \begin{bmatrix} 3 \\ 1 \\ -1 \end{bmatrix}, \begin{bmatrix} -1 \\ 3 \\ 3 \end{bmatrix}\right).$$

Find a basis for S.

Let T be the subspace for $\mathbb{R}^3$ spanned by the list

$$\left(\begin{bmatrix} 1 \\ 2 \\ 1 \end{bmatrix}, \begin{bmatrix} 1 \\ 7 \\ 5 \end{bmatrix}, \begin{bmatrix} 5 \\ 5 \\ 1 \end{bmatrix}\right).$$

Show that $S = T$.

(ii) Show that the set

$$\left\{\begin{bmatrix} x \\ y \\ z \end{bmatrix} \in \mathbb{R}^3 : x + y + z = 0\right\}$$

is a subspace of $\mathbb{R}^3$.

2

(i) Let $A = \begin{bmatrix} 1/\sqrt{2} & 1/\sqrt{2} \\ -1/\sqrt{2} & 1/\sqrt{2} \end{bmatrix}$, $B = \begin{bmatrix} 0 & 1 \\ -1 & 0 \end{bmatrix}$.

With the aid of diagrams, explain in geometrical terms the effects of the linear transformations from $\mathbb{R}^2$ to $\mathbb{R}^2$ which are represented by A and B. Show by means of an algebraic argument that the result of applying one of these transformations and then the other is always the same, irrespective of which one is applied first.

(ii) Show that the linear transformation from $\mathbb{R}^3$ to $\mathbb{R}^4$ which is represented by the matrix

$$\begin{bmatrix} 1 & 3 & 0 \\ 0 & 1 & -2 \\ 2 & 1 & 1 \\ -1 & 1 & 2 \end{bmatrix}$$

is one-to-one. Describe its kernel and its image.

3

Let $\boldsymbol{u}$ and $\boldsymbol{v}$ be eigenvectors of the matrix A corresponding to the eigenvalue k. Show that every non-zero linear combination of $\boldsymbol{u}$ and $\boldsymbol{v}$ is also an eigenvector of A corresponding to k.

Explain what is meant by saying that two non-zero vectors in $\mathbb{R}^n$ are orthogonal. Show that if $\boldsymbol{u}$ and $\boldsymbol{v}$ are orthogonal to $\boldsymbol{x}$ in $\mathbb{R}^n$ then every non-zero vector in the subspace spanned by $(\boldsymbol{u}, \boldsymbol{v})$ is orthogonal to $\boldsymbol{x}$.

Let $\boldsymbol{u} = \begin{bmatrix} 2 \\ -2 \\ -1 \end{bmatrix}$, $\boldsymbol{v} = \begin{bmatrix} 1 \\ -2 \\ 0 \end{bmatrix}$ and $\boldsymbol{x} = \begin{bmatrix} 2 \\ 1 \\ 2 \end{bmatrix}$.

Verify that $\boldsymbol{u}$ and $\boldsymbol{v}$ are orthogonal to $\boldsymbol{x}$, and find a linear combination of $\boldsymbol{u}$ and $\boldsymbol{v}$ which is orthogonal to $\boldsymbol{u}$.

Suppose that a certain 3×3 matrix A has $\boldsymbol{x}$ (as above) as an eigenvector corresponding to the eigenvalue 2, and has $\boldsymbol{u}$ and $\boldsymbol{v}$ (as above) as eigenvectors both corresponding to the eigenvalue 1. Write down an orthogonal list of eigenvectors of A. Find an orthogonal matrix P such that

$$P^{\mathrm{T}}AP = \begin{bmatrix} 2 & 0 & 0 \\ 0 & 1 & 0 \\ 0 & 0 & 1 \end{bmatrix}.$$

Hence find A.

4

(i) The equation

$$9x^2 + 3y^2 + 8xy - 11 = 0$$

represents a conic. Find vectors in the directions of the principal axes of this conic, and decide whether it is an ellipse or a hyperbola.

(ii) Find the general solution to the simultaneous differential equations

$$\frac{dx}{dt} = 8x + 5y$$

$$\frac{dy}{dt} = -4x - 4y.$$

Use this to find particular functions $x(t)$ and $y(t)$ satisfying these equations, and also satisfying the conditions $x(0) = 0$, $y(0) = 3$.

FURTHER READING

There is a very large number of books available on the subject of linear algebra. This book is intended to be different from most of them, in the manner described in the Preface, so reading other books simultaneously could serve to confuse rather than enlighten. But this book is not comprehensive: it does not cover more advanced or abstract parts of the subject, nor does it pursue applications. And of course further examples and exercises are always valuable. Here, then, is a selection of books which might be useful.

The first two are books of worked examples:

F. Ayres, *Matrices*. Schaum's Outline Series, McGraw-Hill, 1968.

J. H. Kindle, *Plane and Solid Analytic Geometry*. Schaum's Outline Series, McGraw-Hill, 1950.

Next, four books with subject matter similar to that of this book, but with different approaches:

H. Anton, *Elementary Linear Algebra*, 4th edition. John Wiley, 1984.

D. T. Finkbeiner, *Elements of Linear Algebra*, 3rd edition. Freeman, 1978.

B. Kolman, *Elementary Linear Algebra*, 4th edition. Collier Macmillan, 1986.

I. Reiner, *Introduction to Linear Algebra and Matrix Theory*. Holt, Rinehart & Winston, 1971.

Now some books which provide either a more abstract approach, or cover more advanced topics:

E. D. Nering, *Linear Algebra and Matrix Theory*, 2nd edition. John Wiley, 1970.

H. Samelson, *An Introduction to Linear Algebra*. Wiley-Interscience, 1974.

G. Strang, *Linear Algebra and its Applications*, 2nd edition. Van Nostrand Reinhold, 1980.

As a reference for the geometrical aspects of the subject:

P. J. Kelly & E. G. Straus, *Elements of Analytical Geometry*. Scott Foresman, 1970.

Last, here are some books which concentrate more on applications:

T. J. Fletcher, *Linear Algebra through its Applications*. Van Nostrand Reinhold, 1972.

F. A. Graybill, *Introduction to Matrices with Applications in Statistics*. Wadsworth, 1969.

C. Rorres & H. Anton, *Applications of Linear Algebra*, 3rd edition. John Wiley, 1984.

INDEX